CONVOY

Martin Middlebrook

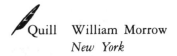

Quill   William Morrow
New York

Library of Congress Catalog Card Number: 86-60101

ISBN: 0-688-06428-0 (pbk.)

Printed in the United States of America

First Quill Edition

1 2 3 4 5 6 7 8 9 10

Contents

List of Illustrations and Maps   vii
Introduction   ix

1   The Battle for Merchant Shipping   1
2   The Convoy Men   19
3   The Germans   56
4   New York   77
5   The Early Voyage   102
6   The Approach to Danger   126
7   The Shadowing   154
8   The Battle of St Patrick's Day   169
9   The Wild Donkey   195
10   Review at Dawn of 17 March   201
11   Six Hearses Bearing 180 Degrees   211
12   The Second Round   234
13   The Final Stage   257
14   Distant Operations   279
15   The Aftermath   286
16   An Analysis   302
17   The Years That Followed   322

Appendix 1   *The Merchant Ships
of Convoys SC.122, HX.229
and HX.229A*   332
Appendix 2   *Local Escort Groups*   340
Appendix 3   *Ocean Escort Groups*   342
Appendix 4   *U-boats Involved in
the SC.122/HX.229 Operations*   344
Appendix 5   *The Roll of Honour*   351

Appendix 6 *German U-boats Destroyed*
September 1939–May 1945   358

Acknowledgements   360
Bibliography   367
Index   368

CONVOY

Illustrations

1 S.S. *Nariva* (Furness Withy Line)
2 S.S. *John Fiske* (U.S. Navy Department)
3 H.M.S. *Highlander* (Imperial War Museum)
4 H.M.S. *Pimpernel* (Lieutenant-Commander H. D. Hayes)
5 H.M.S. *Volunteer* (R. G. Goudy) and Lieutenant-Commander G. J. Luther (Mrs A. Luther).
6 Kapitänleutnant Rudolf Bahr (H. Bethke)
7 Oberleutnant Hans-Joachim Bertelsmann (M. Zweigle)
8 Crew of U.618 (K. Baberg)
9 U-boat pens (Bundesarchiv)
10 Loading aircraft as deck cargo (U.S. Office of War Information)
11 St John's (Public Archives of Canada)
12 Oiling at sea (R. T. Brown)
13 U.S.C.G. Cutter *Ingham* in storm (U.S. Coast Guard)
14 Grossadmiral Dönitz and staff (A. Schnee)
15 Western Approaches Wren (Imperial War Museum)
16 Admiral Sir Max Horton (Imperial War Museum)
17 U-boat torpedo room (Bundesarchiv)
18 Crewmen of U.S.S. *Babbitt* (O. Jensen)
19 Survivors (Imperial War Museum)
20 Survivors (U.S. Coast Guard)
21 Liberator (T. J. Kempton)
22 Fortress (R. V. Gurnham)
23 Sunderland (Squadron Leader D. G. T. Hayes)
24 Aerial attack on U-boat (Imperial War Museum)
25 and 26 *Canadian Star* survivors (D. C. Christopherson)
27 H.M.S. *Beverley* (Dr W. A. B. Campbell)
28 U.S.S. *Babbitt* comes into Londonderry (O. Jensen)
29 American survivors (R. J. Smith)
30 U.338 comes into St Nazaire (H. Zeissler)
31 U.338 crewmen and R.A.F. prisoner (H. Zeissler)
32 Kapitänleutnant Heinrich Müller-Edzards (Frau M. Müller-Edzards)

33 U.618 crewmen receive mail (K. Baberg)
34 Tower Hill Merchant Navy Memorial (Commonwealth War Graves Commission)
35 American Memorial at Cambridge (American Battle Monuments Commission)

Maps

1 The Planned Convoy Routes   91
2 Convoys and U-boats   116
3 The U-boat plan   136
4 Situation at dawn, 15 March   141
5 The U-boat reaction – stage one   155
6 The U-boat reaction – stage two   157
7 The U-boat reaction – stage three   207
8 The disengagement   288–9

Diagrams

1 SC.122's escort and convoy disposition   132
2 HX.229's escort and convoy disposition   133
3 HX.229's movements, Daylight 16 March   161

Maps and diagrams drawn by Illustra Design Ltd from preliminary drawings by Edward Sylvester

Introduction

Convoy: 'A fleet or number of merchant ships under the protection of ships of war, or powerful enough to defend themselves.'

Oxford English Dictionary

At 11.00 a.m. on 3 September 1939, Britain's ultimatum that Germany should cease its attack on Poland ran out and Britain declared war on Germany. Ten hours later Oberleutnant Fritz-Julius Lemp, captain of the German submarine U.30, sighted the 13,581-ton Donaldson liner *Athenia* 250 miles north-west of Ireland. Lemp torpedoed and sank the liner.

On the night of 7 May 1945, three days after the German armed forces had surrendered in the West, Kapitänleutnant Emil Klusmeier in U.2336 sank the small Canadian merchant ship *Avondale Park* off the Firth of Forth. Klusmeier, who was on his first operational patrol as captain, later claimed that he had not received the orders to cease fire broadcast by U-boat Headquarters.

Lemp's sinking of the *Athenia* and Klusmeier's of the *Avondale Park* demonstrate that the German submarine campaign constituted the longest single battle of the war, but neither incident was typical of that campaign. Both victims were sunk within a reasonable distance of land, whereas throughout most of the war the main scene of the German submarine effort was on the ocean trade routes, particularly those of the North Atlantic.

It is well known that the Allied naval escorts and later their air forces held and eventually defeated the German submarine challenge. But there was a period during the winter of 1942–3 when the Germans came close to cutting the North Atlantic lifeline. The crescendo of this crisis was reached in March 1943.

Early in that month two convoys sailed from New York harbour for England. They constituted one 'turn' in what was known as the 'convoy cycle', and their voyage across the Atlantic became one of the turning points of the sea war. This book will describe the ships, aircraft, equipment and tactics of the rival forces, the men who were caught up in the contest, and the resulting convoy battle.

Although I happen to be an Englishman, I have tried to write my story impartially. There will be no 'us' and no 'them'; after more than thirty years, courage and patriotism can surely be admired whichever side a man fought on.

This then is the story of Convoys SC.122 and HX.229, but, in many ways, it could be the story of any of the hundreds of convoys that sailed the oceans in those perilous years.

CONVOY

The Battle
for Merchant Shipping

The only thing that ever really frightened me during the war was the U-boat peril . . . The Admiralty, with whom I lived in the closest amity and contact, shared these fears. Winston Churchill, *The Second World War*, Vol. II, p. 529

The convoy battles of March 1943 took place well past the mid-point of the war, and it will help if a brief description is first given of the course in those earlier years of what may loosely be called 'the battle for merchant shipping', setting out the main aims of both sides, the major turning points of the battle and the development of equipment and tactics. This may be tedious to experts but will enable the general reader to follow more easily the 'action' chapters.

Those who directed Britain's naval affairs in 1939 had abundant memories of how the German U-boat campaign of 1916 and 1917 had nearly cut the supply lanes to Britain's ports.* The U-boats had sunk 2,520 merchant ships up to the end of April 1917 and Britain faced starvation. The situation was saved only by the introduction by the Admiralty, although somewhat reluctant, of the convoy system. The losses fell dramatically. The lesson had been well learnt and, when Britain found herself at war in 1939, convoys were introduced at once.

In one way, this was already a small victory for the Germans. The delays forced upon individual ships while a convoy assembled, the need to take a common route which was not necessarily the most direct for all ships in the convoy, the reduction in speed forced upon the faster ships in a convoy

---

* The term 'U-boat', after the German word *Unterseeboot*, is popularly used in connection with German submarines and will be used in this book.

and, finally, the congestion at ports of destination on the arrival there of a host of ships all requiring unloading – all these factors immediately cut the effective cargo-carrying capacity of Britain's merchant fleet by one third, and this loss would continue for as long as merchant ships had to sail in convoy.

U-boats were not the only danger to merchant ships; mines, aircraft, surface warships and armed raiders all had to be combated. But, although these achieved considerable success early in the war, their effectiveness declined. Mines only affected coastal waters; land-based aircraft had a limited radius of action; *Graf Zeppelin*, the only German aircraft-carrier, never came into service; the Royal Navy soon cleared the oceans of German surface raiders. The German submarine menace, however, would become ever more powerful and was nearly to defeat the Allies.

One basic aspect of the battle for merchant shipping should be stressed. Britain's island position meant that all of her oil, most of her raw materials and much of her food had to be imported. The Germans were never strong enough at sea to impose a complete blockade, and so the campaign between the U-boats and the merchant ships of Britain, and eventually the ships of a growing number of Allies, was essentially a campaign of attrition of shipping tonnage. When a German U-boat captain torpedoed and sank an Allied merchant ship, he had not only destroyed that ship, the cargo it carried and probably some of its crew – the British Empire and America had more than enough men, war material and civilian supplies to crush Germany. What the U-boat captain had achieved was to deny the Allies the opportunity of transporting many more cargoes to the vital war theatres in that vessel on later voyages. The Germans called it 'The Tonnage War'. If more ships could be sunk than the Allied shipyards could replace with new construction, the Germans would inevitably achieve a tightening stranglehold on Britain's supplies.

If this had happened in that period before Pearl Harbor when Britain stood alone in Europe, Britain would have fallen however great the courage of her people. If the U-boats had strangled Britain in 1942 or 1943, there would have been no base for the Anglo-American invasion of Europe in 1944. In either instance the history of our times would have been immeasurably altered.

These were the issues at stake in the battle that would take place on the ocean trade routes. Like Wellington's Waterloo, it was to be a close run thing.

The German Navy had forty-six U-boats available for action on the outbreak of war and sent as many of these as possible to sea several days before Poland was invaded. These operated intensively until the end of March 1940, when the U-boats were diverted to help with the Norwegian campaign, and they sank 222 merchant ships. But most of these sinkings had been of ships sailing alone; very few attacks had been made on convoys. Other successes were the chance encounter south-west of Ireland between Oberleutnant Otto Schuhart's U.29 and the aircraft-carrier *Courageous*, which resulted in the sinking of the carrier, and the brilliant penetration of Scapa Flow by Oberleutnant Günther Prien in U.47 on 14 October and his torpedoing of the battleship *Royal Oak*.* The Germans paid relatively dearly for these early successes, losing seventeen U-boats – a high proportion of their strength.

It would be appropriate to examine here the manner in which the Germans observed International Law on submarine warfare. Germany had signed the 1935 London Submarine Agreement in which all parties undertook to observe the rules of the Hague Convention. These rules laid down that a submarine must first stop its intended victim, then order its crew to take to the boats and, after sinking the ship, ensure that the lifeboats had capacity for the entire crew. The rules would also apply to defensively-armed merchant ships but not to ships in convoy.

In September 1939 Hitler ordered that these rules should be observed, hoping that Britain and France would not be unduly antagonized and might come to terms after Poland's defeat. But it is almost certain that the German Navy had made up its mind to be rid of these restrictions as soon as

---

* The reader may be interested in the subsequent fortunes of the four U-boat commanders mentioned so far. Lemp, who sank the *Athenia*, went on to sink fifteen more ships, including the battleship *Barham*, before dying in U.110 in May 1941, sunk by H.M. Ships *Bulldog*, *Broadway* and *Aubretia* in a convoy battle. Klusmeier, who made the last U-boat attack of the war, is still alive and is a household appliance dealer in Bochum. Schuhart was given a shore appointment in January 1941 and also survived the war. Prien's fate will be described later.

possible. The use by the merchant ship of its radio while the U-boat was giving the required warning would endanger the U-boat, and the responsibility for taking surplus survivors on board the U-boat would be a reduction in its ability to continue operations. When the *Athenia* was torpedoed and sunk on the first day of the war, no warning was given and 110 of her civilian passengers and eighteen crew members were drowned. Oberleutnant Lemp afterwards claimed that he had identified *Athenia* as an armed merchant cruiser; no action against him was taken by his superiors, but the sinking had brought Germany much unfavourable publicity in neutral countries. On the day following *Athenia*'s sinking, signals were sent to U-boats ordering that the agreed rules must be observed and, by Hitler's personal order, no more passenger ships were to be sunk even if in convoy.

This was a good start but, within only three months, the German Navy had persuaded Hitler to drop all restrictions on attacks against vessels identified as hostile; neutral countries were also warned that their ships would be attacked without warning in all areas around Britain. The Official Historian, in detailing the various steps taken by the Germans, comments:

> It is impossible to avoid the conclusion that Admirals Raeder and Dönitz and the German Naval Staff had always wished and intended to introduce unrestricted warfare as rapidly as the political leaders could be persuaded to accept the possible consequences.*

Hitler's eventual approval of the admirals' policy marked the end of the short period during which the agreements on submarine warfare had been honoured and some degree of mercy had been shown to the merchant sailors. For the next five-and-a-half years there would be no such protection for them.

It should be recorded, however, that there were to be several instances of U-boat captains helping survivors when it was safe for them to do so.

> Admiral Dönitz never ordered us to shoot survivors. We had a clear order, which we always complied with as far as conditions allowed, to help survivors but they must not be brought on board. We ourselves certainly met English survivors in the North Atlantic and gave them food and cans of water. We even sent a

* Captain S. W. Roskill, *The War at Sea 1939-1945*, H.M.S.O., 1954-61, Vol. I, p. 104. Hereafter this work will be referred to as the Naval Official History.

signal giving their position. There was always the chance that our own boat would be sunk one day. (Funkgefreiter Werner Hess, U.530).*

Another contributor, a British officer, was on the armed merchant cruiser *Laurentic* when it was torpedoed one night in November 1940. The crew took to the boats and one of these was approached by a surfaced U-boat. A voice called out and asked if the survivors had enough food but, fearing that it might be a trap and hoping to remain hidden in the dark, they refused to answer. A voice called 'Good night British' and the U-boat disappeared. It turned out to have been the legendary U-boat commander Otto Kretschmer. There are many other stories of survivors being offered help in this way. Unfortunately there are also stories of merchant sailors in boats or in the water being machine-gunned, but it is difficult to judge whether these were propaganda-inspired atrocity stories or actual events.

The U-boat men, by nature of their training and traditions, were unlikely to surrender their boats when cornered and, because of the U-boat practice of diving when detected, there were unlikely to be many survivors among their crews. The German sailors called their boats 'iron coffins' and were well aware that they stood every chance of being trapped within them.

Thus, within only a few months of the opening of what was destined to be a long war, the pattern of the U-boat campaign was set. 'Sink on sight' was the U-boat rule for merchant ships. British propaganda would whip up extreme hatred for the U-boat men; their officers were depicted as fanatical Nazis, the crews as vicious murderers. The U-boat would always be hunted to its death.

It would be a fight to the finish with no quarter asked for and none conceded.

The great victories of the German Army in the spring and early summer of 1940 resulted in important changes in many of the factors affecting the U-boat war. The successful

* This and similar quotations are the results of interviews or correspondence with contributors who were participants in the battle for Convoys SC.122 and HX.229, or were at various headquarters at the time of that battle. Ranks and ships, or appointments held, are those of March 1943. Christian names instead of initials will be used with German names because it is the German custom to do so.

invasion of Norway, followed closely by their victories in France and the Low Countries, gave the Germans the valuable ports on the coastlines of these countries. Some of these were rapidly brought into use as advanced bases and enabled the U-boats to extend their patrol limits by 450 miles. At the same time the establishment of airfields and torpedo-boat bases in France and the Low Countries effectively closed the approaches to London, Britain's main port, to large-scale shipping movements. Liverpool, Glasgow and the Bristol Channel ports would now have to cope with virtually all of Britain's overseas trade. Finally, when Italy entered the war on 10 June, Mussolini placed twenty-six modern submarines at the disposal of the Germans.

Britain gained only two advantages in these desperate months. The first was her occupation of Iceland to forestall a German occupation; this move would later give Britain naval and air bases in a vital area. The second was an agreement with the still neutral American Government whereby Britain received fifty obsolete United States Navy destroyers, while America received from Britain naval bases and sites for airfields in Newfoundland, Bermuda, the West Indies and British Guiana.

Although Britain's new bases in Iceland and the ex-American destroyers were to play vital roles in what would soon be known as the Battle of the Atlantic, the changes that occurred in 1940 had tipped the balance decidedly in favour of the U-boats.

The Germans were quick to take advantage of their improved position, and as early as July 1940 made an all-out attack with every available U-boat on shipping in the approaches to Britain's western coasts. Many merchant ships were still sailing independently, and those that were in convoys were escorted only for 200 miles to the west of Ireland, but the U-boats were now operating as far as 25 degrees West, 700 miles out in the Atlantic. They found easy pickings among the independently sailing ships and the unescorted convoys.

In September there occurred a dramatic change in the German tactics. U-boat attacks until then had usually been made in what might seem to have been a conventional manner. The U-boats had operated singly and found their own targets, and had either carried out submerged attacks with torpedoes or, when a completely defenceless merchant ship was found, surfaced and sunk their victims by gunfire. Such attacks were usually made by day.

Admiral Dönitz, commander of the German submarine arm, had discovered while commanding a First World War U-boat in the Mediterranean that a U-boat on the surface at night was almost invisible and, being able to use its air-breathing diesel engines, could travel fast. He had found little difficulty in getting through a convoy's screen of escorts and sometimes right into the convoy itself, calmly aiming and firing his torpedoes at the merchant ships all around and then making his escape at speed on the surface or by sub-merging.

Dönitz had always intended that the night attack on the surface should be the ultimate U-boat tactic. If several U-boats could be gathered together and all strike on the same night, then untold execution could be wrought among the merchant ships, and any escorts present would be almost powerless to prevent this. He even wrote a book on the subject which was openly published in Berlin just before this new war.

Dönitz was always very sensitive to changes in the tactical situation and, with the longer autumn nights, the reduction in the number of ships sailing independently and the gradual extension by the British of escorted convoys, he judged that the time was right to introduce the new tactic. The Germans called it *die Rudeltaktik*, 'the wolf tactic', and the British soon named the groups of U-boats that carried out these attacks 'wolf packs' – a very apt description. These tactics were immediately successful when they were introduced in the autumn of 1940. They were to be the basis of U-boat operations for the next three years.

A watchful U-boat or a long-range German reconnaissance aircraft would spot and report a convoy, and U-boat Head-quarters would order all nearby boats to close in. When sufficient U-boats had assembled, the attacks would begin. The escort vessels could do little to prevent the slaughter that followed. They had no radar, their underwater search equipment, Asdic, was almost useless, and the U-boat's low silhouette could rarely be seen with the naked eye. If detected, the surfaced U-boat could even outrun the slower escort vessels and escape.

Although the Germans now had only twenty-eight operational boats – twenty-one less than on the outbreak of war – and a proportion of these were always in transit or at their home ports, the U-boats sank 274 merchant ships totalling nearly 1,400,000 tons in the five months June to October 1940. In just three nights, in October, three convoys lost thirty-eight ships to wolf-pack attacks. It is little wonder that the Germans called this 'the Happy Time', and it was in these months that the first wave of famous U-boat aces, notably Endrass, Kretschmer, Prien and Schepke, established their reputations. Sources differ slightly on their exploits, but it is probable that these four were responsible for sinking 133 merchant ships totalling over 700,000 tons, plus the battleship *Royal Oak* (by Prien) and the destroyer *Daring* (by Kretschmer). Although all four were to disappear from the Atlantic within a little over a year, the exploits of these men, admittedly achieved in relatively easy conditions, were to prove an inspiration to the great host of U-boat men who would follow.*

The five months of the U-boat men's Happy Time hit the British hard. In addition to the 274 merchant ships sunk during that time by U-boats, a similar number had been lost to other causes, mainly German aircraft and magnetic mines. By the end of 1940 the Germans had sunk 1,281 ships, mostly British, totalling 4,747,033 tons. This was equivalent to more than one fifth of Britain's 1939 merchant fleet or to five years' peace-time construction of new vessels. The implications of the battle for merchant shipping tonnage were becoming apparent.

How had Britain, with the world's most powerful navy, come to be in this desperate position? The Royal Navy appeared to have been unprepared for the U-boat threat and seemed to have no answer to the latest tactics. The U-boats

---

* The four aces were all put out of action by the Royal Navy, three of them in one week. Günther Prien, twenty-nine sinkings, died when his U.47 was depth-charged by the corvettes *Camilla* and *Arbutus* off Rockall on 7 March 1941. Otto Kretschmer, forty-five sinkings, was taken prisoner when the destroyer *Walker* torpedoed U.99, an unusual feat, on 17 March 1941, and Joachim Schepke, thirty-nine sinkings, was lost in U.100 on the same day, depth-charged by *Walker* and another destroyer, *Vanoc*. Engelbert Endrass, twenty-two sinkings, mostly in U.46, died when his new boat U.567 was depth-charged by the sloop *Deptford* and the corvette *Samphire* off the Azores on 21 December 1941.

were now sinking more merchant ships than the rest of the German Navy and the Luftwaffe combined, and they gave every appearance of continuing their success. In the five months of the Happy Time the Germans had lost only six U-boats and only two of these had been sunk by convoy escorts; the escorts would sink only two more U-boats in the next four months. Why was the Royal Navy so unprepared for Dönitz's U-boats? How could it combat the *Rudeltaktik* of surfaced attack at night by the U-boat pack?

It is now freely admitted by naval officers that the threat of the submarine had been underestimated in the inter-war years. The Royal Navy's great tradition of offensive action by surface vessels had somewhat obscured the need to provide a defence against submarine attack. It may have been suffering from what naval historian Arthur Marder calls 'the Jutland syndrome'. Although the Admiralty had instituted the convoy system promptly on the outbreak of war, it had not in the inter-war years provided the necessary anti-submarine vessels to provide adequate escorts for the convoys. Even allowing for the modest level of peacetime naval budgets, the specialist convoy escort had a lamentably low priority. This will be highlighted in later chapters.

The Royal Navy's main anti-submarine weapons were the 1917-type depth charge and the Asdic underwater detection device developed in great secrecy after 1918, but the approved tactics consisted of one or more destroyers hunting the single, submerged submarine.* Although peacetime exercises had shown that British submarines on the surface at night could easily evade screening destroyers, the lesson was never really learnt and Dönitz's pack attacks came as a complete surprise in 1940. All Fleet destroyers were fitted for anti-submarine work but their primary training was for action in support of

---

* Depth charges were 250-pound drums of explosives dropped over the suspected position of a submarine and pre-set to explode at the submarine's estimated depth.

Asdic (now called Sonar) was a submarine detection device housed in a dome under the hull of the anti-submarine ships. It sent out a narrow beam of sound in a series of pulses which would produce an echo or 'ping' from any solid object detected within a maximum range of 3,000 yards from the transmitter. This echo produced an accurate range and bearing of the object. Asdic was less effective against a surfaced object. It could be used passively as a hydrophone to detect propeller noise, when it would give a reasonable bearing but no range. Asdic could be upset by differing density layers in the water and the echoes were drowned by the noise of rushing water if the ship exceeded eighteen knots.

capital ships. The following quotation, by an officer who later became Director of the Anti-Submarine Division of the Naval Staff, illustrates the low priority given to peacetime anti-submarine work.

I was serving in River Gunboats in China in 1930 when I decided to specialize in A/S because I reckoned that we had nearly been brought to our knees in the First World War by the submarine and this must be an interesting line for the future. To qualify in A/S was looked on rather poorly in those days and many friends raised eyebrows that a bright and keen officer should head for such a backwater full of lazy, hard drinkers. I was Staff A/S Officer of the 6th Destroyer Flotilla at the time of the Spanish Civil War and I found that among the four Asdic operators in each of the nine destroyers, half only managed to get in FOUR HOURS operating their Asdic sets in the YEAR! This was mainly due to the non-availability of submarines but as operating the Asdic is like playing the harp – it needs much skill and practice – this was hopeless. (Commander C. D. Howard-Johnston, Staff Officer Anti-Submarines, Western Approaches H.Q.)

When the war did start, the Navy's difficulties were further compounded by a basic error in the employment of the available anti-submarine vessels. There were two schools of thought. Again, the Royal Navy's tradition lent itself far more readily to offensive action in the form of anti-submarine sweeps along the known trade routes than to the dull, defensive, close screening of the actual convoys on those routes. But the sweeps simply did not work. The U-boat never sought a confrontation with warships; if destroyers were seen on the horizon the U-boat simply submerged, went deep and kept quiet. It was rarely detected. When the destroyers had passed on, the U-boat surfaced and resumed its searching for merchant ships. A minority of British officers, usually those with past experience in submarines, urged that every available anti-submarine warship should be placed in the convoy escorts, thus forcing the U-boat to come to the warship; they were not heeded, although it later became obvious that this so-called 'defensive' attitude would, in the end, force the U-boats to fight and then there would be a chance to destroy them.

The surface escort was not the only counter to the U-boat. First World War experience had shown that aircraft could considerably harass and occasionally sink U-boats. Coastal Command of the R.A.F. was well established on the outbreak of war but its patrol aircraft were of only modest range. The movement by the Germans to mid-Atlantic rendered these short-range aircraft almost useless. When the occupation of Iceland gave Britain an advanced air base in the North Atlantic, an Admiralty request to the R.A.F. for long-range aircraft for Coastal Command was refused. The R.A.F. was falling into the hands of those who felt passionately that they could win the war by the determined application of every available heavy bomber against the industrial capacity of Germany's homeland. In this grand aim they were backed up by Churchill, and instead of the available long-range aircraft being shared between Coastal Command and Bomber Command, Bomber Command was not required to part with any, at least for some time. Berlin and the Ruhr were bombed while the U-boats cruised with impunity in the Atlantic. That the bombers may not have achieved their leaders' high hopes is still a subject of great debate for which there is no space here.

The scene was now set for the Battle of the Atlantic. Britain had been forced into a battle for which she was sadly ill-prepared, but it would be just as important to her survival as the Battle of Britain and as vital to the eventual Allied victory as any other aspect of the war.

The year of 1941 brought a steady increase in the tempo of the U-boat war and the merchant ships were again to suffer grievously, but there were to be some developments favourable to the Allied forces engaged. The German successes came only in the face of fiercer convoy defences and at increasing loss to themselves. There were to be no U-boat Happy Times in 1941.

The force of anti-submarine escort vessels available to the Royal Navy was now significantly increasing, and the Royal Canadian Navy was also building up rapidly. The Canadians gradually extended their convoy cover eastwards; the Royal Navy pushed their cover steadily to the west. In May 1941 the two navies met, and the first eastbound convoy received continuous escort all the way across the North Atlantic; in July the first westbound convoy received the same protection. This major achievement by the Allies in closing the 'escort gap' denied the U-boats the opportunity to attack convoys in mid-Atlantic which, until then, had been virtually defenceless.

Unfortunately the 'air gap' was not closed at the same time.

Many of the merchant ships crossing the Atlantic belonged to the still neutral United States. The U-boats were being forced steadily westwards in their search for unprotected shipping, and in March 1941 the United States Navy allocated three destroyer flotillas and five flying-boat squadrons to patrol certain areas of the Western Atlantic, warning the Germans to keep their U-boats out of these areas. Although this 'Neutrality Patrol' was under American control and its ships were not allowed to attack U-boats until themselves fired upon, it was an effective deterrent in the areas patrolled and it freed the British from having to protect their own shipping in those areas. In May, the Americans established a naval and air base at Argentia in Newfoundland and in July sent a military force to Iceland. Then, in September, the United States Navy started to take full responsibility for escorting British organized convoys in the Western Atlantic that contained American merchant ships. This increasing involvement by the Americans had its inevitable consequence in October when their destroyers *Kearney* and *Reuben James* were torpedoed while on convoy escort. The *Kearney* remained afloat but the *Reuben James* went down with much loss of life. It is amazing, after the Germans had brought the United States into the First World War by its unrestricted U-boat campaign, that it should have again antagonized the Americans by torpedoing their merchant ships and now these two destroyers. The Americans, however, must have known that this was likely to happen, for both the *Kearney* and the *Reuben James* were escorting mainly British convoys on regular convoy routes. The *Reuben James*, at least, was well outside the area in which the Germans had been warned that United States Navy ships would be operating, and the position of its sinking was much nearer to England than to the United States.

Not only were the British escorts gaining in effectiveness with more ships and air support, and gaining a new ally, but they had found a partial counter to the surfaced night attack by the U-boat which had led to the disastrous sinkings of late 1940. The answer was the development of radar for the escorts to 'see' the U-boats in the dark and the use of starshell and of 'snowflake' flares to illuminate the U-boat. The surfaced U-boat would still be hard to detect in the dark, especially in rough sea conditions, but at least the escorts now had a chance.

The wolf-pack attack tactic was also found to have one essential element of which the Allies were able to make vital use. For the pack attack to be successful, the U-boats had to use their wireless, first to report the sighting of a convoy, and then to send off shadowing reports to enable other boats to join. The Germans even used their wireless to report the result of attacks, their stocks of fuel and torpedoes and to send weather reports. They realized that Allied shore stations would pick up these signals and, by taking cross-bearings from two such stations, get an idea of the whereabouts of the U-boats, but they believed that such 'fixes' were not accurate and accepted what they thought to be the slight risk of detection. In fact the Allies were to produce in 1942 a ship-borne direction finder called the High Frequency Direction Finder (HF/DF or 'Huff-Duff' for short), and fit these to a proportion of their escorts. The result was that a U-boat signalling near a convoy could be 'fixed' with great accuracy and hunted. This relatively simple device was to cause the destruction of numerous U-boats in the following years.

When this was discussed with Kapitän Hans Meckel, the staff officer in charge of U-boat signals, he stated that one of the German Naval Intelligence branches knew all about Huff-Duff early in its existence and sent in a report warning U-boat Headquarters of its danger. Unfortunately for the Germans, this warning was contained in a routine report, its importance was not appreciated by the officer reading the report and, as the intelligence branch never bothered to follow up its warning, no action was ever taken on it.*

And so the Allied escorts started to sink U-boats in appreciable numbers for the first time. The successes were almost always around the convoys, the previous fruitless, offensive patrolling having ceased. But with each tactical hurdle that was cleared, another obstacle showed itself. To detect by radar a U-boat attacking a convoy at night was not

---

* The matter was also raised with a former U-boat commander and it was suggested that he was required to use his wireless too often. He replied that he, and other captains that he knew, often refused to send weather reports when requested, but if the request was repeated for three consecutive days then they sent the signal on the third day because they knew that three days' silence would result in their relatives being informed that their boats were presumed to have been sunk.

too difficult and the swift approach of an escort was usually sufficient to force the U-boat either to make off at high speed or to submerge. If the U-boat submerged it immediately came under the searching pulses of the escort's Asdic and was attacked with depth charges. But the depth charge was a very imprecise weapon. To sink a U-boat the charge had to explode within twenty feet of the U-boat's pressure hull, and to achieve this with the U-boat able to move in three dimensions was a difficult art made even harder as the U-boat captains became more skilled in evasion and the U-boats themselves were built more strongly and capable of diving to greater depths. Once a U-boat had been 'put down' the ideal solution was to leave two escorts over it to make a long series of careful attacks or, better still, wait for the 36–48 hours necessary for the U-boat's air supply to run out. The U-boat was then an easy kill when it surfaced.

This was fine in theory but was rarely possible for the escorts of a convoy under attack by a U-boat pack. To leave even one escort for more than an hour over a known U-boat left a gap in the convoy screen which other U-boats might penetrate. This was the great dilemma always facing the commanders of escort groups – whether to leave escorts behind to sink U-boats but risk the convoy or whether to be content with forcing the U-boat down, dropping two or three quick patterns of depth charges in the hopes of a lucky kill, and then back to close the ring of escorts around the convoy.

The answer lay in the provision of even more anti-submarine vessels, preferably in the form of Support Groups, sometimes called 'Hunter-Killer' Groups. There was now no argument that the best way for surface vessels to fight U-boats was to have two kinds of escort group: the close escort which remained permanently with one convoy, with other groups of vessels with greater endurance which could remain in the main danger area and reinforce those convoys directly threatened by U-boats. These Support Groups were again fine in theory but the Royal Navy was desperately short of just this type of escort vessel. Some naval officers believe that the necessary ships should have been found to destroy the U-boat arm in 1941 and 1942 while it was still comparatively small, even if at the expense of other theatres. They are convinced that this was the Navy's greatest mistake in the Battle of the Atlantic. This is one of those fine points of history which will long be debated, but the result was that, although U-boats were being sunk in increasing numbers, many more were only shaken up by short, hasty depth-charge attacks and they were able to return to their bases to provide the nucleus of experienced officers and senior ratings for the increasing number of U-boats now being built.

But despite their apparent successes the Germans had already made one vital mistake. During the first year after the war had started, the U-boat construction programme had been almost halted; only thirty-seven new boats were commissioned in those first twelve months, despite earlier plans to build over 100 boats in that time. Hitler had hoped for a quick victory on the Continent and that Britain would agree terms for peace when she found that she stood alone in Europe. It was 1941 before the building programme was really set going again, but a valuable chance had been lost. Although 230 new boats were being built in April of that year, only thirty-two boats were available for operations!

If Hitler had not miscalculated the British mood and had allowed the U-boat building programme to proceed in 1940, Dönitz could possibly have won the Battle of the Atlantic in 1941 or 1942. Instead, Dönitz had to husband his slender strength of boats and constantly shy away from the defended areas, sending his boats further and further afield for easy pickings. Dönitz told me that he considered Hitler's decision to suspend most of the building in 1940 as one of the two decisive setbacks that his U-boats suffered. The Allies were thus given a breathing space in which to build up their strength of naval and air escorts.

And so 1941 ended, almost in a stalemate. The U-boats had sunk 432 ships during the year, a decrease on their 1940 successes, although the capacity sunk remained the same at over 2,000,000 tons. In turn, thirty-five U-boats had been sunk, twenty-seven of them by escort vessels of the Royal Navy.

The story of 1942 can be told more quickly. The year was to see few changes in tactics, and the two main factors influencing the battle for merchant shipping were to be the entry into the war of the United States and a big increase in U-boat strength..

America's participation in the war would eventually bring immense reinforcement to the Allied cause and open the way to final victory but, initially, the effect on the sea war was catastrophic. The Japanese attack on Pearl Harbor surprised the Germans, but they declared war on the United States a few days later without any apparent thought of the consequences. Dönitz was quick to see the opportunity presented by the vast amount of unconvoyed shipping that sailed along the American Atlantic coast and sent six of his few long-range boats to this area in January 1942. The U-boat captains were delighted. The American coastal cities had no blackout; buoys and beacons were still lit; there appeared to be no effective anti-submarine patrols; ships were not sailing in convoy. The second Happy Time had begun.

The next six months were to be disastrous ones. Although the American coast was 3,500 sea miles from the Biscay bases, the Germans were able to assemble and maintain there a force averaging eight U-boats, partly by the use of tanker U-boats. The way the American authorities reacted seems, in retrospect, to be almost criminal. Despite the fact that there had been an American Naval Mission at the Admiralty since August 1940 and that the members of the Mission had studied every development of Britain's U-boat war, the United States Navy seemed determined not to follow British methods. No attempt was made to form convoys, and the few available anti-submarine vessels were sent on the offensive sweeps that the British had found such a waste of time. It was not until April that the civilian authorities imposed a coastal blackout. Not a single U-boat was sunk for three months.

The British were not well pleased by the American failure to profit from the hard-earned British experience. Many of the ships being sunk off the American coast were British ships that had been laboriously escorted by the British and Canadians across the Atlantic only to be torpedoed while sailing as independents in American-controlled waters.

The U-boat men's Happy Time ran to the end of May 1942. There are no separate figures for sinkings off the American coast, but in these five months 362 Allied merchant ships were lost to U-boats and already the tonnage sunk exceeded the U-boat successes for the whole of 1941. The entire effort of the United States Navy off its own coast in these five months had resulted in the sinking of just one U-boat, while a Coast Guard cutter had sunk a second. In fairness to the Americans it must be stated that they were being desperately pressed by the Japanese in the Pacific at the same time, but it was still an unimpressive showing.

In the end the United States Navy abandoned its sweeps, gathered together the available anti-submarine vessels, including some loaned by the British, and set up a convoy system. Immediately, the sinkings slackened off and more U-boats were sunk. Dönitz, reacting as usual to a firm defence, immediately moved his boats on to easier waters – to the Caribbean, to the coast of Brazil, to the South Atlantic.

This ever-widening search by the U-boats for easy hunting grounds obviously could not continue indefinitely. Furthermore, Dönitz's new boats were now coming into operation in large numbers and he felt that at last he could afford to make an all-out effort on the main convoy routes across the North Atlantic and this time accept the casualties that the confrontation would entail. Gradually the main U-boat strength shifted back to the north.

There was every sign that the decisive stage was approaching. By October 1942 Dönitz had no less than 196 front-line boats, compared with under 100 at the beginning of the year. The German shipyards were turning out twenty new boats every month, while U-boat losses from all causes averaged only eleven per month. For the first time in the war Dönitz had enough boats to mount a huge offensive wherever he wished, and could still spare boats to make a showing in other areas and keep the Allied escorts at full stretch. The long winter nights which suited his pack tactics so well were approaching once again. There was still a large gap in the North Atlantic not covered by air patrols. Here would be the winter battleground.

The new phase opened in August with a convoy battle so fierce that the crews of three merchant ships took to their lifeboats even though their ships had not been torpedoed. Two crews did go back but the third refused, a most unusual attitude by the normally steadfast British merchant seamen. Their ship had to be left and a U-boat finished it off later.

All through the autumn and into the winter the battles continued. In the last six months of 1942 the U-boats sank another 575 ships in all areas, totalling 3,000,000 tons. This brought the total sinkings by U-boats for 1942 to a staggering 1,160 ships of over 6,000,000 tons weight, and a further 1,500,000 tons had been sunk by other German forces. This was more than had been sunk in the years 1939, 1940 and 1941 combined. The Allies had now lost 14,000,000 tons of merchant shipping through all causes and had replaced less than half of it with new construction. They were in danger of losing the battle for merchant shipping.

It was not, however, a one-sided battle. The Germans lost sixty-six U-boats in the last six months of 1942, and one significant factor emerges when these losses are analysed. For the first time, the U-boat sinkings by aircraft exceeded those by surface escort vessels. A handful of what were called Very Long Range Liberators were now operating from Iceland and Northern Ireland and, although the Air Gap still remained, it had become significantly narrower.

We are now, in the New Year of 1943, only ten weeks from the battle for convoys SC.122 and HX.229. It is a convenient time to leave this description of tactics and tonnages and to look at the men engaged in this deadly game.

2

The Convoy Men

In each North Atlantic convoy battle there were four major elements involved – the merchant ships that made up the convoy, the naval vessels that escorted the convoy, the aircraft that were sometimes able to give it added protection, and, finally, the German U-boats opposing the convoy's passage. The strength and character of all these changed considerably as the war progressed; this chapter and the next will describe them as they were constituted in those critical months early in 1943. Ships, aircraft, weapons and equipment are all relevant to this story and some space will be devoted to them, but many excellent books on these subjects have been published since the war and this book has little new to offer in these directions. The main emphasis will be on the men who found themselves fighting the convoy battles.

Of the four participants in each engagement, the merchant ship element formed the very essence of each convoy and had the greater number of ships and men involved at all stages. The merchant sailors will take pride of place in this chapter.

The pre-war British merchant shipping fleet was the largest in the world. It contained almost one third of all the world's merchant ships, and almost exceeded the combined total of its next three rivals – the United States, Japan and Norway – although the British fleet had been gradually declining in size for some years while those of her rivals were growing rapidly.* Britain had been a seafaring nation for several

* Merchant fleets exceeding 1 million gross tons in 1939:

	No. of ships	Gross tonnage
British Empire	2,965	17,524,000
U.S.A.	1,409	8,506,000
Japan	1,054	5,030,000
Norway	816	4,209,000
Germany	713	3,762,000

(continued overleaf)

centuries, and her merchant fleet had been a major factor in
the great wealth and power of that country in Victorian and
Edwardian times. But the British fleet had been hit hard in the
trade depression of the 1930s, and the ships on which the
country was to depend for survival in 1939 were, on average,
ageing and dilapidated.

If the ships were a motley collection, so too in many ways
were their crews. The officers were usually men of good
education who went to sea following a family tradition. Their
pay, in the pre-war years, was abysmally low in proportion to
the skills and qualifications these men required. There was
never, however, a shortage of young men anxious to become
apprentices or cadets. The apprentices' indentures were strict,
sometimes giving a small pittance, sometimes even demanding a premium by the boy's parents. The indentures of one
company even laid down that the boy should not 'frequent ale
houses or taverns'.*

The rest of the men fell roughly into three categories –
seamen, engine-room men and the catering staff. Many of the
United Kingdom men naturally came from the great ports –
London, Liverpool, Cardiff, Glasgow, Hull, Southampton.
One engineering officer describes his men, from Liverpool, as
'the *crème de la crème*, the last of, or descendants of, the old
Liverpool firemen – hard cases and hard workers'. As well as
the large ports, several smaller communities also had strong
seafaring traditions. One captain declared that the best seamen he had ever had were men from the Orkneys and Shetlands, although they sometimes picked their voyages to fit in
with the harvest time or the big fishing season at home.

Those shipping companies that sailed regularly to East

	No. of ships	Gross tonnage
Italy	571	3,107,000
France	502	2,639,000
Holland	477	2,651,000
Greece	389	1,663,000
Russia	299	1,048,000
Sweden	259	1,040,000

These figures include all ocean-going vessels above 1,000 tons. More than
90 per cent of the British tonnage was registered in the United Kingdom.
Information taken from C. B. A. Behrens, *Merchant Shipping and the Demands
of War*, H.M.S.O. and Longmans, Green, 1955, p. 23. For convenience, this
publication will be referred to hereafter as the Merchant Navy official history.
   * Alfred Holt & Company's indentures, 1941.

Africa, India or the Far East frequently crewed their vessels with men from those areas, the British Government approving a special type of ship's articles which allowed such employment as long as the round voyage for that crew ended in the men's own country. These men were the Lascars of which there were usually 50,000 from a total at any one time of 190,000 seamen in British ships. Three quarters of the Lascars were Indians, and it was considered that the best seamen came from Bombay and the waterways of the Chittagong area, that the Sikhs from the Punjab were the traditional strong men in the stokeholds of the coal-burning ships, and that the best stewards came from the Portuguese colony of Goa.

A smaller number of Lascars were Chinese from Hong Kong or from China itself, although they were rarely known by the official title of Lascar. The Chinese were invariably clean and efficient, but the big menace was the world-wide organization of their syndicates or 'tongs' with their excellent communications and intelligence. They often engaged in opium smuggling and this caused their British officers much trouble.

The Africans served mainly in the engine-rooms, their fine physique and their upbringing in a tropical climate making light work of stokehold jobs. Many could speak little English and had unpronounceable tribal names. The chief engineer often 'christened' them with his own choice of name based on a whim or on the man's character – Micky Finn, Jawbone, Ben Coffee, John Trybest, Tom Everyday – and this name often stuck and appeared in later ship's articles. It was these African merchant seamen, together with others from the West Indies, who had earlier started settling in some of the English ports and produced England's first coloured communities in places like Tiger Bay, Cardiff, and Liverpool's Toxteth.

The pre-war merchant fleet may have brought prosperity to Britain and to the shipowners and shareholders, but the merchant seamen had seen little of the benefit. There was no guaranteed employment; a man signed articles for the one round voyage and was not automatically re-employed for the next, although the better-class companies kept what was known as a 'following' of carefully chosen, reliable men that it would employ more or less permanently. One method of

getting regular work on the poorer-class ships was to decline
leave at the end of a voyage and sign new articles at once.
There was thus a small floating population, usually very poor
Arab or African, which never came ashore.

In those bad years of the depression, the seaman who was
not fortunate enough to be with a good company had to wait
for a vessel to secure a cargo and then take pot-luck with
many others for the few jobs available.

You had to put your fiver in your discharge book when you
showed it to an officer. If you got the job you were provided with
a bunk and a straw-filled mattress – we called it the 'donkey's
breakfast' – but you had to provide the rest of the bedding and your
cutlery. I was a cabin boy and often worked fourteen hours a day
and Sunday was like a Monday, in fact Sunday was worse because it
was Captain's inspection day, 9.0 a.m. sharp. If everything was not
in order you were 'logged' 5 shillings loss of pay. Then there were
the 'field days' that we hated, when all hands were ordered out to
work extra to their normal watch, usually chipping and painting. It
was very rarely that there was any overtime pay and if you complained you were never employed by that company again. (Cook S.
Banda, S.S. *Zouave*)

The food in many of these ships was atrocious. The
Merchant Shipping Act laid down a minimum scale of fresh
food for each man, but the Act also allowed substitution of
salted or tinned meat 'not to be used without reasonable
cause', and most pre-war merchant sailors became well
acquainted with salt 'junk' (beef) or salt pork while working
for what were commonly called 'Starvation Lines'.

The coming of war made very little difference to the
composition of the merchant ship crews. But the increased
demand for shipping brought back into service many ships
laid up during the depression and delayed the retirement of
others that were due for the scrapyard. All merchant ships
were taken under the control of the Ministry of Shipping and
the Government insurance scheme was not too choosy about
the quality of the vessels employed; virtually anything that
could carry cargo was welcome. The result was full employment for the merchant seamen and good profits for the
owners. The war did bring one new type of man into the
merchant service – the younger man who preferred to do his
war service in the Merchant Navy rather than in the fighting
services. The reasons for this choice could be anything from a
desire to see the world to the attraction of higher pay and less

discipline than in the services. Some of the Lascars stopped coming to sea but most continued as before, although with one big difference. Before the war they had always served on shipping routes in warm or temperate climates. Now, as a ship was likely to be sent anywhere by Government order, the Lascars often found themselves in cold and stormy waters. They suffered severely from this change and did not like the new conditions but, as the Merchant Navy official history somewhat cold-bloodedly comments, 'they were always plentiful, almost invariably docile and they gave no trouble'.*

Although the Merchant Navy started suffering heavy losses in ships as soon as the war started, a proportion of the crews were usually saved. These survivors, in total, were evenly balanced by the new ships being built. Thus, although the overall numbers of ships and seamen declined for several years, there was neither unemployment nor the need to conscript new men, and throughout the war there was always a high proportion of pre-war seamen. It will be seen later that this was an important factor.

In two departments of the merchant ship the war did bring changes of personnel. Each ship now had to have three trained radio officers, whereas before the war smaller ships had usually managed with one. Now the radio had to be manned at all times, and a large influx of hastily trained operators joined for what must have been the most boring job at sea – keeping two four-hour radio watches a day but only allowed to send a signal if attacked by surface vessel or U-boat. The other new faces on board the merchant ship were the gunners. The Admiralty had made preparations to arm all merchant ships even before the war started and, by 1943, every ship had at least one large gun and a variety of smaller weapons. The first gunners had been provided by the Royal Navy and were mainly Reservists, but as the war progressed these were supplemented by younger 'Hostilities Only' naval ratings. Even these could not be supplied in sufficient numbers and the Admiralty soon appealed to the War Office for help.

* Behrens, op. cit., p. 157n. The worst incident for the Lascars occurred on 8 December when the ex-Italian cargo liner *Calabria*, under the management of British India, was torpedoed by U.103 (Korvettenkapitän Victor Schütze, fifth in the list of ace U-boat commanders) 350 miles west of Ireland while bringing spare Indian crews to England. 450 Indians and twenty-two English crew members lost their lives. This tragedy is not mentioned in either the Royal Navy or Merchant Navy histories.

This resulted in the formation of an unusual unit, the Maritime Regiment of the Royal Artillery. By 1944, 24,000 naval gunners and 14,000 army gunners were serving on merchant ships.

The war brought more secure employment. The merchant sailors registered at their local Merchant Navy Pool office which found them a ship and even paid them stand-by money when not at sea. A man could refuse the first two ships offered by the Pool, although he would think carefully about the second because he had to accept the third and this might be a very poor prospect. The same rules operated for officers. Many men liked this haphazard way of getting a ship; it gave them a certain freedom of choice denied to the man who worked ashore, and a man could easily sail on the *Queen Mary* on one voyage and a humble tramp steamer the next. But if he attempted to leave the sea or refused the three ships offered by the Pool he was immediately called up for military service.

So, a man would sign on a ship he had probably never seen, for a voyage that could take a month or a year and with men he had never served with before. His family would have little or no idea as to his whereabouts or when they would see him again. The only route that was really hated was the Arctic run to Russia, with its icy water that would give a man little chance of survival if he was torpedoed and not rescued quickly. The North Atlantic usually meant shorter voyages and a chance to go ashore in Canada or the U.S.A., but the Atlantic's fogs, storms and U-boats meant ever-present tension. The ideal was a long voyage through comparatively safe, warm waters to Australia or New Zealand.

Most of the seamen were probably content with the company and seeming security of a convoy, but the officers usually preferred independent sailing and the reply of this officer, when asked about his attitude to convoy work and the North Atlantic run, was probably typical:

Morale was high and there was invariably a good spirit on board. Attitude to the North Atlantic run? I don't think there was an 'attitude'. You just went out of port, latched on to a convoy and that was that – you just hoped you'd make it back again from wherever you were going. If you've got to go, you've got to go and the sea water you drown in will be S.G. (specific gravity) 1025 wherever you drown – so don't think about it! But the attitude to convoy work? Ah – that's different; we HATED it! The soul-destroying frustration; the physical strain of watch-keeping and station-keeping in convoy was one of the worst aspects of wartime seafaring. (First Officer G. H. Hobson, M.V. *Losada*)

When his ship came back to England, the merchant sailor could draw all his back pay and make a popular return home with the 25-pounds weight of rationed food that the Customs allowed him to bring ashore and the much-sought-after nylon stockings for wife or girlfriend. After home leave equivalent to three days for each month of his last voyage, he went back to the Pool office to sign on a fresh ship and start the whole process again.

Sailing with the Merchant Navy was undoubtedly a dangerous occupation and its men suffered grievous casualties during the war. In this peculiar service which was officially a civilian occupation, approximately one quarter died during the war years, a higher proportion than any of the 'fighting' services, although certain sections within those services sustained higher casualty rates. The chapters covering the passage of Convoys SC.122 and HX.229 will portray this danger and there is no need to labour it here. But there was also a great element of luck; many men never came under attack in six years of war while others might survive three or more sinkings.

The Merchant Navy contained its fair share of heroes – 6,000 awards were made for gallantry – and also of slackers and cowards. Occasionally ships could not sail because crew members who had signed articles failed to appear and the law forbade ships to sail undermanned; statistically, firemen were the worst offenders. But, although there was the National Union of Seamen, there was never any strike action or even threat of such action, at a time when it would have been easy to press for more money and at a time when many unions did call strikes among equally important groups of workers. Nor was there, so far as this writer is aware, any instance of mutiny among British seamen or of crews refusing to sail, however dangerous the conditions.

Although Britain had not treated him particularly well during peacetime, the merchant seaman served his country well in the years of war.

The pre-war experiences of the crews of the American merchant ships had been very similar to those of the British – poor conditions, bad food, competition with cheap foreign labour and uncertain employment with unofficial 'payments' to union dispatchers and ships' officers in order to obtain jobs. Most American merchant seamen came from the major ports – New York and Baltimore for the big passenger and cargo companies trading overseas, Philadelphia for tankers, Boston for coastal ships, New Orleans and Mobile from where the Lykes Brothers and the Waterman ships sailed, in which it is reputed that a man was only employed if he had a southern accent. The West Coast was served by San Francisco, regarded by one man as 'one of the most beautiful ports in the world in the 1930s'. This port had its own union, Harry Lundberg's Seamen's International Union. One of the most popular companies here was Matsons with their service to Hawaii.

The war in Europe brought a big increase in activity and every ship was soon busy. It is said that some of the older seamen went ashore to wait out the war but there were plenty of younger men willing to take their places. The main reason was not support for the Allies, although most were fairly sympathetic, but the higher wages being paid and the bonus for sailing in the war zone. When Russia came into the war in June 1941 both seamen's unions became enthusiastic for the Allied cause and it was only a few months later that America herself was in.

One of the major success stories of the war was that of the American shipbuilding industry, in the production both of warships and of merchant ships. America had built sixty ships, the Ocean-type cargo ships, for Britain before Pearl Harbor, and this was followed by a truly phenomenal effort much of which was devoted to the famous Liberty ships. These were cargo ships of a simple design, originally a British design by J. L. Thompson and Sons of Sunderland but modified by the Americans to allow for a diesel engine and to simplify the construction still further. The finished product was a sturdy but inelegant, 7,200-ton, 11-knot ship designed for mass production. Roosevelt called them 'dreadful looking objects' and their title of 'the Liberty Fleet' was a belated but successful attempt to capture the public imagination.

New shipyards were soon springing up and producing the Libertys all round the United States. They were not ship-yards in the traditional sense but were really assembly plants where the 30,000 prefabricated parts of each ship were pieced together. The great breakthrough which saved so much time on traditional shipbuilding methods was the introduction of welded plates instead of riveting. One special effort by a yard at Richmond, California, in November 1942 resulted in the Liberty ship *Robert E. Peary* being launched only four days and fifteen and a half hours after its keel had been laid. Of the American businessmen who pioneered the building of the Libertys the most famous was undoubtedly Henry J. Kaiser, a civil construction man with no previous shipbuilding experience. Kaiser is reputed to have once placed a single advertisement for 20,000 additional workers for his yards and to have known so little of ships that he referred to the 'front' and the 'back' of the ships he built.

In all, 2,710 Liberty ships were built during the war. Several of the early ones were lost when their welded seams parted in rough weather, and they were always regarded as expendable with no more than a five-year working life. But the Libertys and other American-built ships eventually more than replaced the Allies' losses of merchant shipping. More than 200 Libertys were sunk during the war.*

It was one thing to build ships but another to provide the crews. This massive expansion of the merchant fleet caused many problems which the Americans were never fully able to solve. At the peak of the wartime shipbuilding boom, six new ships were coming into service every day and requiring crews. The Government attempted to solve the problem by setting up three huge training camps to produce merchant seamen – on the East Coast at Sheepshead Bay, Brooklyn, on the Great Lakes at Waukegan, Illinois, and on the West Coast at Treasure Island, San Francisco. These 'boot camps' were run on military lines, the recruits wore uniforms and men were produced in large numbers trained in the basic skills of the merchant seaman. The unions would not allow these men to be employed until every one of their own men had found employment so it was not until early 1943 that the Government-trained men started to go to sea. There were also Government training schools for officers where a man with just one year's sea service could attend an officer's course that was eventually shortened to only four months, and a man could become a master after only two years as a deck

* I acknowledge the use of *The Liberty Ships* by L. A. Sawyer and W. H. Mitchell, David & Charles, 1970, for much of the above information.

officer. By these mass production methods the officers and the crews as well as the ships were produced.

The life and attitudes of the average American wartime merchant sailor were in some ways similar to those of the British but differed in others. He too was allowed to refuse the first two ships offered at his Union Hall but had to take the third, and he was exempt from the armed services draft only as long as he remained a merchant sailor. He was paid roughly double the wages of his British counterpart; an able seaman earned 90 dollars a month plus overtime, plus a war-area bonus of 100 per cent for the North Atlantic and the Mediterranean, 80 per cent for the South-West Pacific and the Indian Ocean and 40 per cent for other areas. Each man also had his life insured by the Government for 10,000 dollars! The American liked to get ashore in English ports and was much impressed by the bomb damage there and by the wartime living conditions of the civilians, but often felt that he was being overcharged, particularly in London. His favourite ports were elsewhere, 'I know that all U.S. seamen raved over their receptions in Australia and Tasmania and just slightly below in popularity was South Africa.'

Every American merchant ship had a large contingent of U.S. Navy gunners called the Armed Guard and commanded by a '90-day wonder' ensign. The Armed Guard men were rarely popular in the merchant ships. The ensigns were regarded as 'pompous, arrogant and secretive as a result of their Navy training', and the naval gunners 'were mainly young farmboys with unbelievable appetites; their eating was continuous throughout the day'. Several merchant seamen commented on their admiration for the relaxed and efficient way the Royal Navy escorts conducted themselves, and the long British experience of convoy work was much relied upon.

Unfortunately the British merchant seaman had a poor opinion of his American counterpart. There was much jealousy over the superior American pay, which left the Briton at a big disadvantage in dockland drinking houses all over the world, and the obvious inexperience of some of the American crews was derided. Several legends still persist among the British: the Americans were always anxious to abandon a ship that had been hit but was not actually sinking in order to collect a huge survivor's bonus; American officers could gain their tickets by correspondence course; there were special rest camps where men suffering from 'convoy fatigue' could recuperate. In fact there was no 'survivor's bonus'; the Americans only received compensation for loss of belongings in exactly the same way as did the British seamen. An American could study by post for an officer's ticket but he still had to put in the necessary sea time. There was, however, at least one small rest camp, provided by President Roosevelt on his estate in one of the New England states, but it was only of small capacity and the average American merchant sailor never went to or even heard of such a rest camp.

The British would have done well to remember that had the Americans not mass-produced their merchant ships and sailed them with their raw and admittedly sometimes panicky crews, Britain might have starved.

By 1943 the British and American merchant ships made up the greater part of any convoy, but there were usually ships of two further groups sailing in the convoys – those of the other Allies and a few from neutral countries.

When Germany cleaned up half of Europe in its Blitzkriegs of 1940 and 1941, a large proportion of the merchant ships of the countries that were occupied were at sea and escaped capture. It is interesting to see what became of these ships when the head offices of their shipowners and the crews' families came under German control. The masters of these ships were in an appalling dilemma, and the first to face an unpleasant choice were the Danes and Norwegians in April 1940. The B.B.C. immediately put out a series of broadcasts offering protection to these ships and payment for their services. At the same time, rival appeals purporting to come from the shipowners, but in reality broadcast by the Germans, urged the masters to return to their home ports. The German broadcasts all began with the words 'My Dear Master' – a form of address that shipowners would not have used – and the masters were eventually persuaded to sail to an Allied-controlled port mainly through the broadcast appeals of a well-respected Norwegian shipowner, Mr Hysing Olsen, who happened to be in England at the time. The same procedure was successfully adopted when Holland and Belgium were attacked a month later, and not one ship of any of these countries went back to the Germans. In 1941 a fresh supply of shipping dropped into Britain's lap when Greece and Yugo-slavia were invaded, the large Greek fleet making a particularly valuable contribution.

The crews of those ships which went to work for Britain were not to see their homelands or families for the next five years. Instead of changing ships every voyage they usually stayed with their ships permanently, the ships becoming their only homes. There were always many welfare agencies in ports to care for them ashore, and they were shown great kindness by many families in Britain and America, but this could never fully compensate for their long exile and the anxiety about the fate of their families under the Germans.

We were approaching Rio de Janeiro when we heard the news. I remember that day as a beautiful, sunny one, overshadowed by the news we had just heard. Cut off from country and family, ignorant about their fate, unable to help, we felt miserable and desperate. In the long run the officers and elderly seamen took their plight better than the younger ones but the fighting spirit never left anyone. We found in Mr Churchill a shining example to keep faith in the future and the courage to carry on. Nowadays, it gladdens my eyes, when I see in so many a Dutch town or village, Mr Churchill honoured by a street or square named after him. (Second Officer J. Wassenaar, S.S. *Zaanland*)

Another Allied sailor remembers that 'the hospitality and guts of the British people during the war helped us all to overcome this long period'.

It may seem surprising that neutral ships were also risked when their owners sent them to work for the British, particularly after 1940 when Hitler declared a total blockade of Britain and warned that any ship of whatever nationality would be sunk on sight and without warning. The main neutral country to be affected was Sweden, which offered 60 per cent of her cargo fleet outside the Baltic on charter to Britain. Thus, Britain received another 480,000 tons of much-needed shipping, and the Swedes were to be found in convoys throughout the remainder of the war. They accepted convoy discipline and agreed to respect British naval secrets.

Britain's astute actions eventually brought more than 700 merchant ships from the occupied countries of Europe into her service; these, with the neutral Swedes, provided Britain with over 3,000,000 tons of priceless shipping capacity, an increase in her effective fleet of 25 per cent.

One small neutral country which had very little choice

about sailing in convoy was Iceland. Her ships could not go about their business without passing through U-boat-threatened waters, although they were not working for Britain but merely carrying civilian supplies to their own country. At first they did not like the convoy system but when one of their ships, the *Hekla*, was torpedoed by U.564 with the loss of fifteen lives while sailing alone between Iceland and Canada in June 1941, the Icelanders felt safer in convoy. They were to lose at least two more ships, the *Godafoss* and the *Dettifoss*, before the war was out, and to see plenty of ships from other countries torpedoed. These men, from a country with a long tradition of neutrality, were shocked by the seeming brutality of the U-boat war and eventually became strongly pro-Ally.

It is time to leave the merchant ships of these many countries and turn to the faithful naval vessels that escorted the convoys back and forth across the Atlantic. A look will be taken at the ships, at the men who sailed in them and the routine of their life, and at the commanders ashore who directed both convoys and escorts.

Earlier in the war there had been the threat not only of German U-boats but also of surface raiders, and a North Atlantic convoy might have had any type of escort from a trawler to a battleship, but by the period in which we are interested, the Royal Navy had largely eliminated the threat of surface ships and it was now only against U-boats that the convoys needed protection. But the Allied navies were still desperately short of small warships; the almost world-wide network of convoys demanded hundreds and hundreds of escorts. Although the North Atlantic was possibly the decisive theatre of the naval war it still had to manage with those ships that could be spared for that theatre, and in early 1943 these were never enough. The task of convoy protection had fallen mainly on two classes of warship – the destroyer and the corvette – with a little assistance from a handful of other types. The main strain was still being taken by the Royal Navy with growing help from the Canadians, and there was a small force of American ships.

The Royal Navy possessed many fine classes of modern destroyers but its main allocation for convoy work were the oldest and least effective destroyers it had. These were the

vessels of the V and W Class and of the Town Class. The
1,000-ton Vs and Ws had been produced in large numbers
during and just after the 1914–18 War and, in their original
form, could not even cross the Atlantic without going into
Iceland to refuel. Since then, twenty-two of the Vs and Ws
had been drastically altered; the main change was that one of
the three boilers had been removed in each ship and replaced
with an extra fuel tank. This left the destroyer with less speed
and necessitated the removal of the forward funnel but it
increased the range and these ships could now cross the
Atlantic without refuelling.

The other main escort destroyers, the Town Class ships,
were none other than the old ex-American vessels given to
the Royal Navy in 1940 in exchange for bases. These had
been taken over by British crews and brought across the
Atlantic, and some effort had been made to fit them out for
modern anti-submarine warfare. They had all been renamed
after towns common to both the United States and the
British Empire and many were now back in the Atlantic as
convoy escorts. These Town Class destroyers were usually
known as 'four-stackers' after their distinctive four funnels.
The Germans knew them too and they were probably the
most easily indentifiable ship in the Atlantic.

Their crews hated them. They were narrow ships which
rolled viciously; their propellor shafts stuck out several feet
beyond the stern and they were very difficult to handle. They
had a huge turning circle which was not of much help when
attacking a U-boat. They were certainly unsuited to Atlantic
weather: at least two had their bridges smashed by heavy
seas, and in one of these the captain and several others were
crushed and killed. This ship then broached to and lost most
of its superstructure. The captain of one four-stacker says
that 'it seemed to me that all my previous sea experience was
just a preparation for my time in H.M.S. *Mansfield*', and that
the fate which befell H.M.S. *Campbeltown* which was packed
with explosive and blown up on the lock gates at St Nazaire
was 'the best thing that could have happened to the Town
Class destroyers'. These sentiments would find much approval
among the sailors who had to weather so many Atlantic
storms in these old ships.

By contrast, the only modern destroyers permanently
allocated by the Royal Navy to the North Atlantic were really

fine ships. These were the six ships of the Havant Class that had been built in England for the Brazilian Navy but were taken over by Britain when war broke out. They were fast, had advanced depth-charge armament and good fuel endurance and, at 1,340 tons, were a little more stable than the Vs and Ws and the four-stackers. Added to this, the Brazilians had specified a higher standard of interior comfort than was customary for Royal Navy ships.*

At this time there were roughly thirty destroyers of these three classes available for North Atlantic escort work, but the main escort strength was still being provided by a remarkable little ship – the corvette. This 925-ton ship was based on a whale-catcher design and had been ordered by the Royal Navy just before the war as a coastal patrol vessel. When the U-boat war had spread to the Atlantic these corvettes, being easy to build, were produced in large numbers to provide convoy escorts until something more suitable could be found. The Royal Navy corvettes were all named after native English flowers, a little incongruously considering the violent lives they led; Canadian corvettes were named after towns in their country.

In many ways the corvettes made useful anti-submarine ships. They were very seaworthy, carried sufficient fuel to make the through crossing, and once a U-boat had been put down the corvette could carry out a perfectly adequate attack on it. The corvette's main disadvantage was its lack of speed; its 16 knots was not sufficient to catch a U-boat running on the surface, and if it remained for long over a submerged U-boat or on survivor rescue work it could take hours to catch up the convoy it was supposed to be escorting.

The winter of 1942–3 saw some seventy corvettes operating in the North Atlantic; they were already becoming obsolete to the specialist and more effective frigates and sloops that were appearing, but the corvettes were still badly needed and much will be seen of them later in this book.

The basic escort organization was that of the group. In theory this was a balanced collection of anti-submarine ships which remained permanently together under one commander

---

* Three of the Havants did not survive the war. H.M.S. *Havant* was sunk by a German bomber off Dunkirk in 1940; H.M.S. *Harvester* was torpedoed by U.432 in the Atlantic in March 1943; and H.M.S. *Hurricane* was torpedoed in the Atlantic by U.415 on Christmas Eve 1943.

and was responsible for the protection of one convoy for all or part of its passage. Looking firstly at the mid-ocean groups which had the task of guarding the North Atlantic convoys at that time, it will be found that this heavy burden was mainly carried by twelve groups, seven British numbered B1 to B7 and five Canadian, C1 to C5, although other groups were sometimes available to help out. The paper strength of each group was nine ships – three destroyers and six corvettes. The group commander's ship was usually one of the group's destroyers, in the British groups usually one of the Havants that had been built for the Brazilian Navy.

The British groups all had their main bases on their own side of the Atlantic, three groups each at Liverpool and Londonderry and the remaining group at Greenock. The Canadians were based at St John's, Newfoundland. The normal routine for a British group was to take one convoy across to a position off Newfoundland, go into St John's for a few days rest, and then bring another convoy back to England for a longer period of rest at its main base. The Canadian groups carried out the same process in reverse. The timetable for each group and the convoys it was to escort was worked out for months ahead; one group commander, Commander Peter Gretton of B7 Group, wanting in February 1943 to fix the date of his wedding in May, was told the exact day on which he would be free and he was. But with only twelve regular groups and up to eight convoys at sea at any one time there was little to spare if anything went wrong with the timetable, and it would be this lack of spare time for the hard-pressed North Atlantic groups that would be a vital factor in the convoy battle to be described later.

These mid-ocean groups were not the only escorts that the merchant ships saw. When the U-boats had struck the shipping lanes off the American coast early in 1942, one result had been the formation of the Western Local Escort Force which escorted North Atlantic convoys between the quaintly named WOMP (Western Ocean Meeting Point) off Newfoundland and the terminal ports – Halifax, Sydney (Cape Breton Island) and New York. The Royal Canadian Navy had assumed responsibility for this work. The smaller ships were all Canadians, usually newly commissioned ships being broken in on this now quiet coast before transfer to the more active mid-ocean routes, but there were also a few old destroyers and their crews on loan from the Royal Navy and the United States Navy.

Among the escorts that were working in the more active part of the North Atlantic in early 1943, there was a handful of escorts that were not members of regular groups. These were mostly stationed at Iceland, where they formed a useful reserve either for escorting merchant ships destined for Iceland or as reinforcements to the escorts of convoys which were under heavy U-boat attack. Most of these spare ships were American and were the remnants of the more powerful United States Navy force of escorts that had helped with North Atlantic convoy protection in 1942 but were now being withdrawn.

The best of the American ships still remaining were five United States Coast Guard cutters – *Bibb, Campbell, Duane, Ingham* and *Spencer* – all named after U.S. Treasury Secretaries. The United States Navy had taken over the Coast Guard in 1941 for the wartime years and these cutters made first-class anti-submarine vessels. Their 22 knots was quite sufficient for anti-submarine work; their displacement of 2,200 tons made them twice as heavy as the British destroyers and they were actually the largest warships operating on North Atlantic convoy work. In a five-month period that winter the five Coast Guard cutters sank three U-boats and shared in the sinking of a fourth.* These successes were greater, in proportion, than those of any other type of escort then in the North Atlantic and show what might have been achieved if other more powerful escorts had been available.

This then, was the line-up of naval vessels facing the U-boats in that crucial North Atlantic winter. The Allies were fighting the U-boats with ships that were virtually the cast-offs and surplus from more glamorous or supposedly more important theatres of war, although history would later look back on this phase as the most critical of the sea war. These tired, rust-streaked, battered, usually obsolete ships steamed back and forth across the Atlantic with their convoys in the worst of winter storms, with occasional but fierce brushes with U-boats, and with only a few days rest between voyages. Official statistics show that one escort in every six was under

---

* *Ingham* sank U.626 on 15 December; *Spencer* sank U.225 on 21 February and U.175 on 17 April, and *Campbell* shared the sinking of U.606 on 22 February with the Polish destroyer *Burza.*

repair at any one time, mostly from storm damage. A group rarely managed to sail with its full strength and the American official history records that so many escorts were out of action in early 1943 that the group organization was in danger of breaking down completely.

The officers and men who manned the escort vessels of the Royal Navy's Western Approaches Command must sometimes have felt that they almost formed a separate navy of their own quite apart from the main body of their parent navies. They rarely saw a naval vessel larger than their own destroyers; the battle they fought out in the Atlantic against the elements and the hidden enemy had its own peculiar rules and lifestyle for which service in the remainder of their navy had not prepared them.

Twenty-one escort vessels will be met later in this book providing the mid-ocean escorts for SC.122 and HX.229. There were fifteen British ships, three Americans, two Belgians and one Canadian. Leaving aside the three American ships but including the Belgians which were administered by the Royal Navy, a survey shows that 127 officers sailed in these ships. Of these only twenty-four were regular officers; half of these were commissioned gunners or engineers and four more were either commanders of groups or members of their staffs. This leaves just eight regular Royal Navy deck officers in these eighteen escorts compared with 103 reserve officers. Nine of the reserve officers were Canadians and two were Australians.

This lack of regular officers was unusual, even considering the rapid expansion of the wartime navy. There are at least two reasons for it. The pre-war professional naval officer was essentially a Fleet man and when war broke out he wanted service either in the Home Fleet or with the Mediterranean Fleet. Since 1939 the 'little ship' officers in the two Fleets had seen plenty of action in the Norwegian and French campaigns and in the Mediterranean. It was in these waters that the pre-war officer found decisive action, promotion, decorations and the fulfilment of his long peacetime training. There was little appeal in the escorting of trade convoys in the cold and stormy Atlantic in a twenty-five-year-old destroyer. The first regulars to go to North Atlantic convoy work were often those who could not obtain interesting positions elsewhere.

There were a few exceptions, officers who in peacetime had followed the unfashionable trade of anti-submarine work and would now find the fulfilment of their career. The most notable of these was the legendary Captain Frederick Walker. This brilliant officer commanded two different escort groups which sank thirty U-boats during the time he was in command. Poor Walker died of overstrain before the war finished.

The second reason for the lack of regular officers in the Atlantic was that there were such seemingly good alternatives in the form of the Royal Naval Reserve (R.N.R.) and Royal Naval Volunteer Reserve (R.N.V.R.). The Navy had long relied on the R.N.R. as a reserve of experienced officers. They were merchant service officers who had served a year with the Royal Navy in peacetime and then been granted an R.N.R. commission. They were liable for recall in time of war and were paid an annual retainer of £25, although one admitted that the chief attraction was the prestige of his R.N.R. commission.

The R.N.V.R. was composed mainly of younger civilians with a shore-based occupation but a love of the sea; it contained a high proportion of amateur yachtsmen. There were R.N.V.R. training ships in most of Britain's ports; the standards required were high and a commission in the R.N.V.R. was highly prized. The better educated wartime naval ratings could also apply for commissions in the R.N.V.R. and, of the three services, the Navy was probably the hardest in which to get a wartime commission. The average level of potential ability was, therefore, very high.

A doggerel of that time said that 'R.N.R.s were sailors trying to become gentlemen and R.N.V.R.s were gentlemen trying to become officers'. Although this is somewhat unkind, it probably contained an element of truth.

The Navy sent large numbers of both types of reserve officers into anti-submarine escorts. These ships required less sophisticated training than a ship destined for Fleet work, and the seamanship and navigation of the R.N.R.s were invaluable in the rough weather that the convoys encountered far more often than U-boats. As the escort force grew, the R.N.R.s were given commands, especially of the corvettes and older destroyers. The R.N.V.R.s provided the great mass of junior officers. They had little experience and had to learn many of

their specialist duties – navigation, gunnery, radar and communications – as they went along, but they also found a
tremendous responsibility and self-satisfaction in being
trusted as Watchkeeping Officers. By 1943 a few of the best
R.N.V.R.s were beginning to take command of corvettes.

But the high quality of the R.N.R.s' seamanship and the
enthusiasm of the R.N.V.R.s should not be allowed to cloud
the fact that the Battle of the Atlantic would have been fought
better if the Navy had made available even a few more of its
skilled regular officers in the early years of the war. A second
phase came in 1942 and the number of groups started to
grow. When the first wave of group commanders moved on
to more restful positions there were not enough experienced
regular officers to come up from the escorts and take over.
Many in Western Approaches thought that at this stage the
R.N.R.s should have been appointed but they were not,
partly because the R.N.R.s themselves did not push for command of groups. Instead, with the Battle of the Atlantic
suddenly becoming 'fashionable', a fresh wave of regular
officers was attracted and, as one R.N.R. officer put it, 'starting politicking to get commands in Western Approaches'.
These officers sometimes went straight to the command of an
escort group without any North Atlantic or even anti-
submarine experience.

They were conscientious officers who did their duty to the
best of their ability, but there were few Walkers among them
and the convoys might have fared better in that vital winter
of 1942 if the Admiralty had earlier directed just a few more
of its talented regular officers to the Atlantic.

Whatever the merits of the escort officers, they all depended
on the ordinary sailors who formed the crews of their ships.
The men who served in these small escort vessels were in
most respects a typical cross-section of their wartime navies.
For the British at least there was no question of volunteering
for a particular ship. The manning depots simply produced
the correct proportions of trained men required for each
newly commissioned ship and these men stayed with that ship
indefinitely. Those who found themselves in the North
Atlantic were destined to spend several years of boredom,
fatigue, discomfort and some danger. But most sailors are
'little ship men' at heart; these escort men were young

enough to recover quickly from hardship. They could see
with their own eyes how great a contribution they were making to the war effort, and many tell of the immense satisfaction of bringing a convoy safely across the Atlantic. Once
they had got beyond the seasickness stage many of the
younger men were thrilled with life at sea and, as one said,
'were rather fascinated by the raging seas and howling gales'.
Another says, 'I relished gales and never minded how much
green water, rain or hail came over me. This, I felt, was what
the Atlantic was supposed to turn on and I'd have been
disappointed if it were different.'

Statistically the casualty rate of the escorts was low and
they certainly suffered far less than the merchant sailors. On
the other hand, merchant sailors spent much of their time in
safe areas while the escort sailor was rarely free from the
knowledge that a torpedo could strike his ship at any time,
despite the mathematical odds that it was unlikely to do so.

At the end of each westbound convoy there were a few
days rest in Newfoundland (not then part of Canada), either
at the Canadian Navy base at St John's ('Newfiejohn' to the
sailors) or the American base at Argentia. The latter was
reckoned the best for ships' repairs but was otherwise un-
attractive. St John's did have a little to offer the off-duty
sailor.

St John's had the appearance of an outpost in the Canadian North.
The streets were steep and narrow, large drifts of snow everywhere,
the locals wore snow boots, fur caps etc. In nearly every shop
window was a card advertising that their wares were 'just landed'
or 'just unpacked', frontier phrases that seemed very appropriate.
Many of the buildings which were built of wood had a makeshift,
impermanent air about them. There were few cars and some
sleighs. There was a railway station, American style – no plat-
forms and the train had a snowplough and bell. The cinema was an
all-wood affair – I remember seeing *The Glass Key* with Alan Ladd
there – no smoking of course. There were no pubs as we knew them;
the beer was a light lager. No spirits were sold. To obtain spirits
you first had to obtain a liquor licence and then line up at a Government liquor shop to purchase the spirits. (Signalman A. H.
Dossett, H.M.S. *Havelock*)

These liquor stores would sell nothing less than a bottle and
sailors ashore often bought a bottle each. This could not be
drunk in public, nor could it be brought back aboard ship. So
the sailors finished the equivalent of a bottle of spirits each,

either in a back alley or a hotel bedroom. Another source of drink was the bootlegger. There was an alleged alliance between the bootleggers and the powerful Nonconformist Church, which was officially against all drink but preferred illegal drink to be available by the glass rather than the sale of full bottles. Other relaxations at St John's were a roller-skating rink and long walks along the cliffs overlooking the marvellous natural harbour.

On his return to Britain, the escort sailor docked either at Londonderry, Greenock or Liverpool, and half of each crew could usually get a few days leave. For those who did not, Londonderry offered a good selection of pubs and cinemas, a friendly population and no hint of the religious dissension it would later see. Greenock was not so much used, being the home of only one escort group. Liverpool was easily the most popular. It was reckoned to be 'the heart of Western Approaches' because the headquarters of that command were nearby with all the latest news of events in Western Approaches and, also, a large contingent of attractive Wrens. One of Liverpool's attractions was the new Philharmonic Hall where Dr (later Sir) Malcolm Sargent conducted the Liverpool Philharmonic Orchestra. Members of the forces had free entrance to rehearsals there and could sit on the stage for sixpence.

The escort vessels were based at Gladstone Dock and one attraction here is described, again by Signalman Dossett:

My memories of Gladstone Dock must, of course, be mainly of the Flotilla Club. This was on the quayside for lower-deck personnel of the escorts. A very exclusive club, crew members could go in overalls and a naval cap so it was a favourite spot for those of us a little short of the ready and those on duty who thought they would not be missed for a short while. From the dark of the quayside, with the shaded gangway lights of the destroyers, frigates and corvettes being the only illumination, you stumble through the door and the blackout curtains to find yourself in a long wooden shed, with a refreshment counter at one end and a bar at the other. The canteen was manned by women, the wives of Liverpool businessmen who did this work voluntarily. Everywhere there were sailors, no women except the staff, the noise, the fug, the pilchard sandwiches and, of course, the beer – all made a haven of rest where we could forget our troubles for a short time.

By 1943 a growing proportion of the Atlantic escorts were provided by the Royal Canadian Navy. There was no question here of failing to send enough regular officers or the best

ships to the Atlantic, because in 1939 Canada had virtually no
navy to send anywhere. Its active fleet had then numbered a
mere six ships. Canada had from the first assumed a share in
the Battle of the Atlantic, although it left much of the overall
direction of that battle to the Royal Navy. Canada set out to
build up a large front-line navy. The ships came entirely from
her own shipyards, and the men, except for a few key officers
borrowed from Britain, also came from her own resources,
many being from inland provinces and having never before
seen the sea. By the end of the war Canada had a navy of
90,000 men and 400 ships. It was a fine achievement by a
country with no well-established maritime tradition.

The five Canadian mid-ocean escort groups used the bases
at St John's and in Britain just described. The Canadian-
controlled Western Local Escort Force contained both
Canadian and British ships and the crews of these had the
opportunity to see a series of ports not often visited by the
mid-ocean escort men. Their main base was at Halifax
('Slackers'), Nova Scotia, with its huge harbour and fine sur-
rounding scenery, but the city itself was not popular. The
expansion of the Canadian Navy had placed too great a strain
on the shore facilities and there was too much naval presence
and discipline for complete relaxation.

The American ports were far more desirable. Rationing
was almost non-existent; there was no blackout and the
people provided abundant hospitality for wartime sailors.
The escort base at New York was at Staten Island and repairs
were carried out at the Brooklyn Navy Yard where 'absolutely everything was laid on'. Staten Island was only a
ferryboat trip and a short subway ride from all the delights
of New York. There is not room here to describe adequately
what the city offered to its visitors. One man said 'I could
write a book about the wartime delights of New York',
another that 'I couldn't keep up with the pace in New York
and perform my duties properly', a third said that 'New York
was just plain exhausting'. Although New York gave its
visitors a gay time many men preferred Boston, although it
was not visited so often. Sailors say that 'its hospitality was
abundant even by American standards and more personal than
in New York', and that it had 'a more English character' and
that 'its appearance was delightful in the autumn'.

Turning back for the last time in this survey to the mid-ocean escorts, there are two more smaller groups of sailors who should be described – the Americans and the men from the occupied Allied countries.

The Americans were unlucky in that most of their remaining ships were now based in Iceland, not too bad in summer but a tough place in winter.

During the summer it was delightful. We would even go swimming from the side of the ship and go over on the land and try to find a place flat enough and rock-free enough to have a baseball game or indulge in some other sport. I even recall one night we were over the side painting the ship at midnight. It was really a beautiful sight. I also recall one evening being at anchor in Reykjavik harbor and I decided to try my hand at fishing. I put over a hand line with a little dough that I secured from the galley and much to my amazement I hardly had my hook on the bottom before I got a nibble and pulled in what we call a flounder. You possibly refer to them as sole. As a matter of fact, we caught so many that day we had enough to feed the entire complement and it was a real treat to eat fresh food for a change. (Lieutenant L. D. Hirschler, U.S.S. *Babbitt*)

I remember (wish I could forget) Hvalfjordur – that God-forsaken snow-capped, wind-swept fiord, surrounded by abruptly rising shale mountains and jagged rocks which reached out for our thin-skinned hulls in the everlasting winter gales, particularly in those long, black nights when you weren't sure your anchor was holding in the silt bottom and there was nothing on which to check one's bearings. There were no facilities ashore and there existed very little feeling of security which might afford rest while at anchor. I felt safer at sea. (Lieutenant Commander E. B. Ellsworth, U.S.S. *Upshur*)

The crews of the American ships in the Atlantic were in no way different from those of hundreds of other ships in the massive navy that the Americans were putting into other theatres of war, but the men on the five American Coast Guard cutters were, in one respect, quite unique. These ships still had their original pre-war Coast Guard crews and thus were probably the most experienced seamen in the Battle of the Atlantic. One of their officers says that 'this was the only way we survived the winter of 1942–3 intact'.

Some Americans were loath to admit any British superiority but a Coast Guard officer pays his allies a small tribute.

We did not mind being in a mainly British area of operational control. I think the majority of us felt that the British had something to offer in the years of experience they had already gained in the war. The training sessions on the British naval vessel in Hvalfjordur were invaluable in ASW (anti-submarine warfare) matters. There was no doubt in my mind that everyone fully recognized the vital importance of the North Atlantic area in the early period of the war. During the winter of 1942–3, just being with a couple of convoys was very convincing. (Lieutenant Commander J. D. Craik, U.S.C.G. Cutter *Ingham*)

But another American officer admitted, 'I did not enjoy convoy work in the North Atlantic. It was cold, miserable duty under the very worst possible weather conditions, most of the time with very little feeling of accomplishment for our efforts. I would gladly have accepted duty elsewhere – in fact just about anywhere. But someone had to do it.'

Finally there were the men of the occupied countries who, like some of the merchant sailors, had escaped to fight for the Allies. Escort vessels manned by men of four of these countries – France, Belgium, Norway and Poland – were serving on the North Atlantic convoy routes at this time. It is impossible to go into the background of all these groups, but two Belgian-crewed corvettes will be met later in this book and a little of their story can be told.

After the fall of Belgium in 1940, a large number of Belgian seamen had found their way to England. The merchant sailors continued to serve in their own ships but the remainder had nothing to do. One of these was Victor Billet, formerly the second officer of one of the Ostend–Dover ferry-boats. Billet wanted the Belgian Government-in-exile to form a Free Belgian Navy but initially he received no support; the Belgian leaders in England considered that their country was now out of the war and that Britain too might shortly collapse. So Billet joined the Royal Navy and asked that other Belgians willing to join should be allowed to serve together. This was agreed and so the Section Belge of the Royal Navy was formed.

The first members of the Section Belge served in coastal minesweepers at Harwich but in 1941 the Government-in-exile, now established in Eaton Square (sometimes known as 'Shake Hands Square' because of the continental habit of shaking hands on every meeting), had become more active in the war and it appealed to Belgians of military age all over the world to volunteer for the Free Belgian Forces. In this way the Section Belge grew to a strength of 450 men and the

Royal Navy was asked if the Belgians could have larger ships.
They were allotted two new Flower Class corvettes, H.M.S.
Buttercup and H.M.S. *Godetia*.*
Initially each ship had a British captain and a few British
technical ratings but later they were completely Belgian-
manned. The officers came mainly from the Ostend ferries but
were made up with Scheldt river pilots, officers of the
merchant service, and from the Belgian training ship
Mercator, which was stranded by the war in the Belgian
Congo. The ratings were mostly fishermen from Ostend,
Nieuport, Zeebrugge and La Panne, or seamen from the
Channel ferries or the tug and pilot-boat service, although a
few had been brought in by the world-wide appeal, including
a handful of Belgians from the Foreign Legion. Some of their
officers say that the Foreign Legion men were the worst of
the lot, others that their discipline was superb.
   Buttercup and *Godetia* were to spend three years in the
Atlantic. Some of the Belgians had their families in England;
others were found '*marrains de guerre*', ordinary English
women who were prepared to provide a home for the stranded
Belgians when on leave. Political intrigue at Eaton Square
led to the Section Belge being neglected by the officials there.
One result of this was that no one bothered to see that the
officers of *Buttercup* and *Godetia* were promoted or appointed
to bigger ships. *Godetia* had three officers with Master's
Certificates and the Belgians claimed that there was more sea
experience among the officers of either of their corvettes than
in the average Royal Navy cruiser. Neither of the Belgian
corvette captains was ever promoted beyond the rank of
lieutenant. Poor Victor Billet, who had started it all, never
managed to serve with the Belgian ships; he was killed in
H.M.S. *Dinosaur*, a support ship at the Dieppe Raid in
August 1942.
   Buttercup and *Godetia* performed sturdy service with B5
Group. There were occasional difficulties over language;
both French and Flemish were spoken and an order would
sometimes be issued in English, passed on in French and
again in Flemish. The men, especially the fishermen, made
excellent seamen and these ships were often detailed by the

* The Belgian-crewed *Godetia* was the second corvette of that name; it
replaced another *Godetia* sunk by collision off Northern Ireland on 6 September
1940, the first R.N. corvette to be lost.

group commander to cope with the rescue of survivors or other tasks that posed difficult seamanship problems.

The ships and the seamen of the North Atlantic will be left for a while, and a look taken at the various naval head-quarters that controlled the movements of the merchant ships and those that trained, administered and directed the escort vessels. The overall picture is again one of British involve-ment but with an important American and Canadian interest.

There still exists in London, between Horse Guards Parade and The Mall, an ugly fortress-like concrete building which during the war was the Admiralty's Operational Intel-ligence Centre, known to those who worked there as The Citadel. It was from here that much of the Royal Navy's war-time operations were directed, including the control of convoys.

Captain B. B. Schofield's Admiralty Trade Division looked after all British merchant shipping throughout the world, but its Mercantile Movements Branch under Captain C. M. Leggatt was responsible for day-to-day movements of con-voys and independent ships at sea. The really important place was Room 12, deep underground, for it was here that the Trade Division's Master Plot was kept and the main decisions on convoy operations were made. The officer in charge of the Master Plot was the bluff Commander Richard Hall, son of a famous Director of Naval Intelligence; he had retired from the peace-time Navy to go into business but had come back into the Trade Division on the outbreak of war. Commander Hall's staff watched over all convoys for which the Admiralty had responsibility, and every independently sailed British merchant ship anywhere in the world was also plotted, much of this work being carried out by a team of retired merchant shipmasters. A convoy or an independent could be diverted at any time and for any reason.

Next door to Hall's Master Plot was Room 8, the home of the Submarine Tracking Room, part of the Admiralty's Intel-ligence Division. In charge here was Commander Rodger Winn, in civilian life a barrister and later destined to be a Lord Justice of Appeal. Rodger Winn had suffered polio as a boy and this had prevented him following the naval career he had wanted. He had been brought into the Submarine Tracking Room in 1939 as a civilian and so suited was

his keen, analytical mind to this work that, in 1941, he had
been given an R.N.V.R. commission and put in charge of the
Room.

The story of how information on the whereabouts of
German U-boats reached the Tracking Room is a complicated
and interesting one, but will be best left for a later chapter. It
is sufficient to say here that Commander Winn did receive
regular information of a varying degree of accuracy. The
Submarine Tracking Room's plot was marked up daily with
his estimate of the disposition of the U-boats. No convoy or
independent ship was routed by the Trade Division before the
Tracking Room was consulted and, once at sea, nothing was
diverted from its route against Winn's advice. Commander
Winn and his staff did valuable work in saving merchant ships
from U-boat attack, but unfortunately some of this fine work
was cancelled out by a German radio intelligence unit which,
during certain periods of the war, was able to decipher many
of the diversion signals sent to convoys at sea and based on
Winn's warning information. This German success will also
be described later.

A daily estimate of U-boat dispositions – called 'the
Admiralty Guess' – was sent to all operational headquarters
and broadcast, in code of course, to all naval ships at sea. But,
again, the Germans were able to decode some of these signals
and this daily report came into their possession. This cat-and-
mouse wireless game then took yet another turn when the
Germans redirected the operations of their U-boats accord-
ingly.

The British had operational control over the North
Atlantic convoys only in that area east of the 35 degrees West
line of longitude, roughly two thirds of the way from Iceland
to Newfoundland. This line was known as the CHOP
(Change of Operational Control) Line. Beyond this line convoy control was in early 1943 exercised by the United States
Navy, more specifically by its Convoy and Routing Section,
situated in Washington under Rear Admiral Martin K.
Metcalf. The British did not like this divided control. It
meant that convoys with a majority of their merchant ships,
escorted mainly by British and Canadian escorts, were subject
to control for half of their passage by a navy that had withdrawn almost all its own escorts from the fight in this area.
But Admiral Ernest King, Commander-in-Chief of the United

States Navy, insisted that he should maintain control over the Western Atlantic. The arrangement was particularly disliked by the British on account of the confusion that divided control could bring when convoys were under U-boat attack near the CHOP Line.

The Americans had a Submarine Tracking Room next door to Convoy and Routing at Washington. The American Tracking Room was organized by Captain Kenneth Knowles, a regular officer who had retired with poor health before the war but had been recalled to duty. There was yet a third Submarine Tracking Room at Ottawa. Any U-boat information that became available to the British, the Americans or the Canadians was immediately passed to the other two so that all three Tracking Rooms shared much the same information, although the way this was interpreted depended on the skill of the staffs in each Room.

The routes chosen for Convoys SC.122 and HX.229 by the Convoy and Routing Division in Washington and the subsequent diversions from those routes on the advice of the American and British Submarine Tracking Rooms will form a major part of the story of those convoys.

There was one more headquarters and one more commander whose influence affected the fortunes of the convoys. The headquarters was that of Western Approaches Command of the Royal Navy and its commander at this time was the legendary Admiral Sir Max Horton.

Western Approaches originally had its headquarters at Plymouth, but when London was virtually closed as a major port in 1940 a new headquarters was chosen and prepared in Liverpool and the move was made early in 1941. The Royal Navy set-up was in Derby House and the Merchant Navy part was at the nearby Liver Building. There were many other shore establishments as far afield as Greenock and Milford Haven and also in Northern Ireland; it is said that Western Approaches eventually became responsible for more personnel than had been administered by the entire Admiralty before the war.

But the shore establishments existed only to maintain at sea the escort vessels which were the Navy's front line in the Battle of the Atlantic. Western Approaches' task was, in that time-honoured naval phrase, 'to ensure the safe and timely arrival' of the convoys. The escort vessels of the North

Atlantic and their crews have already been described. Western Approaches trained these ships, developed new equipment and tactics and sent the escorts out over and over again to do battle with the U-boats. Although the protection of North Atlantic convoys was its most important responsibility it also had to provide escorts for convoys to Gibraltar, to West Africa and to Iceland. It had no say over the number of convoys it had to protect; officially it had no say over the routes those convoys took, although when a convoy was attacked or known to be threatened the Admiralty often agreed to diversions suggested by Western Approaches.

From early 1941 until November 1942 Western Approaches had been commanded by the much-respected Admiral Sir Percy Noble who, despite a quiet and fastidious manner, had built much of the basis for Western Approaches' later success, but he was replaced by Sir Max Horton, chosen presumably because he was more forceful than Noble. It should be stated that despite the praise later heaped upon Horton some officers remained convinced that Noble was equally sound, and that he would have achieved in a less rumbustious manner all that Horton achieved, but the majority would say that Western Approaches really started to hum when Horton took over and it is certain that nothing was lost by the change.

Admiral Sir Max Kennedy Horton, known sometimes as 'Max K', was a successful First World War submarine commander and his sinking of the German cruiser *Hela* on 13 September 1914 was actually the first-ever sinking of an enemy ship by a British submarine. Before he came to Western Approaches he had been Flag Officer, Submarines, and his deep knowledge of submarines undoubtedly helped him to counter the U-boats. He was of Jewish extraction and one of his staff described him as 'absolutely ruthless but so astute that he could see through a brick wall'. He certainly thought big, was not overawed by his task in the Atlantic, and is reputed to have turned down the command of the Home Fleet because the strength of that Fleet was so small compared with 1914–18 and because its operations were controlled too closely by the Admiralty. If this is true, it was as well for Britain that he went to Western Approaches instead.

Admiral Noble had been a 'charming and popular C.-in-C. who chose a good staff and left most of the work to them', but that was not Max Horton's style. He lived in at Derby House, did routine work in the morning and played golf at Hoylake most afternoons. He would then have dinner and play bridge until late in the evening before coming down to the Operations Room. His poor staff officers, who had already been working hard, were then driven even harder, especially if a convoy battle was developing that night. If this was the case, Horton might sit up all night in pyjamas and dressing gown helping with the more serious decisions which had to be taken.

Horton's Chief of Staff had fallen foul of his boss and had been relegated to the administration of the shore establishments. The officer in charge of day-to-day operations was the Deputy Chief of Staff, Captain R. W. (Bill) Ravenhill, described by one who worked with him as 'a man who was so intelligent that he would not get far in the Navy'. Ravenhill's staff consisted mainly of regular officers who had retired before the war and been recalled to duty, R.N.R.s and R.N.V.R.s, all well backed up by Wrens. Once again young active-service officers were almost non-existent.

The Wrens had liked Admiral Noble. 'He was always polite and courteous with us, not a bit grand; he knew all of our christian names.' But Max Horton usually only spoke to the senior officer present and, if he addressed the girls at all, he simply bellowed 'Wren!' – 'He treated us as something the dog brought in.' It is unfortunate that there is not space here to tell of the marvellous work that the Wrens did in Western Approaches. The girls who worked there would say that their worst times were when news arrived that one of their escort vessels had been lost in the Atlantic.

It is quite common for fighting men to criticize their commanders and even more so their staff officers. In the course of the research for this book I never met one Western Approaches sailor who had a serious criticism of Max Horton or any aspect of the shore organization.

There remains one last group of Allied participants in an early 1943 convoy battle – the aircraft and airmen who provided such effective protection for the convoys, at least in those areas of ocean that the aircraft could reach. As with the Allied naval forces operating in the Battle of the Atlantic, the air effort was made up of men from many countries operating from widely separated bases; but the greatest strain was being taken by the men and aircraft of the long-range squadrons of 15 Group, R.A.F. Coastal Command, and these will be dealt with first.

Whether a convoy could receive air cover or not depended on how far it was from an air base and what flying range the aircraft at that base possessed. As the U-boats usually operated in or on the edges of the Air Gap, it was the aircraft with the longest range which saw most action. Of all the aircraft operated by the Allied air forces it was a lumbering, ungainly-looking, American-built bomber that was performing this function. This was the four-engined Consolidated B-24, known as the Liberator. Coastal Command had been flying Liberators supplied under the Lend-Lease Agreement since the summer of 1941 but for the first year and a half there had been only enough Liberators to equip just one of the squadrons operating over the Atlantic. This was 120 Squadron, now based partly at Aldergrove in Northern Ireland and partly at Reykjavik in Iceland. For most of this time 120 Squadron's Liberators had been the Mark I and Mark II versions but, latterly, it had started receiving the first Mark IIIs. This version had extra petrol tanks fitted into its bomb bays and it became the V.L.R. (Very Long Range) Liberator with a total fuel capacity of 2,900 gallons, a flying time of up to twenty hours if pushed to the limit and, with an extreme range of around 2,300 miles, it could fly to the middle of the Atlantic and back from either Northern Ireland or Iceland. The intervention of one of these aircraft in a convoy battle was frequently the turning point in that battle.

120 Squadron had flown for over a year before getting its first U-boat 'kill' – U.597 – in October 1942; this long wait was partly due to the policy of flying the 'offensive' patrols and then of attempting to escort every convoy whether it was threatened or not. This policy had recently been changed by Air Marshal Sir John Slessor, the new C.-in-C. of Coastal Command, who ordered that convoys under attack or threatened were to have absolute priority. By March 1943, 120 Squadron had sunk five more U-boats and would sink or share in the sinking of thirteen more before the end of the war. Little wonder that 120 Squadron was called 'conceited' by the rest of 15 Group, but its crews considered that their long experience and accumulated skill justified their pride. In March 1943 a second Liberator squadron, 86 Squadron, was just coming into action. An old 86 Squadron had been cut

to pieces flying Beauforts from Malta in 1942 and disbanded. The squadron was reformed at Thorney Island with V.L.R. Liberators and had just moved to Aldergrove, but had not yet gained its first U-boat kill. It would, however, go on to rival 120 Squadron and would sink almost as many U-boats in a much shorter period.

The next aircraft able to reach a convoy in trouble was another American-built bomber, the Boeing B-17E, the Flying Fortress, officially the Fortress Mark IIA to the British. These were more Lend-Lease aircraft and there were now two squadrons of them, 206 and 220, both based at isolated Benbecula in the Outer Hebrides. The Fortress performed a valuable function, taking over from the Liberators as soon as a convoy had come within range and releasing the Liberators for more distant work. By March 1943, the two Fortress squadrons had sunk seven U-boats.

These American-built aircraft, especially the Fortress, were quite popular with their crews and, for comfort, compared well with British types.

I did like the Fortress. It was double-glazed and soundproofed, didn't let the rain in, was well heated, had good windshield wipers; one could see out of it quite well. It was easy to fly on four, three or two engines. It flew like a bird with artless ease. The whole aircraft was designed, constructed and assembled with great solidity and was completely reassuring. It never frightened me. (Sergeant T. E. Kynnersley, 220 Squadron)

The Liberator did not handle quite as well as the Fortress and one man described it as 'a cold, draughty plane, not a good machine for crashing; it broke up too easily'. But most men were quite satisfied with the Liberator and they rarely had to try out its crashing qualities.

Usually the last aircraft to become involved in a convoy battle was the British-built Sunderland flying boat. Sunderlands had been flying against the U-boats since 1939 but their usefulness was now much hampered by their slow speed and limited range of around 1,300 miles, although they too released the longer range aircraft and often came into action if the U-boats stayed with a convoy long enough. The U-boat crews knew these slow, lumbering giants, calling them '*müde Bienen*' – 'tired bees'. The Germans had become so used to seeing the Sunderland that they often assumed that any aircraft forcing them to dive was a Sunderland and would report it as such even though they might be well beyond the Sunderland's range. U-boat men of the March 1943 era, when asked about Fortresses and Liberators, hardly knew of their existence, but all knew the Sunderland.

The Atlantic Sunderland squadrons were 201, 228 and 423 (Canadian), all stationed at Castle Archdale on Lough Erne in Northern Ireland, with 246 Squadron further back at Bowmore. The Sunderland carried a large crew of ten, and was liked for its huge interior which allowed rest bunks and a galley to be provided, but ground duties such as mooring the aircraft to buoys and putting covers over engines in the middle of winter on an icy lough were not so popular.

The men who flew with the Atlantic squadrons came from the United Kingdom, Canada, Australia and New Zealand. They were all volunteers for flying duties although they had been allocated to Coastal Command without any choice. Their tour with a front-line squadron ended only when 800 hours of operational flying time had been completed, which usually took about a year. Losses were not frequent; promotion was slow; medals were not plentiful. Their biggest enemy was boredom – boredom at their isolated bases, boredom on their long flights over the ocean. Many crews never saw a U-boat throughout their tour and few received the satisfaction of a confirmed kill. In a six-month period from September 1942 to March 1943, 15 Group's aircrews had flown 12,387 hours and attacked only twenty U-boats, sinking eleven. This gave an average of 642 hours flying for each attack and well over 1,000 hours for each kill! This situation was soon to change.

Pilots and navigators were specially chosen for this work: usually they were those who had topped their training courses. The navigators, in particular, had the difficult task of guiding their pilots across the unmarked wastes of the ocean to find a convoy up to 1,000 miles from the last landmark, and then, fifteen hours or so after take-off, they had to make a precise landfall in order to get their crew comrades home again. It is not surprising that they regarded themselves as 'the cream' of the R.A.F.'s navigators. In particular they regarded the navigation of Bomber Command with its bomber stream and Pathfinders as 'a follow-my-leader type of navigation'. One young Sunderland navigator, after his first operational flight, brought his crew back to a good landfall but before landing calmly walked up to the front of the flying boat, opened a door and stepped out. It is believed that he could not face the strain of having to guide his crew home from every flight. A note found on his navigation desk explained that he took full responsibility for his action.

These men lived on some of the most isolated stations of the Royal Air Force. The Northern Irish bases of Aldergrove and Castle Archdale were the most civilized, with the added attraction of unrationed food in neutral Eire a few miles away. Castle Archdale is described as 'always raining although there was good fishing in the Lough', and the Nissen huts inhabited by the men were 'always swarming with rats'. Benbecula was pleasant if you liked shooting, fishing or walking; if not, 'you just lived from leave to leave'. The nearest pub was eighteen miles away. One feature of Benbecula was a constant stream of visiting aircraft, often pleased to have found an airfield in this remote region.

Reykjavik was a big airfield with nearly 4,000 men, more than half of whom were Americans. There was little contact with the Icelanders but the camp had a good social life and it was not as bleak as might be imagined.

It was not a bad place – we got home on leave every three months; the climate was pretty grim, not so much the cold as the wind. The Nissen huts had to be anchored down. Food was rather monotonous, it put me off fried fish and tinned peaches for life. Aircraft coming back from Aldergrove always loaded up with fresh vegetables. We had a very passable Mess, bingo most nights and a Sunday Spectacular once a week – Sam Costa, Denny Dennis and Cyril Stapleton were in the band. Beer was rationed but with non-drinkers and fiddling the ration cards I never went short of a bottle; it was Canadian, strong and with a peculiar taste but we forced ourselves. The whisky and gin were so foul we used to flog it to the Yanks – they would drink anything. (Sergeant B. H. Harvey, 120 Squadron)

For the pilots Iceland was a testing place from which to operate, with plenty of bad-weather flying experience, radio blackouts, big compass variations and snow landings. The airfield had no proper snow-clearing equipment but clinker from a local power station was spread on the runway snow. When the snow melted, the clinker was swept aside into big heaps ready for use the following winter.

Bored though they often were, the men of 15 Group knew they were playing a useful role in the war. They envied the glamorous fighter pilots and admired the courage of Bomber Command. They may not have seen U-boats on every flight, but sinking or burning merchant ships and the survivors in lifeboats they sometimes came across made a deep impression, and their war was certainly fought with some feeling. One 120 Squadron crew sank a U-boat and then machined-gunned the survivors, excusing themselves afterwards on account of the scenes of shipwreck they had seen. The Operational Research Section of the Admiralty calculated that one V.L.R. Liberator on convoy escort, by direct action in attacking U-boats, by forcing the U-boats to lose contact with convoys, or simply by denying the Germans the ability to operate over large areas of ocean, saved at least six merchant ships during its flying lifetime. There is no way of proving this estimate but it does show how the Liberator crews and indeed all the Atlantic squadrons were performing a worthwhile duty.

There were many other aircraft involved in the Battle of the Atlantic. Squadrons of 18 Group patrolled between the Hebrides and Iceland in an attempt to catch new U-boats coming out into the Atlantic from Germany, and 19 Group sent their aircraft down into the Bay of Biscay to catch more U-boats travelling to and from the main U-boat bases on the French coast. These were a constant danger to U-boats in transit and they scored steadily. In Iceland there was an R.A.F. Hudson squadron and a United States Navy squadron of Catalinas, in Newfoundland the American 20th (Anti-submarine) Squadron with Fortresses and two Canadian squadrons with Canso flying boats and Hudsons. These had an occasional success if the U-boats ventured within their radius of action. Then there were a whole host of Canadian and American squadrons covering the coastal waters of those countries; they rarely saw a U-boat but their deterrent effect meant that convoys were reasonably safe for that part of their voyages.

The men who flew these aircraft were as brave and skilful as those who operated on the edges of the Air Gap but, as long as the Air Gap existed, it would be the men of the long-range squadrons who would meet the Germans more often and be in the limelight.

If this chapter has been long and detailed, the very length of it illustrates the diversity of the Allied forces deployed in a typical Atlantic convoy – the merchant ships of so many countries which were the *casus belli* as it were of the whole struggle; the corvettes, destroyers, frigates, sloops and cutters of the escort groups; the Liberators, Fortresses, Sunderlands, Catalinas and Hudsons of three Air Forces and a Navy. In every case it seemed as if the forces involved had to fight this crucial battle not with the best that their respective services could provide, but with what was left over after more glamorous theatres of war had taken the best.

And in the end it all depended on the men – United Kingdom and Empire men, Americans, exiled Europeans, Lascars, neutral Swedes and Icelanders. A few were professional fighting men but the vast majority were civilians at heart, the ordinary men who turn out in every war, do their duty and then disappear, those who have survived, quietly back into civilian life.

3

The Germans

Of all the millions of men who fought for Hitler's Germany in the Second World War, two groups – the aircrew of the Luftwaffe and the sailors of the U-Bootwaffe – were regarded by their fellow Germans as élite bodies of fighting men. Both took only men of the highest physical and mental standards, both operated sophisticated, modern weapons of war in the most dangerous conditions and often far away from any friendly base. Their courage and skill were rightly recognized as deserving public adulation and this they received. Although the honour accorded to the Luftwaffe waned somewhat as the war years passed, the U-boat men were heroes from beginning to end.

To the British, however, and to most of the rest of the world, the U-boat man was far from being a hero. The apparently callous sinking without warning of the liner *Athenia* on the first night of the war and the harrowing nature of the effects of further U-boat operations against merchant ships were a godsend to Allied propaganda, which was at that time fighting the U-boat threat by every means available. The U-boat captains were portrayed as fanatical Nazi killers and their crews as vicious, murdering bullies. The U-boat man rapidly became both hated and feared.

Is it possible to turn back the years and take an impartial look at the U-boat men? Were they gallant heroes or were they callous murderers?

After the First World War the German Navy was forbidden by the Treaty of Versailles to have any U-boats, but this was repudiated by Hitler in 1934 and in the following year a new agreement, the Anglo–German Naval Agreement, allowed Germany to build U-boats once more. In the interval, German naval architects had kept their hand in by designing modern submarines for foreign countries and, although not all the work was carried out by German shipyards, the expertise so gained allowed a swift start to be made with Germany's new U-boat fleet. In September 1935 the first flotilla of six new U-boats, the famous Weddigen Flottille, was formed at Kiel.* Its first commander was a Kapitän Karl Dönitz. By 1939 five more flotillas were in service.

The successes of these early U-boats have already been described. By 1943 the men who served in them were mostly dead, prisoners-of-war, or had gone ashore. The 1943 U-boats, which are the particular objects of these notes, were almost entirely crewed with new men who had only come into the U-boat arm since the outbreak of war. The traditions and standards they followed had already been set by others.

The majority of the 1943 U-boat captains were men who had been junior naval officers on the outbreak of war and had served on all types of naval vessels from torpedo boats to battleships; many had been seconded to the Luftwaffe as observers in maritime reconnaissance squadrons. These men had then been drawn into the expanding U-boat arm in the heady days of 1940 and 1941, had sailed on at least two operational patrols as watch officers in other U-boats, and then attended a captain's course before taking over their own newly commissioned boat with a new crew. A few of the captains were older men who had been brought in from the merchant service or had been instructors at U-boat schools; these last were reckoned by their crews to be good men to serve under, as they had managed to amass a considerable submarine experience in safe waters. But most of the captains were in their twenties, were tremendously loyal to Hitler and to Dönitz, felt greatly honoured to be in U-boats, and were delighted with the responsibility of command in a ship whose operations allowed so much scope for individual initiative. They wore with pride the white cap cover allowed only to captains of ships and thus giving them a status that could not be claimed by many officers far senior to them in rank. The U-boat captains were the élite of the German Navy.

When you were on a big ship you were a nobody but when you were captain of a U-boat, if you had the confidence of your crew, you were almost a God. (Kapitänleutnant Kurt Neide, U.415)

The average Atlantic U-boat had a crew of fifty: captain,

* Named after Otto Weddigen, captain of U.9, which sank the British cruisers *Aboukir*, *Cressy* and *Hogue* off the Dutch coast on 21 September 1914.

two watch-keeping officers, one engineering officer, seven petty officers – two of whom were the boat's navigators – and thirty-nine ratings. The claim is often made with great pride by senior U-boat officers that every man who served in U-boats was a volunteer and, that however severe the losses, there was never a shortage of fresh men volunteering to man new U-boats. In the course of the research for this book I interviewed or corresponded with Admiral Dönitz, five staff officers and thirty-nine operational U-boat officers and men. It became obvious that the 'all-volunteer U-boat crew' claim was not quite justified, or at least did not quite coincide with the concept of a volunteer in other countries. Because the motives for which men undertake exceptionally hazardous duty are always of interest, this aspect of U-boat life was pursued a little further than others. The investigation threw up much useful light on the U-boat men's attitudes.

Taking the officers first, it must be borne in mind that a German naval officer was not expected to volunteer for anything. It was assumed that having become an officer he would at all times be ready to perform whatever duty his superiors considered him suited for. Several ex-U-boat officers, when asked why they had volunteered for U-boats, expressed surprise that the question should even be asked and one, comparing the German naval officers' complete accept-ance of orders, quoted Nelson's 'England expects every man to do his duty'.

So the U-boat officers were not and could not be volunteers, and numerous examples can be quoted of how they came to be in U-boats. The following are the views of officers who joined in each of the first four war years.

I was an adjutant ashore at Wilhelmshaven and tried to get a job as a *Wachtoffizier* in torpedo boats, but when I applied and had a medical, I was passed fit for U-boat service. I didn't worry too much; it was 1939 and I felt the war would be over before the six-month U-boat training was finished. (Oberleutnant Siegfried Koitschka, U.616)*

A career officer in the German armed forces was not asked to volunteer but accepted the duties he was called for. After I lost my

---

* Koitschka was to spend the next four and a half years in U-boats. He carried out sixteen patrols, was awarded the Ritterkreuz and was finally taken prisoner when his boat was sunk in the Mediterranean in May 1944 after a three-day hunt by R.A.F. aircraft and eight American escort vessels; this was the longest individual U-boat hunt of the war.

position in the destroyer *Erich Giese,* sunk at Narvik, I was ordered to the Submarine School at Pillau in July 1940. I always was and still am very proud to have been a member of the German U-boat force (in spite of the regrettable fact that my skipper and I did not get along too well). (Oberleutnant Rudi Toepfer, U.406)

When the U-boat arm started to grow it was particularly short of engineering officers. At the end of 1941 all destroyers had three engineering officers and the 'Thirds' were all taken out and put into U-boats. I had mixed feelings; on the one hand I was sorry to lose the large and powerful machinery on the destroyer but on the other I was pleased to get a position as Chief Engineer so quickly. Also, at that time the U-boats were going well and there was much prestige to be with them. (Oberleutnant Walter Lorch, U.523)

I was a midshipman at the Naval Air Service School; we were training for eventual duty on the aircraft carrier *Graf Zeppelin.* I had once seen a U-boat and was horrified at its small size and technical complexity. I had been glad then that I was with the air arm and not with U-boats. When Admiral Dönitz took over the Kriegsmarine early in 1943 he drafted 125 of us straight into the U-boat service. I arrived at Kiel one day, joined my boat and we sailed for the Atlantic the next day. (Fähnrich Wolfgang Jacobsen, U.305)

The experiences of the ratings were slightly different. A large number had responded to appeals and were genuine volunteers for the U-boat service. When those who served in the 1943 U-boats had joined earlier in the war there had been many attractions – the evident fame and glamour of the job, an opportunity for active service when most of the German Navy's large surface ships were virtually confined to harbour, good pay, and, for the technically minded, the chance to work with advanced types of machinery and equipment.

One man, however, had motives connected with political persecution at home.

I went to the U-boat arm of my own free will. The first reason for this was that one could earn a lot of money. The second reason was that in this way I could help my father. He was a well-known Social Democrat and had enormous difficulties under Hitler. Life on a U-boat was hard and primitive but after I became a U-boat man my father was left in peace. (Funkgefreiter Werner Hess, U.530)

But as the easy U-boat victories began to fall off and the Allied defences became stronger, so the rush of eager volunteers slackened.

I estimate that 80 per cent were genuine volunteers but from the end of 1942, when radar-equipped aircraft began operating against

us, there weren't so many volunteers. I remember an almost complete absence of married volunteers after this time. This was sad
because a U-boat wanted a few older men in the crew to give an
example of steadiness and to hand on experience. (Bootsmaat
Hermann Lawatsch, U.530)

The U-boats were particularly short of technical men and
suitable men were often 'invited' to volunteer. For a German,
with his attitude of complete obedience and loyalty, there was
no question but to comply.

When our *Lehrgang OF* (a course for Oberfeldwebel machinists)
passed out at Kiel we went for a medical examination and sixty-four
of us were sent to U-boats. We weren't too pleased; we had been
trained for big ships with plenty of room and the open air above;
instead, we had these tiny ships with foul air and no sun. They were
like prisons, although we weren't too worried about the danger at
that time. (Obermaschinist Artur Kolbe, U.406)

In quoting these examples the reader must not get the
impression that men were driven unwillingly to the U-boats.
A large proportion were genuine volunteers and even those
that were given little say in the matter accepted their lot
without protest. All but a handful were proud to belong to
the U-boat arm and, in true German manner, they served
there with absolute loyalty and devotion to duty. But to say
that every U-boat was manned only by volunteers was, by
non-German standards at least, a myth.

By the end of 1942 the German shipyards were producing
five new U-boats every week. The crews for these, with the
exception of the captain and one or two petty officers, were
entirely new entrants to the U-boat service. These new men
had been given their individual submarine training on the old
boats of the various U-boat schools in the Baltic and had then
met for the first time as a crew at the shipyard where their
new boat had been built. There was a formal '*in Dienststel-
lung*' ceremony when the Kriegsmarine officially took over
the boat, followed by a celebration lunch given by the
shipyard for the new crew. In the normal course of events
most of these young men, average age twenty-one, would
stay together for the remainder of that boat's career.

There was some talk among U-boat men concerning which
shipyard produced the best boats. Opinions were coloured by
affection for a man's own boat, especially if it had brought

him safely home after suffering severe depth-charge damage, but many say that the boats built in the Hamburg shipyards were the most soundly built – although opinions are divided on whether the Blohm and Voss boats or the Deutsche Werft boats were the best of the Hamburg boats. Blohm and Voss certainly had the reputation of providing small extra items that the standard boats did not have and of finishing off their boats well.

One day while we were at Kiel two men suddenly appeared from Blohm and Voss and insisted on stripping down the exhaust. One admitted that he may have accidentally left a screwdriver in it while the boat was being built and his conscience had troubled him ever since in case the rattling of the loose screwdriver might one day endanger the boat while we were making a quiet underwater escape from a British escort. They found the screwdriver, re-assembled the exhaust and went back to Hamburg. We were full of admiration for the man who had admitted his mistake and for Blohm and Voss for following it up so thoroughly. (Leutnant Claus von Egan-Krieger, U.615)

AG Weser at Bremen specialized in the building of larger boats and are considered to have built the best of the Type IXCs. The German Naval Dockyard at Wilhelmshaven built comparatively few boats and are reputed to have given their boats greater strength by using special alloys not available to other yards; their boats are believed to have cost 25 per cent more than others. The average cost of a Type VIIC boat was 3 million Reichmarks (£300,000 or 1.2 million dollars) and of a larger Type IXC about 4 million Reichmarks.

The newly commissioned U-boat then sailed for its six-month working-up period with one of the training flotillas in the Baltic – the 4th Flottille at Stettin, the 5th at Kiel or the 8th at Danzig. In these safe waters the men would learn to work as a crew. Many say that this time was far more strenuous than actual operations; it culminated in eight days of continuous diving exercises and mock torpedo attacks. After completion of this training there was another ceremony at which the boat was declared *Frontreif* – 'ready for front-line operations'. It was at this stage that the captains were allowed to have an insignia painted on the conning tower. These were of great variety, but one of the most coveted was the five-circle Olympic symbol which was carried only by those boats whose captains had been members of the German

Naval Academy class of 1936, the year of the Berlin Olympic Games.

The U-boat next went back to the yard in which it had been built for final repairs to any faults discovered, then to Kiel for loading with food and operational stores. Here there was one more ceremony at the quayside, attended by many of the crew's families, and the boat sailed off for the hard life of a *Frontboote*. 'We knew we must all be at our best from now on.'

The Germans had operational flotillas based in Norway, in the Mediterranean, and even a small flotilla, the 30th, in the Black Sea: the latter was made up of a few small 300-ton Type IIB boats which had been taken in sections by Autobahn to the Danube then reassembled and sailed into the Black Sea for operations against the Russians. But the main body of U-boats were those whose bases were in the French Biscay ports. It was these boats that fought the Battle of the Atlantic.

Before reaching its permanent base the new U-boat topped up with fuel at a Norwegian port, then sailed into the Atlantic and had to carry out a full operational patrol before coming into Biscay. The direct route through the Straits of Dover was effectively closed by a British minefield which had claimed several U-boats early in the war, and no attempt was now made to use this. The route out from Norway lay between the Faeröes and Iceland, this route being known to the U-boat men as 'Reich No. 1' or the 'Iceland Passage'. There was another minefield covering this, the British having laid 90,000 mines there. The minefield was known to the U-boat men, who named it 'the Rose Garden', but this massive effort claimed only one U-boat for certain during the whole of the war. The Iceland route was, however, also patrolled by Allied aircraft and several new U-boats were sunk by these.

Having arrived in the Atlantic the U-boat was now controlled by U-boat Headquarters and could be sent into a convoy battle within a few days of the crew saying goodbye to their families at Kiel. The U-boat men knew all about that part of the North Atlantic which the British called the Air Gap; they called it '*das Todesloch*', 'the death-hole', for this was where the fiercest convoy battles took place and where so many of their comrades in earlier boats had been lost.

To the U-boat man his crew was everything. By the very

nature of the submarine's construction and its operations, the crew were closely confined together for long periods, saw little if anything of other boats or crews, and depended absolutely on each other for survival. It was in these days that a crew really settled down.

The life on board our Atlantic operational boats was very hard because of the constricted space and the proximity of the sea; even on the bridge we were only five metres above the water. As every man on board was visible to everyone else and regardless of rank and position exposed to the same hardships, sacrifices and dangers, there had to develop quickly a strong feeling of togetherness, of sharing the same fate. It fulfilled us completely even when we were not at sea. It was our whole life. We had been put into it with all its glory and terror and we accepted it, often with joy and enthusiasm, often with anxiety and fear. (Oberleutnant Helmut Dauter, U.448)

Another man says that 'a well-trained and happy crew was half of a life insurance'.

The German Navy had a saying – 'to live like God in France'; it signified that it would be the ideal way of seeing out the war, and the permanent staffs of the U-boat organization on the Biscay coast enjoyed this for several years. The U-boat man arriving at his new port was able to sample something of this idyllic existence.

The first use of a French port by a U-boat was on 5 July 1940 when Kapitänleutnant Lemp, sinker of the *Athenia*, called in at Lorient with U.30 for fuel and stores and resumed his patrol four days later. Within a month Brest and La Pallice had also been brought into use and St Nazaire and Bordeaux followed later. For the next four years these ports were to be the main bases for the Atlantic U-boats. By early 1943 no less than eight operational flotillas, each with a nominal strength of twenty-five U-boats, were using these ports. The 1st and 9th Flottillen were at Brest, the 3rd at La Pallice and the 6th and 7th at St Nazaire; all these were equipped mainly with the Type VIIC, the standard North Atlantic boat. Lorient was the home of the 2nd and 10th Flottillen, both made up mainly of the larger Type IXC boat used for operations further afield, and Bordeaux contained the 12th Flottille, composed of a variety of types for mine-laying, supply and refuelling and for operations at extreme range.

These bases existed to service and repair their U-boats

between patrols and administer to the needs of their crews. The whole organization came under Kapitän Hans-Rudolf Rösing, Flag Officer U-boats West, with his headquarters in a château at Angers on the Loire. The only operational control Rösing had was that of planning the U-boats' passage across the Bay of Biscay; beyond 12 degrees West they came under the direct control of U-boat Headquarters.

The importance of the Biscay bases soon became obvious to the Allies, but the story of the action taken is an unhappy one. In July 1941, R.A.F. Coastal Command had urged that Bomber Command should bomb each port in turn 'to the limit of its resources', but nothing at all was done in that year. In April 1942 Sir Arthur Harris, the new commander of Bomber Command, offered to bomb the ports but the Air Ministry for unknown reasons would not allow this. When the U-boat danger had become apparent to all, at the end of 1942, both the R.A.F. and the U.S.A.A.F. were ordered to make U-boat construction yards in Germany and bases in France their first priority. But as far as the Biscay ports were concerned it was too late.

The Germans had realized the danger of their valuable port facilities in France being bombed and in 1941 had embarked upon the immense task of putting the essential parts of the U-boat organization under bomb-proof shelters. Work was started at once by the Todt Labour Organization and was completed by mid-1942. The Germans had been very worried that these shelters would be bombed during the vulnerable stages of their construction but they were not – an omission for which the Allies were to pay dearly.

The big Allied bombing offensive finally got under way on 21 October 1942 when a force of American Fortresses and Liberators based in England bombed Lorient. The inhabitants of Lorient turned out in force to watch the accurate bombing of the U-boat shelters. There was little sympathy for the forty French workmen killed; they had gone to work for high German wages. Five direct hits were scored on the roofs of the concrete shelters by 2,000-pound bombs but these did not break through the twelve feet of reinforced concrete.

A general warning was given to the townspeople in all the ports, and many more raids followed over the next few months. The Americans precision-bombed the shelters by day, the R.A.F. 'area-bombed' the towns by night. Brest,

Lorient and St Nazaire in particular were laid waste and became almost ghost towns with the civilian populations fleeing the towns. It had been hoped that the bombing, if it could not hit the U-boats, would at least force the German shore organization to abandon the ports. In fact the Germans moved everything into the concrete shelters; everyone whose work was not essential moved out of the towns before dark, while round-the-clock work on the U-boats continued unhindered inside the shelters.

Reports vary, but it seems that up to 20,000 tons of bombs fell on the Biscay ports and over 100 Allied bombers were lost on these raids, well over half of them American. It is not known how many Germans were killed, but the municipal records of the Biscay towns reveal that 617 civilians were killed – 455 in St Nazaire, 83 in Brest, 55 in Lorient and 24 in La Pallice. Lorient at the conclusion of the bombing is described as 'a desolation in which only a few Germans were moving about and an army of monstrous rats'. Some disruption must have been caused but it was minute in comparison to the effort. Not one U-boat was hit. The money spent by the Germans on these concrete shelters must have been one of their finest investments of the war. By enabling U-boats to be serviced and sent back to sea without returning to Germany, ten days were added to the time a U-boat could spend in its operational area. This was equivalent to an increase in the U-boat strength of approximately 40 per cent!

When a U-boat arrived in its Biscay port its stay there depended largely on its mechanical and structural state. A boat in good condition could be made ready for sea again in as little as ten days, although a month was the more usual time and a delay of two months or longer for a damaged boat was not unusual.

For the crew the length of the stay was all-important. One half usually got away to Germany on leave for a week, but if their boat was to be in port for several weeks the entire crew could get two weeks at home. The U-boat men received extra pay which roughly doubled their basic naval pay; this accumulated during their voyage and was paid out to them in French money when they landed. This is when the good times began. Those going on leave were able to purchase scarce goods in special shops reserved for U-boat men and then were off home as soon as possible.

All U-boat men will remember with affection the *B.d.U.*
Zug (U-boat Headquarters Train). This express train was
reserved exclusively for the needs of the U-boat organization
in France, partly for leave men and partly for the dispatch
from Germany of urgently needed stores and spare parts. The
main train ran from Nantes through Le Mans, where a section
from Brest joined, Paris, Charleroi, Rotterdam and on to
Bremen, Hamburg and Flensburg. The German sailors who
did not live in these towns could easily join a civilian train and
all could be home within two days of leaving their U-boat.
One springtime feature of the B.d.U. Zug was the huge
display of flowers always available at Rotterdam station for
the U-boat men to buy for their families.

Those men who could not get leave were sent off to the
U-Bootsweiden, U-boat 'pastures'. These were various French
premises commandeered as rest camps. Some of these,
especially those for the officers, were in châteaux or luxury
hotels; all were well away from the ports so that Allied
bombing would not interrupt the U-boat men's recuperation.
In the 'pastures' every effort was made to rest the crews and
get them fit for their next patrol.

Even after the bombing of the French ports, the U-boat
men still spent some of their leisure time in the towns. By
common consent Brest, 'a dirty place', was the least popular.
Lorient, a smaller and more intimate town, had been considered '*à la mode*' until the bombing. St Nazaire was popular
because of its superb 'pastures' for both officers and men in
the hotels and luxury villas of La Baule. Further south, La
Pallice had not been affected so much by bombing, but the
best of all was undoubtedly Bordeaux which was not bombed
at all and was a beautiful city compared to the bleaker
Brittany towns. Its inhabitants, too, were reckoned to be far
friendlier than the Bretons. 'You could always enjoy yourself
in Bordeaux.'

Relations with the French civilians were mixed. The
U-boat man ashore had plenty of money and after his long
spell was anxious to spend it and to seek human companion-
ship at the same time. The French restaurant owners and
shopkeepers were happy to oblige, as were some of the
ordinary townspeople; several U-boat men say they were
treated 'like sons' in French homes. In general many of the
French were more tolerant of the Germans than might have

been expected, and British servicemen who had fought in France in 1940 would not like this comparison.

I liked St Nazaire, particularly the rest area at La Baule, very much. We got along fine with the French; they admitted that they preferred our polite disciplined attitude to the stuck-up and demanding ways of the British Expeditionary Force (no hard feelings please). Two tailors made a pre-paid uniform and two suits for me between patrols and got in touch with me even when we evacuated the town after the heavy 'Allied' bombing of St Nazaire. (Oberleutnant Rudi Toepfer, U.406)

And so the U-boat man relaxed between patrols. When his boat was ready he came back from leave or the 'pastures' and helped load his boat for the next patrol. The loading of the food was always done by the boat's own crew, mainly to prevent anyone stealing the splendid rations being put aboard. It took two days of careful work, packing the stores into every spare corner of their boat in such a way as to make sure that they could be reached in the correct order. At the same time torpedoes, ammunition, fuel and water were all loaded. The officers and men had one last night out when they spent every last franc of their money, and the U-boat and its crew were ready once more to face the rigours of a North Atlantic patrol.

The U-boat men soon settled down to the hard life of an operational Atlantic U-boat. The average patrol lasted at least two months. The boat was ordered here and there by wireless; it might occasionally see a friendly boat and exchange greetings; it might be flung into a series of convoy battles or spend weeks patrolling an empty ocean looking for independent ships. The crew grew pale with lack of fresh air; no one shaved and water was strictly rationed. There was one small toilet for fifty men.

The Germans produced nine main types of U-boat during the war, and twenty-two variants of these. But in early 1943 it was only the Type VIIC and the Type IXC that operated on the Atlantic convoy routes. The famous Type VIIC boat, which had come into service in 1941, quickly became the standard North Atlantic U-boat and was the ideal boat for convoy operations in this area. Its low silhouette and small conning-tower enabled it to be almost unseen on the surface at night; it could fire a salvo of four torpedoes plus a fifth from its stern tube; with a surface speed of 17 to 18 knots, it was fast enough to outrun a corvette in any weather and, if it could adopt its favourite escape tactic of making off on the surface into a heavy sea, it could even outrun a destroyer. If attacked and forced to dive it could run at 7 knots, but only for a limited period.

The IXC was a much larger boat, 1,120 tons compared with the VIIC's 769 tons. It was only about 2 knots faster on the surface but one Type IXC man says that 'those 2 knots were as good as three times the speed of a Type VIIC when a destroyer was behind you'. It had no underwater speed advantage but it did have an extra stern torpedo tube. The IXC's great advantage lay in its cruising range of up to 16,000 miles compared with the smaller boat's 9,000 miles. This enabled the IXC to reach the Caribbean and the South Atlantic and this long-range role was exactly what it had been designed for.

The crews of each type were usually convinced that their own boat was the best, but some comparisons of the factors affecting the crew can be made. The IXC had more living space and carried a doctor, but its voyages were longer, although there was a saying '*die länger war die Fahrt, um so besser war die Kameradschaft*' – 'the longer the voyage, the better the crew spirit'. So, while the VIIC was confined to the cold and stormy north, the IXC men might be operating in warmer waters further south and were often able to maintain their physical condition by taking regular swims as long as they were not in danger of attack; but when air cover reached those regions, interior temperature became a great trial, with up to 55 degrees Centigrade in the engine-room if they were forced to remain submerged throughout the daylight hours.

The most important comparison, however, was in the relative chances for survival if attacked and forced to dive, especially by an aircraft. In an 'Alarm' dive the VIIC could be down in about twenty-five seconds, but the IXC, because of its greater length and bulk, took thirty-five seconds even in a calm sea and more if it was rough. This slow-diving characteristic was a serious drawback of the IXC of which its crews were well aware. Once down, a U-boat's captain frequently sought safety by going as deep as he could, 'into the cellar'. The VIIC was designed to be completely safe to 100 metres and later to 125 metres, while the IXC's design was for 105 metres, but both types, when under depth-charge attack, were often taken much deeper. It became a topic of conversation at U-boat bases as to how far each type could actually be taken before the pressure hull was crushed. After many questions, the greatest depths I heard of were 270 metres (885 feet) for a Type VIIC (U.618) and 310 metres (1,020 feet) for a IXC, but it is doubtful if many U-boats returned from this depth. The maximum setting for British depth charges early in 1943 was only 550 feet!

The Germans would have been happy if only the Type VIIC had taken part in the North Atlantic convoy battles, but the IXCs had to do one North Atlantic patrol while making their maiden voyage from Germany to Biscay. On subsequent voyages they were usually sent south.

The danger of detection was never completely absent. If spotted by an aircraft or warship, the U-boat immediately became the hunted; it could not fight on equal terms. Aircraft attack was feared far more than that of an escort vessel; the main factor here was the speed at which an attack from the air developed, especially if the aircraft came out of cloud or was equipped with radar and caught the U-boat at night.* Although the attacking aircraft usually had time to drop only one stick of depth charges, the U-boat really needed to be 80 to 100 metres down to be completely safe and this could rarely be achieved in time. There were a few captains who favoured a 'stay up and fight' policy when attacked by aircraft and this was officially encouraged at one stage. Many Coastal Command aircraft were shot down by this tactic, but the Allies were quite prepared to lose a few aircraft in exchange for the U-boats that were inevitably sunk sooner or later if they decided to stay on the surface and fight it out. An attacking escort vessel could usually be spotted some way off and its slow approach gave the U-boat time to submerge. A good captain could often evade the resulting depth-charge attacks and many U-boat men survived numerous such brushes with escorts. Those who did not come back to tell the tale were those unfortunate enough to be caught by one of the more skilful escorts or by an escort group that could afford to leave two or more ships behind for a prolonged hunt.

---

* There is not space here for the complex story of the tactical and electronic developments of this type of encounter. The reader may like to consult an excellent book, by Alfred Price, *Aircraft Versus Submarine*, William Kimber, 1973.

The U-boat men had mixed feelings towards their enemies. The Germans all knew of and were either impressed by or jealous of the Royal Navy's long history compared with that of their own Navy. One officer commented, 'For were there any amongst us who had not read Forester's *Hornblower?*' Many speak of their respect for the fighting qualities of the British seamen, although one said that 'we did not always trust the so-called English fairness'.

Some of the U-boat men had visited England before the war.

I had several links with your country. My great-grandfather, James Egan, was a Scotsman and in 1938 I had an exchange visit with a very nice family in Salisbury. At the Portsmouth Navy Week I bought a small doll dressed as an English Admiral and it was my mascot all through the war. Only on my last patrol did I have to sail without this mascot because someone had stolen it. (Leutnant Claus von Egan-Krieger, U.615)*

To quote average U-boat loss rates in the rapidly changing conditions of the Battle of the Atlantic is dangerous. The Germans knew that the *'fette Jahre'*, 'the fat years', of the easy successes were gone but, in early 1943, losses were still reasonable compared with later standards. Even so, it would appear that between 20 and 25 per cent of all U-boats that left for a North Atlantic patrol at this time never returned. The greatest losses were in two groups, those boats on their first patrols and those commanded by bold aggressive captains. There are numerous instances of these latter having a brief, glorious career in action, but invariably they were caught and sunk. In the boats that were sunk the crew's chances of survival were slim. Of forty-three U-boats that took part in the convoy battles of mid-March 1943 and were later sunk, twenty-nine were lost with all hands and from the other fourteen boats roughly half of each crew were saved. If the total survivors are averaged over all the lost boats, then only eighteen men out of a crew of fifty or more could expect to survive if their boat was lost.

But the average U-boat man never really knew the losses. He was not at his home port long enough to realize how many boats sailed and did not return; overall losses were never

---

* This officer cannot blame the English for the sinking of his boat. Off Curaçao on 6 August 1943, U.615 had a long fight with aircraft from no less than four American squadrons, shooting down one aircraft and damaging another, but the U-boat was damaged and could not dive. When an American destroyer turned up, Kapitänleutnant Ralph Kapitsky scuttled his boat and went down with it himself, although forty-three of the crew were picked up.

revealed. Few men realized that even in the comparatively safe times of late 1942 and early 1943 the law of averages would sink their boat on its third or fourth patrol and that when it was lost their chances of being saved from it were about one in three.

Was the picture painted of the cruel, fanatical Nazi a true one? This is one of the more difficult questions to answer so many years later. Most of the U-boat men had been born just after the First World War. They had grown up in the politically turbulent and economically derelict Germany of that era. One man had arisen from among a mass of indifferent politicians. Adolf Hitler, leader of the National Socialist Party, had promised to restore Germany to an eminent position in the world, to give her men work and her children food. He was elected to power and he kept his promises. This writer is not an apologist for the means Hitler used to fulfil these promises and certainly not for the many foul deeds done in Hitler's name, but it can be understood how a young man growing up in the 1930s would have responded when Hitler said 'Follow me!'

The 'fanatical Nazi U-boat officers' were, in the main, career naval officers often following in a father's footsteps. Certainly they admired Hitler, sometimes with all of youth's enthusiasm, but only a handful were signed-up members of the National Socialist Party. The German Navy and the U-boat arm in particular never put pressure on its officers to become Nazis. Were they fanatics? There is only a thin demarcation line between patriotism and fanaticism. These men were fighting to avenge the defeat of the First World War and to redress the humiliating terms of the Versailles Treaty which had reduced their beloved country to such misery and chaos. They were professional officers and their country was at war; for a German that would produce a devotion to duty that another might call fanatical.

Were they cruel? Certainly the effect of their actions produced bitter suffering and, when their efforts were successful, the U-boat men rejoiced. But the whole war was cruel and the U-boat campaign was no worse in its effects than many other means of waging war. Dare one mention the Allies' own submarine operations or the bombing offensive against German cities?

The average U-boat man, officer or rating, did no more than his duty as he saw it. His first loyalty was to the remainder of his crew, then to the U-boat arm. U-boat officers called themselves the '*Freikorps Dönitz*' and the U-boat arm regarded itself as 'a Navy within the Navy'. Hitler and National Socialism were a long way from a North Atlantic U-boat. True, there were men like Fritz-Julius Lemp who had so callously torpedoed the *Athenia* on the first night of the war. British propaganda successfully painted all U-boat men the same colour as Lemp, but the average U-boat man reacted to the war in much the same way as the young men of all the warring nations.

Once an Atlantic U-boat had put to sea its operations were controlled directly by the Operationsabteilung of the Befehlshaber der Unterseeboote – the Operations Division of the Commander-in-Chief, U-boats. This was, of course, Grossadmiral Dönitz and his operational staff. This organization will play a large part in later chapters of the book. To the Germans it was always 'B.d.U.' but it will be more easily followed here if it is called U-boat Headquarters.

Karl Dönitz was the man who absolutely dominated the U-boat scene from beginning to end. A successful U-boat captain in the First World War, he had become a prisoner of the British when his boat had to be abandoned through a technical failure while attacking a convoy in the Mediterranean. Dönitz had commanded the new U-boat arm since its rebirth in 1935 but had been much dismayed by Britain's entry into the war in 1939, partly because he recognized that Britain would be a tough enemy at sea and partly because the U-boat arm was being built up on the assumption that Germany was not likely to be involved in a general war until 1943 or 1944. It was for this reason that he must have supported the pressure on Hitler to abandon the rules of submarine warfare in 1939. This deliberate breaking of an agreement made in his country's name is the main wartime charge that can be made against Dönitz's behaviour, but it is a considerable one and he was to face it at the Nuremberg Trials.

Because of Germany's unpreparedness in 1939 for a major war, Dönitz had to fight the submarine war initially with what was little more than a nucleus, and had to balance the risk of losing his few experienced crews against the need to expand quickly. This was why new U-boats received no more than one officer and only one or two senior ratings with previous U-boat experience. It was one of Dönitz's many achievements that he was able to produce crews of such high calibre from such slender resources.

In January 1943 Admiral Raeder, Commander-in-Chief of the German Navy, was dismissed and Hitler appointed Dönitz in his place. But Dönitz continued to command the U-boats at the same time. This was an amazing set-up, almost as though Admiral Max Horton was appointed to the position of First Sea Lord and continued to command Western Approaches at the same time. This move highlights two aspects of the German naval position in early 1943 – firstly, the decline in the importance of surface vessels in general and of capital ships in particular, and secondly, the priority in resources now to be given to the U-boat effort. It was the greatest of good fortune for the Allies that the change of emphasis had not been made two years earlier.

What sort of man was this fifty-two-year-old Berliner who led the U-boat arm? At first sight, Dönitz's appearance was not impressive. He was not a tall man; his bearing was always formal and correct and he is reputed to have shown no emotion whatsoever when his two officer-sons were killed in action – one in a U-boat, the other in a torpedo boat. But underneath this outward appearance lay two great qualities.

The first of these was a great concern for the welfare of his crews. When his headquarters were in France, he regularly met boats returning from patrols and greeted and talked with every single member of the crew. A shake of the hand and a few kind words were enough to make an enormous impression on a young U-boat rating. He was compared to 'a father with his children' and was known as 'Onkel Karl'. Even after he left France his legendary concern with the welfare of his men kept U-boat morale at a high pitch even in the worst of times.* Everything possible was done for the welfare of a

---

* Dönitz sometimes sent personal wireless messages to his captains while they were on patrol. When the famous Günther Prien's wife gave birth to a daughter the following signal was sent: '*Ein U-Boot ohne Sehrohr ist heute angekommen*', 'A submarine without periscope arrived today.' Other signals allowed were those which arranged marriages by proxy, usually of crew members whose girlfriends were in trouble. The U-boat captain would perform one part of the ceremony at the exact time that the girl attended a similar ceremony in Germany.

boat's crew and for the mechanical state of the boat before it sailed but, once at sea, that boat's crew knew that they would be driven by Dönitz to a point just short of their limits. This was the unspoken bargain that existed between Dönitz and his men. They rarely let him down. There was hardly a U-boat man who did not follow and admire him absolutely.

Dönitz's second great quality was his operational ability. He already had a personal knowledge of what a U-boat and its crew were capable of and he kept in close touch with tactical developments. Captains returning from patrol had to visit U-boat Headquarters for an interview. In this way he was able to follow each Allied move. Dönitz never had the U-boat strength he felt was necessary for complete success but he certainly handled those boats that he did have brilliantly.

There is no doubt that both as an inspirer of men and as a submarine strategist Karl Dönitz will go down in history as one of the great commanders. One more of the names by which he was popularly known amongst the men he led was '*der Löwe*', 'the Lion'. What a tragedy that Karl Dönitz's loyalty to Germany should have also demanded that he saw it as his duty to serve the régime of Adolf Hitler.

The first wartime location of U-boat Headquarters had been a wooden shed at Wilhelmshaven, but there had been several moves since then. For a few months in 1940 it had occupied a building on the Boulevard Suchet in Paris, until it moved to what was intended to be its permanent location at Kerneval near Lorient. Kerneval proved most convenient until March 1942, but in that month there took place the British Commando raid on St Nazaire, and Hitler ordered Dönitz away from the coast in case U-boat Headquarters should be the target of a future raid. Dönitz did not want to move away from the U-boat bases but he had to comply. As a temporary measure Dönitz returned again to a Paris hotel while a new headquarters was prepared in a château at Angers, but before this was ready Dönitz became commander of the entire German Navy and had to move to Berlin.

At the time of the SC.122 and HX.229 convoys, U-boat Headquarters had only just settled into another hotel, the Am Steinplatz in the Charlottenburg district of Berlin. Dönitz divided his time between this and the Kriegsmarine Headquarters at the Tirpitz-Ufer which was five minutes away by car.

Considering the scope of U-boat operations, Dönitz's immediate staff was very small. It was headed by Konteradmiral Eberhard Godt (sometimes called the 'Godfather', but not to his face); Godt had been Dönitz's right-hand man since before the war. In early 1943 the two Operations Officers were Fregattenkapitän Günther Hessler, Dönitz's son-in-law, and Kapitänleutnant Adalbert Schnee, both of whom were younger men who had successfully commanded U-boats earlier in the war. These officers, with a few watchkeepers and specialist officers, formed the brains of U-boat Headquarters. This small staff served Dönitz well. It was a big advantage to have one body to control all U-boat operations from Iceland to Cape Town; many of their opponents would dearly have loved their forces to be so unified. Dönitz's staff were sorry to have to leave Paris with its superb food, freedom from bombing and its closeness to the U-boat bases, and their apprehension about moving back to wartime Berlin was fully justified only nine months later when they were bombed out by the R.A.F. and had to move yet again.

One more German naval department should be described – the Beobachtungs-Dienst (Observation Service), commonly known as the B-Dienst, which had such success in deciphering British naval signals, especially diversion signals sent to convoys at sea.

The B-Dienst was established before 1939 and had already broken some of the Admiralty codes before war broke out. By the summer of 1940 the B-Dienst was 'reading' up to 2,000 messages each month but then there was a big setback in August 1940 when the Admiralty changed all its ciphers, a massive world-wide operation. It took the German cryptoanalysts until well into 1942 to break through again, but by the end of that year many of these new codes were also broken.

The B-Dienst was part of the main German Naval Headquarters at the Tirpitz-Ufer. Here, in closely guarded rooms, worked over 1,000 people. A battery of forty-five teleprinters received signals picked up by listening stations all over Europe, from as far north as Finland to a clandestine one near

Seville operated with the connivance of the Spanish Government. Here also was the massive collection of catalogues
containing all known combinations of ciphers and codes used
by the Admiralty.

We knew which of the British messages were in secret ciphers
and codes and had named the four main ciphers *Köln*, *Frankfurt*,
München Blau and *München Rot*. *München Blau* was particularly
important as it referred to U-boat positions and course changes for
the convoys. In all, there were forty-five different circuits changing
on different days and at different times. It was a massive job to
decipher all this; some of the signals were done at once, some took
longer, and some never did get done.

The Admiral at Halifax, Nova Scotia, was a big help to us. He
sent out a Daily Situation Report which reached us every evening
and it always began 'Addressees, Situation, Date', and this repetition of opening style helped us to select very quickly the correct
code in use at that time. When we found the right group we called
it *'ein Biß'*, 'a bite'. If all went well, we could then have a complete
message decoded within two hours and if it was one of the *München
Blau* series it was immediately sent by teleprinter to U-boat Headquarters. (Kapitän Heinz Bonatz, Head of the B-Dienst)

There is no need to labour here the use Dönitz and his
operations officers were able to make of Bonatz's messages.
This will be amply demonstrated later in the book.

These then were the means by which the Germans fought the
convoys early in 1943 – sound operational U-boats with well-
trained and devoted crews; a brilliant commander with his
tiny staff lodged in a Berlin hotel; a steady supply of decoded
British signals.

This concludes the survey of the four participants in any
convoy battle. It is time to return to the unfolding story of the
Atlantic contest.

4

New York

Thus was the stage set for Germany to fling into the Atlantic struggle the greatest possible strength, directed by the man who had from the beginning of the war controlled the U-boats and had always been their protagonist. It was plain to both sides that the U-boats and the convoy escorts would shortly be locked in a deadly, ruthless series of fights in which no mercy would be expected and little shown . . . In all the long history of sea warfare there has been no parallel to this battle.

Captain S. W. Roskill, *The War at Sea 1939–1945*, Vol. II, p. 355, referring to the situation early in 1943.

On Monday 1 March 1943 the stone eyes of the Statue of Liberty looked out over the great anchorage of New York's Upper Bay. Here and in the Hudson River there were well over 100 merchant ships, fully loaded and low in the water, swinging first one way and then another as the tides turned. These ships were waiting to be formed into convoys and sent to England. It was bitterly cold; there were thick sheets of ice coming down the rivers, banging and scraping against the steel sides of the merchant ships. Snow squalls swept through the anchorages and away out to the cold Atlantic.

Before describing the birth of Convoys SC.122 and HX.229, it might be useful to review briefly the situation on the various war fronts at that time. In most parts of the world the war was going well for the Allies in those first days of March 1943. In November 1942, American and British forces had landed in North Africa in the successful Operation Torch. Now, with the British Eighth Army closing in from the east and the Torch forces from the west, the Germans' days in North Africa were numbered, although they were having a last fling with a counter-attack which rocked the American troops holding the Kasserine Pass. But preparations were going

ahead for the next stage in the Mediterranean war, the
invasion of Sicily in five months time, and in England the
huge build-up of forces continued for Operation Overlord,
the invasion of Normandy planned for early 1944.

At sea the Royal Navy had survived the desperate convoy
battles on the Malta and Russia routes in 1942. The long
agony of keeping Malta supplied was ending with the land
successes in North Africa and there had been no repeat of the
disastrous PQ.17 convoy to Russia of July 1942; the long
nights of the winter months had allowed recent convoys run
to Russia to get through with only light losses. A *Daily
Express* report of 1 March 1943 had just announced that the
Royal Navy had lost a total of 417 ships since the outbreak of
war, 'a proud and terrible roll'.

In the Pacific the great American counter-offensive against
the Japanese was gathering momentum. The Japs had finally
quit the bitter island battlefield of Guadalcanal and on that
very day, 1 March, a Japanese convoy of eight supply ships
and eight destroyers was sighted on its way to reinforce
army units in New Guinea; in the resulting Battle of the
Bismarck Sea, aircraft of the American Army Air Force and
the Royal Australian Air Force sank all the supply ships and
four of the destroyers, and over 3,000 Japanese were
drowned. In Russia, too, the tide had turned with the great
defeat of the Germans at Stalingrad just a month earlier. The
Russians were now streaming forward and recapturing large
areas of their country.

It was at this time that some Germans must have realized
that they could no longer win the war. Their armies were in
retreat on every front in which they were in action and the
Allied invasion of mainland Europe was now inevitable. The
R.A.F. was going into top gear with its bombing of Germany's industrial cities and R.A.F. Bomber Command was
about to start what would later be called the Battle of the
Ruhr, in which the industrial cities of the Ruhr would be
smashed one by one. The American bombers of the Eighth
Army Air Force had still to solve the problem of daylight air
escort and could not yet fly further than targets in France or
on the German coast but, with the arrival of the long-range
Mustang fighter, they would later join in laying waste the
German homeland. Little wonder that the first attempt by
Germans to remove their leader would take place on 13

March, but the time-bomb placed in his aircraft would fail to explode.

The world scene at that time showed the war-weary people of the Allied countries being heartened by solid progress in every theatre of war except one. The North Atlantic was the only place where the Allies were still on a strict defensive and where the outcome was still in doubt.

January 1943 had been a quiet month in the North Atlantic, although the Germans had not intended that it should be so. In the first two weeks of the month not one convoy was discovered by the U-boats, partly because the winter storms hampered the U-boats and partly because the British and American route planners were sending the convoys by ever longer evasive routes. Another factor was that the Admiralty codes had been changed again and the B-Dienst was unable to read the signals diverting convoys away from known U-boat concentrations. Convoy after convoy slipped past the waiting U-boats; only one North Atlantic convoy was attacked in the whole of January and it lost only one ship.

February was different. An eastbound convoy, HX.224, was caught and attacked but, although only two ships were sunk, a British merchant seaman was picked up from one of their lifeboats by U.456 and he disclosed that a slow convoy, SC.118, was following only two days behind. It is not known how the British seaman had obtained this information or why he disclosed it to the Germans. The U-boats were deployed in a line across SC.118's path but the convoy sailed through this line unseen in the night. But then a second piece of misfortune finally disclosed the convoy's presence; a merchant seaman accidentally fired a Snowflake flare which a U-boat, twenty miles behind the convoy, spotted. Twelve merchant ships and three U-boats were sunk in the resulting battle.

By now the B-Dienst had started to break through the new codes and three consecutive westbound convoys were all found and attacked. Merchant ship losses in all areas rose to sixty-three in February compared to thirty-seven in January; nineteen U-boats were also sunk in the fierce convoy battles of that month.

The U-boats were now managing to intercept and attack nearly half of all the North Atlantic convoys and Dönitz could maintain enough U-boats on station to make a decisive breakthrough. Britain was using 750,000 tons of imports more per month than were arriving in the battered convoys, and reserves of some items would be exhausted in two months time. The crisis of the Battle of the Atlantic was near.

Some of the ships anchored in New York harbour had loaded their cargoes at the New York wharves and others had loaded at ports on the East Coast or in the oil ports of the Gulf of Mexico. Such ships were usually Americans – tankers, pre-war freighters and a handful of Liberty ships which were still new enough to be a novelty. There were eight Libertys, all but one of which were fresh from the builders' yard and were about to commence their first ocean voyages.*

When the Americans had finally introduced convoys off their coast after the heavy sinkings in the early months of 1942, the convoys had been organized into what was known as the Interlocking Convoy System. This was a complicated network of convoys starting right down at Rio de Janeiro, coming up through the Caribbean and the Gulf of Mexico and then along the American East Coast to New York. Regular 'trunkline' convoys now sailed up and down this system and were linked with smaller subsidiary convoys into the American ports. It was by this efficient and safe method that many of the American ships had come to New York.

But two thirds of the ships in that anchorage were either British or controlled by the British, and for these ships New York was not the normal convoy assembly port. For three years, their convoys had always started from the Canadian ports of Sydney or Halifax. It was only when the United States Navy took over control of convoy operations in the Western Atlantic in 1942 that New York was designated as the main departure port for convoys to England. The concentration at New York of ships that had previously used two ports was one cause of the overcrowding in the New York anchorages.

There was another reason. When the American and British troops had landed in North Africa for Operation Torch in November 1942, it had been decided that nothing should

---

* The eight Libertys were *Daniel Webster*, *James Oglethorpe*, *John Fiske*, *Pierre Soule*, *Robert Howe*, *Stephen C. Foster*, *Walter Q. Gresham* and *William Eustis*. *John Fiske* was the only one that was not new.

interfere with these first major Allied landings and that the convoys involved should be adequately escorted. Up to that time, ships bound for England from any port in East, South or West Africa and those from South America had sailed to Freetown, Sierra Leone, and there been made up into convoys for England. The sailing of these SL. (Sierra Leone) convoys had been stopped just before Torch, and their regular escort groups sent to protect the Torch assault and follow-up convoys. The British merchant ships from Africa and South America had now to make a long, tedious journey to join the American Interlocking Convoys and then wait at New York for the regular North Atlantic convoys. This major change resulted in a great wastage of shipping capacity, further congestion at New York, and overloading of the regular Atlantic convoys, but this was all acceptable to the Allies in return for the success of the North African landings.

These British and Allied ships, representing many of the well-known shipping companies of that era, had often sailed thousands of miles since they had loaded and New York merely represented one more intermediate stopping place on their long journey to England. The Glen Line's *Glenapp* had loaded cocoa, palm oil and copper at Lagos, but instead of going home with a Sierra Leone convoy had been sent across to Cuba to join the American convoy system, by which means she had arrived in New York at Christmas, no nearer to England than she had been at Lagos. A Lloyd's surveyor had then declared *Glenapp's* port engine unsafe and, while waiting for repairs, she had been berthed at one of the Manhattan piers close by the hulk of the burnt-out French liner *Normandie*.

The Royal Mail Lines *Nariva* and Donaldson Line *Coracero* were both refrigerated ships which had come from Buenos Aires loaded with meat. Due to pressure on shipping space, the meat had been boned and the carcases 'telescoped' to pack in the maximum amount of meat possible. Instead of coming up past Brazil, where U-boats were operating, the River Plate traffic was being sent round Cape Horn into the Pacific and then through the Panama Canal to New York. Neither the *Nariva* nor the *Coracero* was destined to reach England.

Several ships had come from Australia and New Zealand, also through Panama. Among these were the Rotterdamsche

Lloyd *Terkoelei*, loaded with Australian wheat and zinc. She
had once been the German ship *Essen*, one of three German
merchant ships taken as prizes by the Dutch at Sourabaya
when Germany invaded Holland in 1940. The New Zealand
Shipping Company's *Tekoa* had also come from Australia. It
was the first voyage in command for her master, Captain
Albert Hocken. The master of another of the company's ships
had died at Sydney, *Tekoa*'s master had been transferred and
Hocken had been promoted and had taken over. *Tekoa* had
been allotted an anchorage near the George Washington
Bridge but had dragged her anchor one night in the swollen
North Hudson River and drifted downstream a mile but
without hitting any other ships. *Tekoa* and her new master
were to play a vital role in coming events.

Another ship from Australia had never been intended for
convoy at all. She was the fast Blue Star Line ship *Canadian
Star*, which already had a patch on her funnel as a result of a
gun duel with a U-boat near the Azores in 1941. The
Canadian Star normally sailed as an independent but, while
coming across the Pacific, a shell had exploded at gun drill,
killing two gunners, badly injuring several others and wreck-
ing the gun. Fortunately a colonel in the Indian Medical
Service was among the passengers and he looked after the
injured until they were landed at Panama. American naval
workshops at Panama could neither repair nor replace the
gun. *Canadian Star* resumed her independent voyage to
England, but in the Caribbean a message was dropped by an
American airship ordering her to New York, presumably
because she could not be risked as an independent with her
gun out of action.

There was a great variety of ships. There were two brand
new Fort-type ships built in Canada for Britain – *Fort Cedar
Lake* and *Fort Anne*. The two masters had been in Canada for
three months watching their ships being completed. They
were very impressed with the great care taken in the
Canadian yards with materials; not a nut, bolt or piece of
wire seemed to be left lying about spare and they compared
this with the waste often seen in British yards. The crews
gathered together in Britain and sent out to collect the *Fort*
ships were a rough lot. There had been many desertions
while travelling across Canada and, when the ships had
sailed from Vancouver, two men of the *Fort Cedar Lake* were

left behind in jail for insubordination. The *Fort Cedar Lake* had then gone to Portland, Oregon, to load timber at a beautiful up-river wharf, but the crew were so far gone with cheap local wine that Royal Navy men waiting there for their own new ship had to be called in to raise steam for the winches. At Panama a donkeyman came aboard from shore leave drunk and attacked the chief steward, breaking his jaw. The captain debited the donkeyman's pay with the resultant medical expense but, on arrival in England, the Board of Trade did not uphold the master's judgement and the man was not penalized in any way. The *Fort Cedar Lake* and the *Fort Anne* were almost identical in design but the *Fort Cedar Lake*'s engineers could not raise the required steam for a fast convoy and she had to leave her sister ship and sail with the slow convoy.

Another vessel that would sail with the slow convoy was typical of the many old tramp steamers pressed into service for the war.

I had been sailing in the *English Monarch* for eighteen months at that time and she was by present standards 'Rather Tatty'. She was a three-island steamer of about twenty years old, previously owned by a tramp owner who spent little on her. We were inclined to cover up her shortcomings and put up with her discomforts because a few of us were convinced she would survive. Our speed was slow and difficult to maintain as we were a coal burner and always ended up at the tail end of the convoy. On one occasion the Commodore asked if we were 'jet propelled' as our station-keeping was so erratic. (Chief Officer W. D. Morton, S.S. *English Monarch*)

This ship, typical of so many in the convoys, set out from New York but had to put into Halifax for repairs. She did survive the war and was sold to a Japanese company, but sank in a Pacific storm in 1960. Another old ship brought into service was the *Carso*, an ex-Italian ship scuttled at Mogadishu in Somaliland. She had been repaired at Mombasa and Cape Town and then given officers who had survived the sinkings of other ships, and a Lascar crew. *Carso* loaded steel and food at Baltimore and now waited at New York for a convoy.

By contrast, there were two new Royal Navy L.S.T.s (Tank Landing Ships) built by American shipyards and now waiting with their naval crews to cross to England. They would sail in convoy with the merchant ships and each had been filled with cargo – twenty Sherman tanks in the tank deck, tinned milk and food in all other spare space below, and steel rails stacked on the upper deck and welded down to give greater strength to these comparatively small ships of 1,650 tons for their ocean crossings.

There were so many ships at New York that there were at least three collisions. Here is a description of the difficulties that could be encountered in the anchorages.

One evening the *San Veronico* was being anchored ahead of us when she dragged and struck my ship which in turn caused us to drag into mid-stream. Fortunately, a tug was passing at the time with a pilot and we were able to obtain his services before any further damage could be done. We then moved and anchored further up river just below the George Washington Bridge. Whilst at anchor we encountered more ice on the ebb tide and at times it was impossible to get ashore in the launch.

Another evening, my Third Officer informed me that the Greek cargo vessel who was anchored ahead of us was dragging her anchor. We tried every way possible to attract her attention but without success and we were struck on the port bow. The cargo vessel was dragged the length of our ship before she was clear. During this time we had our engines going to keep the weight off our anchor chain otherwise we would have been dragged downstream also. I shall never know how the anchor chains were never fouled. The starboard lifeboat of the cargo ship was smashed between the ships and, at daybreak, when we were able to assess the damage a case of brandy was found on the foc's'le head; evidently it had fallen out of the lifeboat. (Captain W. Luckey, M.V. *Luculus*)

Two more ships that collided were the Dutch ship *Zaanland* and the Norwegian *Elin K*, which had both come from Australia but had got into trouble while anchored off Liberty Island when the *Zaanland*, swinging on the turn of tide at high water, got her stern stuck on the bottom. She managed to get off but in doing so struck the *Elin K. Zaanland*'s forepeak developed a serious leak and had to be filled with cement at the shipyard of the Bethlehem Steel Company; a Dutch officer was very impressed with the urgency shown in the yard to complete the repair. The *Elin K* had also needed repairs to a large hole in her bow. Both ships had missed convoys because of this incident but would be ready in time to join HX.229 and would actually sail together in the same column of the convoy.

These ships were just a small cross-section of the hundred or more ships waiting at New York for the powers-that-be to decide in what manner and by which routes they would attempt to make the North Atlantic crossing.

The crews and passengers aboard the ships in New York harbour knew nothing for certain of their future movements. Every merchant ship was subject to naval control at all times and, although most of those aboard had a good idea that they would soon be bound for England, they could still be diverted at naval whim to another destination. But this had been the merchant seaman's routine for several years and it was accepted quite fatalistically.

The crews of some ships had the great misfortune to miss the chance of savouring the delights of New York. If their ship was fully loaded, had no need to take on fresh provisions or bunker supplies and did not require repairs, then these men had to pass the days of waiting in great boredom and disappointment on their ships in the bleak anchorages with the skyscrapers and bright lights of this great city tantalizingly beyond their reach.

But most ships had the need to tie up at one of the wharves. Many did so in response to a recently issued British naval order that no ship in a convoy bound for Britain was to sail without being loaded with extra deck cargo, presumably if the ship was not already down to her wartime load-line marks. This measure is an indication of the extreme shortage of shipping that existed at that time. The usual deck cargoes loaded in New York were crated American Army vehicles and aircraft, although several ships found themselves carrying such bulky items as infantry invasion barges and railway engines. This unwieldy top cargo, loaded onto ships that had to face winter gales, was not popular.

The American authorities were very strict in all security matters. There had been a wave of German-inspired sabotage in the United States during the First World War and, with a large immigrant population in New York, no chances were being taken now. Ships alongside wharves had to mount special 'anti-sabotage-watches' throughout their stay, and the whole dock area was heavily patrolled.

On one occasion when we first anchored at the Red Hook Flats I had to go to the rescue of the ship's postman – a Leading Seaman who had landed at the nearest jetty to take mail to the Fleet Mail Office in Manhattan and had been arrested by the U.S. Coast Guard as an unauthorized foreigner in a restricted area. When I arrived, a threatening situation was developing as the postman was preparing to fight in defence of his mailbag. The Coast Guard gave grudging permission for me to use the telephone to speak to both the British Naval Liaison Officer and the Duty Officer of the U.S. Navy District. Both asked me to recall the postman and take the ship's boat to Brooklyn Yard, a considerable distance, as they had no control over the U.S. Coast Guard! (Sub-Lieutenant J. H. H. Bayley, L.S.T. *365*)

The security men were particularly careful with sailors from the Allied ships whose homes were in German-occupied countries. There was always the danger that the Germans might infiltrate an agent into such ships or persuade crew members to cooperate by threatening their families.

On arrival at New York, security officers came on board; one of them spoke Dutch fluently. They were very strict. They carried big books and, as soon as you gave your name, they started looking to see if they could find the name in the book. Fingerprints were taken, also photographs with numbers in front of the 'victim'. The only items I remember from the identity card I got afterwards were my criminal picture and my occupation indicated as 'mariner'. (Second Officer J. Wassenaar, S.S. *Zaanland*)

Once out of the docks, the seamen could relax and enjoy the hospitality of wartime New York. The city was well geared to looking after visiting seamen or, indeed, any servicemen, and many speak of the remarkable generosity shown to them by the city, always given 'in such a way as not to be charitable but genuine friendship to visiting sailors'. There was the U.S.O., the Eagle Wing Theatre, and the famous Stage Door Canteen which was the great meeting place off Broadway for seamen, where, it was said, 'the top brick of the chimney is not good enough for the serviceman'. In all these there were lists of cinemas, theatre shows, sports events and various outings. The visitor had only to ask and free tickets were available at once. The Empire State Building with its marvellous view over the city and harbour, Carnegie Hall where Bruno Walter conducted the New York Philharmonic Orchestra, Radio City where the sailors could watch world-famous orchestras recording the Big Band Concert programmes – all these were great attractions.

There were several well-established clubs for Merchant Navy men. Not far from Times Square was the British Merchant Navy Officer Club among whose helpers were a

Miss England and a Miss Ireland! There was a Dutch Club at the Aston Hotel and a Norwegian Seamen's Mission. The British Apprentices Club in a suite of the Chelsea Hotel on 23rd Street was presided over by a Mrs Spaulding, a very popular lady who made 'a real home from home with loads of food, pretty girls to dance with, but no alcohol except on special occasions'. Other men speak of the great hospitality shown by the American Legion and of the many New York families who invited them to their homes.

Then there were the New York night clubs which were certainly not free but were still a great attraction for those who had seen these famous establishments so often on the cinema screen.

There was a little night club where several of us used to go, the name of which I forget, but it was popular in that there was no cover charge and the drinks were not too pricey and the floor show used to change regularly. In the floor show there was invariably a strip act. In those days, our uniform shirts had detached collars and, one night, the Third Engineer, who was a bit of a wag, took an old collar along. When the strip act was on and the girl was going down in the splits the Third took his collar out and tore it sharply lengthways and the resulting rip could be heard all over the room. The poor girl on stage collapsed in a heap, really thinking that her tights had given way. It's the only time I've ever been bounced out of a night club, all six of us. Those bouncers really knew their stuff; they bounced us so quickly that we didn't get a chance to pay our bill! We were fortunate that there was a stack of snow on the sidewalk, so we fell pretty lightly. (Apprentice C. H. F. Hill, M.V. *Glenapp*)

Another *Glenapp* man, Chief Engineer R. W. Douglas, found another attraction.

I remember on one occasion I was called to an engineering workshop to inspect the repair they were doing for us. On my way back to the ship, walking along Broadway, my attention was drawn to a cinema where a stage show with a well-known band and an added attraction of a new discovery was highly advertised. Being Saturday afternoon I decided to go in. Although there were a number of vacant seats I was surprised to see two or three dozen young girls standing in the side aisles against the walls. After the band gave its opening number, the new discovery appeared and was greeted with piercing shrieks from the girls, and thus was Frank Sinatra launched upon the public. The girls, I have no doubt, were admitted free to boost the show.
(Frank Sinatra's New York debut did take place on 30 December 1942 at the Paramount Theatre with the Benny Goodman Orchestra.)

The *Glenapp* men had ample opportunity to study New York life, as their ship had been delayed with engine repairs for two months. The few passengers aboard also had to wait. Two Catholic priests said Mass daily and soon attracted a regular congregation, including the divers working on the burnt-out *Normandie* nearby. Two R.A.F. pilots, who were passengers, disappeared for long periods; they had taken jobs driving explosives lorries from a factory in up-state New York to the docks and were believed to have earned 'a fortune'. Several of the *Glenapp*'s crew worked on part-time jobs ashore – dishwashing at 50 cents an hour or snow clearing for 60 cents being popular. An apprentice took a job in a mail-order store but was sacked for his inefficient packing of parcels.

And so New York played host to hundreds of merchant sailors, most of whom responded like the man who 'made the best of every free hour we had, losing as little time as possible sleeping'. But there were a few discordant notes, one from an American sailor who found city life strange after his native countryside in Georgia.

I did not like New York. It was completely different from my life in the country. It seemed like the sidewalks were about 40 feet wide and full of people, half going one way and half going the opposite way. Where they were going and why they were walking so fast was a puzzle to me. (Able Seaman T. Napier, S.S. *James Oglethorpe*)

And Apprentice Hill of the *Glenapp*:

Personally I had gone off the New York police when, during the first few days we were there, I had asked a policeman directions and he simply said 'buy a map!' A little while later I saw a couple of policemen in Times Square hit a man who had crossed against the traffic lights, hit him at the back of his legs with their sticks and, as he collapsed, roughed him up.

Apropos of this type of thing, I soon learnt that it simply wasn't done to offer your seat to a lady in a crowded bus. It was accepted if the lady was obviously halt, sick or lame, and in her seventies, but otherwise you got a remark like 'what's wrong with you, got ants in your pants?' It wouldn't have been so bad if it was said quietly to you but it was almost shouted out in such a way that it said to the other passengers 'here's a guy trying to get fresh with me'.

Five years before, New York was a city of hard times and depression. As a boy in 1936/37 one could step on shore with one dollar, see a movie-cum-variety show and for supper have a hamburger and egg, sunny side up, followed by apple pie with ice cream and coffee, and return on board with change from your dollar. That afternoon in 1943 New York was all prosperity and go. I took a cab from Battery Park to 42nd Street and bang went $1.40. New York was having a good war. (Third Officer R. McRae, S.S. *Coracero*)

These comments should not obscure the fact that the majority of the men who could get ashore in New York had a thoroughly good time and took with them happy memories of the city's hospitality. For too many of them, New York would be their last experience of human conviviality.

The convoys that would eventually take the merchant ships which were waiting at New York had, in some ways, been planned many weeks before, being part of the long-established 'cycle' of regular North Atlantic trade convoys. There were two types of convoy, quite simply 'slow' and 'fast' convoys.* Ships that could not maintain a steady speed of 10 knots had to sail with the slow convoys, and ships capable of between 10 and 15 knots sailed with the fast convoys. As any convoy could only steam at the speed of its slowest ship, an average slow convoy would make 7 knots and a fast convoy 9 knots. If a ship could make more than 15 knots it was considered fast enough to sail as an independent. The intervals between the departures of each type of convoy varied from time to time but, in March 1943, the slow convoys were sailing at roughly eight- or nine-day intervals and the fast convoys roughly once a week.

The British and Canadians had organized the sailings of North Atlantic convoys until July 1942 when the Americans assumed responsibility for all convoy control functions on the western half of the convoy routes, including the planning and dispatch of eastbound convoys. The American organization responsible for producing the plan for each convoy was the United States Navy's Convoy and Routing Section, which

* The regular North Atlantic trade convoys were code lettered as follows:
HX. – fast eastbound, named after Halifax, Nova Scotia, the original assembly port for these convoys;
SC. – slow eastbound, which had originally sailed from Sydney, Cape Breton Island;
ON. – fast westbound, from Outward North Atlantic;
ONS. – slow westbound, from Outward North Atlantic Slow.

There were dozens of Allied convoy routes all over the world each with their own code; perhaps the most original was the WS. route for fast troopships from England to Suez by way of the Cape of Good Hope. The WS. stood for Winston's Specials.

came directly under the Navy's Commander-in-Chief, the redoubtable Admiral Ernest J. King.* Convoy and Routing was housed in the huge Navy Department main offices in Washington, and had only two weeks earlier moved from the Second to the Third Floor; its hundred or more personnel now occupied Rooms 3517 to 3536. It was in these unwarlike surroundings that the next two convoys in the cycle were born – the slow SC.122 and the fast HX.229.

One thing must be emphasized here. Although these convoys later achieved their own place in history, in their planning stage they were completely routine. Their very numbers, showing that they were the 122nd and 229th convoys of their respective series, indicate that they would have received no more nor less attention than hundreds of other wartime convoys. But, because the initial planning for SC.122 and HX.229 will be important both to the fortunes of these particular convoys and to the easy understanding of later chapters, these plans will be examined in detail.

Convoy and Routing had one paramount function at this stage – the selection of routes. The officer in charge of this was Captain Archer M. Allen, the Assistant Head of Convoy and Routing but, in fact, the Americans were happy to allow their own naval formation, known as Eastern Sea Frontier, whose headquarters were in New York, to suggest the routing details from New York to Newfoundland and the Admiralty in London to suggest the remainder of the route and a convenient CHOP time when the convoys would reach the vicinity of 35 degrees West. Several signals from 'COMEASTSEAFRON' and 'ADMIRALTY' give the details of SC.122's and HX.229's planned routes; the only reply made by Captain Allen's office being a brief 'AFFIRMATIVE' sent out by 'COMINCH C & R'.†

The routes from New York to Newfoundland presented little problem because the convoys would be well within range of ample air cover and the direct route could be followed. The great difficulty lay in choosing routes from

---

* The code name for the Commander-in-Chief's headquarters was COMINCH, an earlier suggestion of CINCUS not being deemed appropriate. Convoy and Routing later came under the Tenth Fleet when that purely antisubmarine organization was formed on 20 May 1943, although Admiral King remained in direct control.

† I am indebted for copies of these and other signals to the Naval Historical Center, Washington.

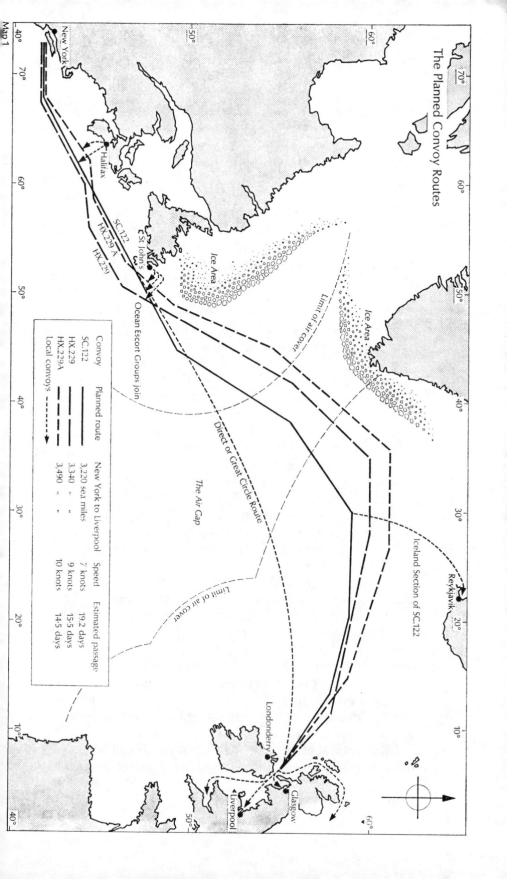

The Planned Convoy Routes

Convoy	Planned route	New York to Liverpool	Speed	Estimated passage
SC.122		3,220 sea miles	7 knots	19.2 days
HX.229		3,340 " "	9 knots	15.5 days
HX.229A		3,490 " "	10 knots	14.5 days
Local convoys				

Map 1

Newfoundland to England that stood a good chance of avoiding U-boats. The shortest route across any ocean is that which follows the Great Circle, and any variation by the convoys from that route caused a loss of time and a waste of shipping capacity. In theory the convoys could take either a southern or a northern diversion from the Great Circle route, but in practice the southern route was a great risk; a glance at a map displaying air cover available in March 1943 shows that the Air Gap was fairly narrow in the far north while it widened considerably with every mile that the convoys strayed to the south. The southern route was also much nearer to the Biscay bases of the U-boats.

The map on p. 91 shows the routes selected by the planners at Eastern Sea Frontier and the Admiralty for the two convoys. The slower SC.122 was to have a shorter route across the Atlantic than HX.229 and, although both routes represented a substantial diversion from the Great Circle, neither was anywhere near the iceberg areas off Labrador and Greenland. But another basic fact should be stressed here. Initial routes selected for convoys were never considered as sacrosanct and could be altered at any time to avoid danger. The routes in the potential U-boat areas for SC.122 and HX.229 allowed plenty of sea room to the north for further diversion, while any southerly diversion would shorten the voyage, although at the expense of air cover. These initial routes were a sound compromise.

While the planning of these routes was taking place, the actual composition of the two convoys was being debated. In theory this was another Convoy and Routing function, but in practice most of the detailed work was carried out at the headquarters of the U.S. Navy's Port Director at New York. This hard-worked officer was Captain F. G. Reinicke and his offices were in the well-known shipping building at 17 Battery Place on the Manhattan waterfront. Captain Reinicke was a man with many problems, for New York's harbour had become more and more congested as each month passed since the SC. and HX. convoys had started assembling there the previous September.

Up to the end of 1942 each of these convoys had taken only thirty merchant ships, but the Royal Navy was in the process of being convinced by a certain Professor P. M. S. Blackett, Head of the Admiralty Operational Research Section, that much larger convoys were practical, provided an increased escort was available, and convoys had recently been increased to sixty-four ships. Faced with the build-up of ships at New York, the Americans had tried to solve the problem by sailing eighty-one ships in the last convoy, HX.228, which had sailed at the end of February. But an order sent to the convoy five days out from New York, an order from New York but almost certainly demanded by the Admiralty in London, had resulted in twelve ships being detached from the convoy and ordered into Halifax. The Royal Navy was not yet prepared to escort such a large convoy across the Atlantic. (When these ships were suddenly detached from HX.228, there developed a prolonged exchange of flag signals between them as to which master should act as Convoy Commodore for the short voyage to Halifax. This was never resolved and, when Sable Island had to be passed, half of the 'convoy' steamed round one side of the island and half the other.) With a few stragglers also dropping out, HX.228 had carried on with sixty-four ships. Even though 128 ships had been sent out with the last two convoys, there were now on Captain Reinicke's desk lists containing the names of at least 160 merchant ships which were waiting at New York or Halifax to join SC.122 or HX.229, when the Admiralty in London would apparently take no more than 128 ships in the two convoys.

There must have been a spate of signals between New York, Washington, Halifax and London at that time; their exact content is not known but the result of them is. A decision was taken to split the fast convoy into two sections – HX.229, which would sail as planned on 8 March, and HX.229A, which would sail one day later. This sailing of an extra convoy in the cycle had occasionally happened before but it was an unusual step and reflects the great pressure to clear the shipping out of New York and get it to Britain. The extra convoy was approved by the British but only because Western Approaches had calculated that it could produce an additional escort group at St John's for HX.229A's ocean crossing. And so the decision was made for SC.122 to take sixty-four slow ships, including nine ships that were destined for Iceland, and for HX.229 and HX.229A to take half each of the eighty faster ships; this would still leave twenty or more slow ships which would have to wait for the next

SC. convoy.* A study of the lists for the two parts of the HX. convoy reveals that the most valuable of the fast ships were allocated to HX.229A. There were no less than thirteen fast tankers, eight large refrigerated ships and four cargo liners among the forty ships in this convoy. HX.229A would cross the Atlantic, therefore, with the fastest, largest and most valuable ships of the three convoys.

The last few details were settled in the planning of the convoys. The Admiralty and Eastern Sea Frontier chose a route for HX.229A (it is included in the map on p. 91) which would take a more northerly diversion than either of the other convoys but would be under air cover for a greater part of its passage. Stragglers' routes were also prepared for all three convoys so that ships which could not keep up the convoy speed could take a common route to England; they could not be escorted on the stragglers' route but at least there would be some idea of where they could be looked for if they became overdue.

Convoy and Routing cooperated with the Canadian naval authorities at Halifax in charge of the Western Local Escort Force over the provision of escort vessels to take the convoys as far as WOMP, the Western Ocean Meeting Point, off Newfoundland. These escort vessels did not operate in regular groups, and it is probable that the convoys had to sail with whatever was available on their separate sailing days. Fifteen escort vessels would take the three convoys out of New York and five more would appear during various stages of the convoys' voyage to Newfoundland. Sixteen of these ships were Canadian, two were British and two were Americans. More details of these will be given in a later chapter, but the choice of the commanders of the escorts of the three convoys as they left New York reflected the mixed bag of vessels making up these escorts. SC.122's senior naval officer was Lieutenant-Commander E. G. Old, R.C.N.R., Captain of the Canadian corvette *The Pas*; HX.229's escort commander was Lieutenant-Commander J. E. R. Wilford, R.N.R., a Union Castle officer in peacetime, in the British destroyer H.M.S. *Chelsea*, and HX.229A's escort would be led by a regular U.S. Navy officer, Lieutenant-Commander C. J. Whiting, in the Benson Class destroyer *Cowie*.

Convoy Commodores were chosen for the three convoys.

* Appendix 1, p. 332, lists the ships in the three convoys.

This would be done by the British Naval Liaison Officer at New York, Captain H. Auten, R.N.R., who had won a Victoria Cross in a celebrated Q-ship action in the First World War.* The Commodores had a simple rule, however, that whoever had been in New York the longest always took the next convoy. In this way, Captain S. N. White, R.N.R., in peacetime a Booth Line officer, would take passage on the British merchant ship *Glenapp* in SC.122; Commodore M. J. D. Mayall, R.N.R., a peacetime Canadian Pacific officer, sailed in the Norwegian ship *Abraham Lincoln* with HX.229, and Commodore D. A. Casey, R.N.R., a Royal Mail Lines officer, with the cargo liner *Esperance Bay*, would look after HX.229A. All three ships were 'reefers' – fast refrigerated ships with comfortable passenger accommodation. Of the three, Commodore Casey would believe that he had the best of the choice; most Commodores liked fast convoys. Each convoy would also have a Vice- and a Rear-Commodore chosen from the more experienced shipmasters of the convoys; they had no extra signalling staff and, if the Commodore's ship was lost, the Vice- or Rear-Commodore would have to carry on as best he could.

All of these details – main routes, stragglers' routes, lists of ships, escorts, Commodores – were concentrated into the Sailing Telegram for each convoy. This was then sent by Convoy and Routing to every headquarters concerned with the passage of that convoy.

While the plans for the convoys were being drawn up, final preparations for sailing were being attended to in the dozens of ships at New York. Bunkers were topped up, food and supplies taken aboard, cargo loading completed. There were many ships requiring repairs, particularly the older British ships that would have been scrapped if the war had not given them a new lease of life. Work was intensified on those which could be got ready for the coming convoys, but others would not make it in time and would be left behind.

There were, as always, a large number of servicemen and a few civilians on essential war business waiting to get across

---

* Lieutenant Auten had been in command of H.M.S. *Stockforce*, a heavily armed vessel disguised as a merchant ship, which had a long battle with U.80 in the English Channel on 30 July 1918. Auten's ship was sunk but he received his Victoria Cross for a determined fight.

the Atlantic. The allocation of the scarce passenger accommodation on the ships in the convoys was strictly controlled by the authorities. Many of the better-class merchant ships had a few passenger cabins; in peacetime such ships had been allowed to carry up to twelve passengers without having to be officially licensed as passenger ships. Every spare cabin was now packed tight with passengers – there were a batch of American naval officers posted for duty in London, a party of ten Australian sergeant-pilots who had completed their flying training and were on their way to serve with the R.A.F. in England, Catholic missionaries trying to get back from Africa to England, civil engineers, businessmen. The United Fruit Company ship *Cartago* embarked ten lady welfare workers of the American Red Cross for passage to Iceland; the Red Cross ladies were not happy when they later found that *Cartago* was in an exposed position in one of SC.122's outer columns. Among the passengers on the *Canadian Star*, the fast independent which had been forced to come to New York for a convoy after her gun had been damaged, were several children who had been born and brought up in India and Malaya. There was much excitement among these when they saw their first snowfall at New York. A year-old baby boy on this ship, Noel Wright, son of an R.A.F. wing-commander, would probably be the youngest passenger to have to brave the Atlantic crossing.

Not all of these passengers were looking forward to the voyage. Ensigns Frank Pilling and Boyce Norris, newly commissioned in the United States Navy, had already made one start. These were Ensign Pilling's feelings:

Our ship, the S.S. *Irénée du Pont*, had put to sea in a convoy at the end of February. She had been assigned 'Hell's Corner' as they called it – last ship in the first line to starboard. None aboard had felt cheered by this, so we considered it a piece of luck when the *Irénée*'s after deck cargo began sliding overboard in heavy seas. Captain Simonsen, the skipper, after he had seen several hundred thousand dollars' worth of crated aircraft preparing to plunge into the Atlantic, decided to bow out of Hell's Corner with a signal to the Convoy Commodore that the *Irénée* would be putting back to New York for re-stowage.

There were nine of us aboard the *Irénée du Pont* and we were happy to be back. 'Maybe,' we were saying in subconscious unison, 'we'll have a better position in a later convoy.' Some of us even made a special trip to the Operations Room of the Commander of the Eastern Sea Frontier to see what the odds were on a safe crossing. This room was evidently a naval version of a Hollywood version of what a naval operations room ought to look like. The officer who arranged our trespass may well have been a psychologist. His prediction that we had 'nothing, no nothing' to worry about, if based upon the convoy statistics of the day, could have been only a misreading amounting to illiteracy.

When we passed this officer's news to our friends, all were in favour of retirement to a cocktail lounge and a great forgetting about Hell's Corner and submarines. All except Ensign Norris. He decided to make one last, all-out effort to get a seat on a plane, or a 'shelf' on the *Queen Mary*. But we had all been through the static dynamics of naval transportation before. Most of us had spent two months in New York awaiting transportation and even Norris had to cave in finally to the transportation officer's negations. Sadly he returned to our linen closet on the *Irénée du Pont* with a foreboding so grim that the rest of us suspected that he had had a more recent interview with the operations fellow.

The *Irénée du Pont* would sail with HX.229 but, this time, she would be in the fourth column of ships from the starboard side of the convoy.

Another group of passengers were those who came under the traditional heading of D.B.S. – Distressed British Seamen. These were men who, for one reason or another, had been stranded ashore. A few had been landed sick from their ships or had got into trouble ashore but, at this time, most were seamen who had been rescued from merchant ships sunk in previous convoys and landed at American ports. Such men were always packed off back to Britain as soon as room could be found for them, and many of the British merchant ships carried a few of these. One ship, however, came in particularly useful. The biggest ship waiting at New York was the 14,795-ton *Svend Foyn*, at one time a Norwegian whaling factory ship but now converted for use as a tanker. The *Svend Foyn* still had its extensive whaling factory crew accommodation and over a hundred survivors of previous sinkings were sent aboard. Many of these had only just come out of hospital and a high proportion were Lascars. The *Svend Foyn* would sail with HX.229A with 195 men aboard and a cargo of 20,000 tons of high flash-point fuel.

There remained one last ritual before the convoys sailed – the Convoy Sailing Conference. On the morning of 4 March, the masters of the merchant ships allocated to SC.122, due to sail the following morning, were called to their conference held at

17 Battery Place. The notification of this conference to the
masters was the first indication to a crew that their ship was
about to sail. When dealing with some of the events connected with the three convoys covered by this book, it will be
more convenient on several occasions to deal with all three
convoys together, even though there may have been several
days between the incidents being described. The description
of the Convoy Conferences will be one of these occasions.
Questions to several of the participants resulted in the
impression that there was little difference between all three
gatherings.

The conferences were all chaired by a senior American
Naval officer, possibly Captain Reinicke. One English master
remembers:

> There was a considerable number of 'brasshats'. It was typical of
> all U.S. conferences which, in my opinion, were always overloaded
> with officers each having something to say. There was always an
> atmosphere of tension as though something sensational was about
> to happen at any moment. At the U.K. conferences the atmosphere
> was that of calm; you were told the facts without any window-
> dressing and you left the conference feeling that all would go well
> during the voyage. (Captain W. Luckey, M.V. *Luculus*)

A young British naval officer attended the conference of
HX.229 and his description may be typical of all three.

> I was fortunate enough to go to the conference with the captain
> (really just to carry his bag). The New York Harbour Master
> addressed the company at considerable length. I remember that it
> took about half an hour to get us down the water and clear of the
> swept channel. One had a feeling that most of the masters were
> impatient to get on with the ocean procedure. I remember wonder-
> ing whether the Harbour Master had ever left the safety of New
> York Harbour himself. As far as I can remember, the Commodore
> of the convoy spoke next and went over the daily routine in detail.
> Next came the Senior Officer Escort; I remember him saying how
> important it was for ships to keep well closed up and for each ship to
> maintain the designated speed of the convoy. Stragglers would be a
> menace and it was quite likely that they would have to be left to
> fend for themselves as escorts could not be spared.
> At the end of the meeting the Commodore asked whether anyone
> present had never been in a convoy before. I remember there were
> two American masters who put their hands up. One, I think, had
> never done any ocean sailing; his experience was confined to the
> Great Lakes. (Sub-Lieutenant A. D. Powell, H.M.S. *Chelsea*)

The masters were told what position they would take when

the convoy assembled, a most important piece of information, bringing pleasure to those not allocated to a vulnerable outside column. Each master was given a sealed envelope containing only the stragglers' route and he was not allowed to open this until twelve hours after he had put to sea. The main route was not disclosed to any of the masters for security reasons.

Although the conferences were largely routine, those who attended these particular ones do remember some of the specific items raised. The current shortage of escort vessels was stressed, although this was something most masters had heard before. Details were given of the estimated strength of U-boats believed to be in the North Atlantic and the fact that the U-boats were operating further west than usual. Masters were warned to be particularly careful about the accidental firing of Snowflake flares; no doubt this warning followed the disclosure of SC.118's position to a searching U-boat only a few weeks earlier by a carelessly fired Snowflake. One officer says that at this time there was always a discussion, 'sometimes an acrimonious argument', over whether naval guncrews on merchant ships could open fire without waiting for the master's permission.

At the HX.229 conference there was a little trouble over the times to be used during the convoy's seventeen-day voyage. Western Approaches instructions stated that Convoy Times would be used (Convoy Time was a time that changed as the convoy moved across the Atlantic). The Canadian Navy had asked the Americans to use Greenwich Mean Time but the sailing orders were actually in B.C.T., the meaning of which is not known. Commodore Mayall had much difficulty in sorting this out at his conference and was not entirely successful because when he issued flag orders to his ships during the convoy's passage he often received a number of queries on what time was being used. When he reached England Commodore Mayall included a sharp complaint on this in his report to the Trade Division of the Admiralty.

It would be an exaggeration to say that all of the merchant seamen were happy to learn that they would soon be sailing. Many of the British seamen had not been home for up to a year and, if they had to sail at all, were pleased at the prospect of seeing their families again. But the Americans were not all keen to leave their homeland for the dangerous Atlantic while, for the men from the occupied countries, it would be just one more voyage in a seemingly endless war.

You would be in New York with all the bright lights and you would think 'I've got to go back on that ship, everything dark and clamped down and sailing tomorrow in that convoy'. You would often think about deserting but you never did. (Cook S. Banda, S.S. *Zouave*)

But a few men did desert – drunks who couldn't find their ships, men whose courage had run out and couldn't face the next voyage, men who had taken up with women. Out of twelve Ships Articles studied, all of British ships, four men were found to have deserted at New York on this occasion – a radio officer from the *Southern Princess*, a fireman from the *King Gruffydd* and two men from the *Fort Cedar Lake*, which had already lost several men at various ports since leaving the Canadian shipyard in which she had been built. If other ships lost men at the same rate, it can be assumed that about forty men might have deserted at New York from the three convoys.

On one British ship, a messroom boy was put ashore ill. The British Consul sent for a 'distressed' able seaman who had left his own ship at Philadelphia with appendicitis several weeks earlier and, now recovered, was waiting for a passage home. The Consul packed this man off in a taxi and he was signed on as the messroom boy's replacement at the last moment. His service on his new ship would last a mere nine days.

One American ship was short of three seamen and the Union could not provide replacements. Three newly-trained men from the U.S. Maritime Service's camp at Hoffman Island were sent for, hurriedly taken to the Union Hall to become Union members, then to the Custom House where they signed articles before a Shipping Commissioner.

After signing on, the captain cautioned us to get our gear aboard immediately as the ship was all loaded and awaiting sailing orders. After a tiring struggle with crowds, subways and taxis, we arrived at Pier 59 on the Hudson River where the *Mathew Luckenbach* was berthed. I will never forget my first impression of the ship. I had never really been aboard a merchant ship before. At Sheepshead Bay (his training camp) they had led us to believe we would be placed on the new Liberty ships that were being built. As we stepped from the warehouse onto the pier, my heart sank. There, lying low in the water, vintage World War One, a mass of rust, booms in disarray, lashings running everywhere and toilet shutes flying in the wind, was the *Mathew Luckenbach*. I felt like turning around and running; only pride kept me going. Standing at the head of the gangway was the ship's crew. They stared at us as we came aboard and asked us who we were, for we had our Maritime uniforms on and they had never seen them before. (Able Seaman P. Civitillo, S.S. *Mathew Luckenbach*)

The men of the British ship *Glenapp*, who had been in New York for over two months, had formed several attachments ashore. Apprentice Hill managed to contact his girlfriend.

Shortly before the great moment of departure arrived, I nipped ashore to telephone Winnie that I would be unable to meet her that night as arranged. I didn't say why, just that I had a job on, and that was the last time I spoke to Winnie. Things didn't seem quite real to us junior members of the crew. It is something I can't really explain but somehow we suddenly felt a sense of loss, and for my own part I can only say that I never was bugged by homesickness but I felt I was losing something by departing from New York in a way that I never felt before, or since. Of course, it didn't take me long to get over this, for youth is pretty well resilient anyway.

With the thirty-one ships at Halifax, a total of 141 ships finally stood ready to sail across the Atlantic with convoys SC.122, HX.229 and HX.229A. There were cargo ships, tankers, refrigerated ships, bulk carriers and the two tank landing ships. They came from ten different countries – eighty-one were British ships, twenty-nine were Americans and the remainder were Panamanian (eleven), Dutch (seven), Norwegian (four), Icelandic (three), Swedish (two), Greek (two), Belgian (one) and Yugoslav (one). Their combined gross weight was approximately 860,000 tons and they carried 920,000 tons of cargo – 170,000 tons of various petroleum fuels, 150,000 tons of frozen meat, 600,000 tons of general cargo such as food, tobacco, grain, timber, minerals, steel, gunpowder, detonators, bombs and shells, lorries, locomotives, invasion barges, aircraft and tanks.

And there were the humans who would sail in the ships, possibly 9,000 merchant officers and seamen and 1,000 passengers of various kinds.

The convoys were ready to sail.

5

The Early Voyage

At 07.18 local time on 5 March 1943, the Glen Line refrigerated merchant ship *Glenapp* cast off from her wharf,
moved out into the North Hudson River and then steamed
slowly down towards New York's Upper Bay. On *Glenapp*'s
bridge stood her master, Captain L. W. Kersley, together
with Captain S. N. White, R.N.R., the Convoy Commodore,
and one of New York's best pilots. As *Glenapp* steamed past
Manhattan's skyscrapers and the terminals of the famous
peacetime transatlantic liners and out between Governor's
Island and Liberty Island, other merchant ships fell in behind
her. While this was happening four Canadian escort vessels
left Staten Island Navy Base and joined the merchant ships as
they passed through the Narrows between the Upper and
Lower Bays. Lieutenant-Commander E. G. Old, R.C.N.R.
('Pop' Old was believed by his crew to be the oldest sea-
going officer in the Royal Canadian Navy) placed his ship
H.M.C.S. *The Pas* ahead of *Glenapp*; *Blairmore* and *Rimouski*
fell in alongside the twentieth and fortieth ships to pass and
New Westminster followed on behind the last ship. Eventually
there were fifty merchant ships with these four escorts in a
column several miles long. The one-hundred-and-twenty-
second convoy in the SC. series was on the move.

The ships continued in single file, firstly south and then
south-east along the Ambrose Channel which had been swept
earlier that morning for any mines; there were none. When
the Ambrose Light Vessel was reached the pilots were
dropped and there was then enough really open water to
commence the complicated process of assembling the ships
into the planned convoy formation.

This was the responsibility of the Convoy Commodore.
Captain White requested the master of *Glenapp* to proceed at
a steady speed of 5 knots, two knots slower than the speed

eventually proposed for the convoy. The merchant ships had been ordered to leave New York in such an order as to make the forming-up as simple as possible. *Glenapp* was to be the leading ship in the centre column. The two ships immediately following her, the Harrison Line cargo ship *Historian* and the Bowring tanker *Benedick*, steamed up and took up their positions as leaders of the columns on either side of *Glenapp*; eight more ships came up and also became column leaders. The remaining merchant ships then fell in behind the leaders, each ship in its correct column and position; some columns had five ships, some had four.

The position a ship found itself in the convoy pattern was obviously a vital factor and one of much interest to the crews, the exposed flanks being reckoned the most dangerous positions. The policy adopted was that ships were allocated positions in the convoy according to their eventual destinations. In SC.122's case, the columns on the port side of the convoy were destined for Iceland, next came the ships which would go to Loch Ewe to join small coastal convoys going round the north of Scotland for East Coast ports, then came the Clyde and Mersey ships and, finally, in the extreme starboard columns, were the ships bound for Belfast, South Wales and Bristol. Some attempt was made to protect tankers and ships with passengers by placing these in the centre of the convoy but not at the expense of the 'port-of-destination' pattern described above.

The merchant ships had been the object of much attention all the time since they had left New York, the United States Navy putting on a good show of strength with numerous patrol vessels and Blimps (naval airships). One British merchant officer remembers that 'there was much more bull about sailing from New York. Despite ensigns and pennant numbers (flags denoting a ship's position in the convoy) being flown, there were always numerous challenges from zealous Morse fanatics.'

By late afternoon the convoy had formed into eleven columns numbered 3 to 13; columns 1 and 2 would be made up on the convoy's port side when the ships from Halifax eventually joined. Another British officer remembers that 'by then it was late afternoon and the weather had turned a little rough. The scudding cloud and hazy yellow sun low in the south-west, and the sea which was being whipped into spray seemed a dirty brown.' *Glenapp* increased speed to 6 knots and, when all ships in the convoy managed this without difficulty, went up to 7 knots ten minutes later. It is not recorded whether Captain White tried the trick of pushing up the speed again by half-knot increases until he found that the slowest ship could no longer keep up and then dropped back the last half knot and so established the maximum speed of his convoy, but it is recorded that the average speed SC.122 would make over the next 3,068 sea miles would be 6·94 knots.*

The sailors caught their last glimpse of land in the deepening gloom, Long Island on their port hand and the New Jersey coast to starboard; no more land would be seen by most of the men of this convoy for another eighteen days. The local patrol vessels and the Blimps having left, one escort officer comments that 'it always seemed a bit odd to us that once one got clear of the land, when you really needed anti-submarine cover, the local escorts and airships left you and all we had was the very small convoy escort which could never be termed adequate'. Lieutenant-Commander Old with his escort group of four tiny ships was now the sole protection these fifty merchant ships had. Old's own ship, *The Pas*, swept back and forth across the front of the convoy, *Rimouski* and *New Westminster* were stationed on the flanks, and *Blairmore* covered the rear. These Canadian ships were responsible for the protection of a convoy whose frontage covered five miles and whose perimeter measured more than thirteen miles.

For those on board the merchant ships who were on their first convoy it had been an interesting day with an impressive display of naval efficiency but, for most, this was all pure routine after more than three years of war. Convoy SC.122 settled down for its first night at sea.

That first night of the voyage passed without incident. The convoy ploughed steadily on taking an easterly course to get well clear of the Nantucket Shoals and to gain plenty of sea room before turning north-east towards Newfoundland. The weather on the next day was fair and the convoy made reasonable progress, covering about 160 miles in its first twenty-four hours out from New York.

Lieutenant-Commander Old was pleased with the be-

* All miles quoted will be sea miles (2,000 yards) unless otherwise stated. A knot is a speed of one sea mile per hour.

haviour of the merchant ships; he noted in his report that there were 'no bad stragglers and very little smoke'.* One patrolling aircraft put in a brief appearance during the day; it was probably a B-25 Mitchell from either the 3rd or the 11th U.S.A.A.F. Anti-Submarine Squadrons based at Fort Dix, New Jersey. There was nothing much for the Mitchell to do but it was a comforting sight. One merchant ship had to put back during the day. The Greek ship *Georgios P*, loaded with sugar, was built in 1903 and was the second oldest ship in the convoy. When she could not keep up, Captain White was not prepared to slow the convoy down for her and ordered her back to New York.

During the late afternoon of that first full day at sea the weather began to deteriorate. Squalls of heavy rain drove at the convoy from the south and soon the rain was continuous and the wind had reached gale force. The merchant ships started to roll heavily with the wind and seas pounding them from the beam and the columns of ships began to lose their neat formation. When night fell the gale was still increasing.

There was little the four escorts could do to hold the convoy together in the dark. The heavy rain and then the sheets of spray from the heavy seas made conditions miserable for the Canadians on their open bridges and look-out positions. One of their officers describes the winter storms which these local escort groups often encountered.

It was sheer unmitigated hell. She was a short-foc's'le corvette and even getting hot food from galley to foc's'le was a tremendous job. The messdecks were usually a shambles and the wear and tear on bodies and tempers was something I shall never forget. But we were young and tough and, in a sense, we gloried in our misery and made light of it all. What possible connection it all had with defeating Hitler none of us bothered to ask. It was enough to find ourselves more or less afloat the next day and the hope of duff and pudding and a boiler clean when we reached port. (Sub-Lieutenant T. C. Marshall, H.M.C.S. *New Westminster*)

On the merchant ships there was some seasickness among those who had not yet regained their sea legs and those who

* British and Canadian ships submitted a Report of Proceedings at the end of every voyage; for convenience these will simply be referred to as 'reports' in the text. American ships and German U-boats kept War Diaries. All ships also kept logs but these contain very little useful information. Copies of the local escort's Reports of Proceedings were provided by Canadian Forces Directorate of History, Ottawa.

had enjoyed themselves too much in New York only two nights ago. Some of the ships were being badly affected by the storm, suffering damage to upperworks or shifting cargo, and several hove to in order to save their lifeboats from being smashed by the seas. The two Tank Landing Ships had been chosen as column leaders, possibly in deference to their Royal Navy status, but they proved very difficult to handle in the gale. L.S.T. 365 had much trouble with her steering which broke down in the storm – the first of fifteen such break-downs she would suffer before the voyage was over.

As Officer of the Watch, the steering failures were particularly to be dreaded. Visibility without lights was bad enough at night but to have the ship go out of control close to other ships and in heavy seas and gale conditions was particularly nervewracking for a junior sub-lieutenant who was keeping a watch on his own for the first time. Because of the steering breakdowns the Commodore later told us to leave our position as column leaders and take station at the rear of the columns. I read a newspaper report after the convoy and one merchant sailor had described us as 'those funny looking ships astern of the convoy'. (Sub-Lieutenant J. H. H. Bayley, L.S.T. 365)*

When dawn broke on the following day SC.122 was scattered over many miles of ocean. The storm abated and the escorts started the tedious task of shepherding the merchant ships back into order. They were helped by the fact that, before the storm, the Commodore had signalled a rendezvous position for noon the next day, so that within a few hours the convoy had once more resumed formation. But eleven of the forty-nine merchant ships were missing.

It should be stressed that this type of experience was almost routine for convoy sailors. Convoys were scattered by storms and reformed, scattered again and reformed, often several times during a voyage. This one-night gale experienced by SC.122 and the subsequent depletion of its numbers was nothing out of the ordinary compared to the experiences of some convoys and, while the passengers might have been sick and terrified, the merchant and escort sailors were quite used to such weather.

None of the missing ships had sunk. Two returned to New

---

* Both of the L.S.T.s arrived safely in England. L.S.T. 365 was fitted out with radar and wireless equipment and became a Fighter Direction Ship. She took part in the landings on Sicily, Salerno and Normandy, and survived the war. L.S.T. 305 was torpedoed, probably by an E-boat, off the Anzio Beachhead and sank while under tow to Naples.

York, six more failed to reach the rendezvous and put into Halifax, one of them, the *English Monarch*, had to have repairs to a 15-foot crack in the plate of her deep tank. These ships all caught later convoys. Two more of the missing ships continued to sail along the convoy's track and caught up two days later. These were the *Vinriver* and the *Kedoe*. The rejoining of the *Kedoe* caused some interest on the bridge of the Commodore's ship.

A couple of days later this small Dutch freighter turned up. No one could believe their eyes at first and Captain White came on the bridge and watched for about an hour as this ship battled against the elements to get into her correct station again. All the time he was muttering words of encouragement and then, before he left the bridge, he told the signaller to send 'Well done. Look up Luke 15–6'.* A couple of hours after this we got a frantic call up from the Dutchman and the message 'I cannot find it in my Confidential Books'. We never did find out whether he had, in fact, been looking in his Confidential Books or whether he just had a sense of humour too. (Apprentice C. H. F. Hill, M.V. *Glenapp*)

There remained one more ship missing. This was the British cargo vessel *Clarissa Radcliffe*, which was due to become the mystery ship of this convoy. She was seen on 9 March, two days after the storm, by Lieutenant-Commander Old just after *The Pas* had left SC.122 and was putting into Halifax. Old gave her the convoy's position, course and speed, and advised her to try and catch up again. At this time the *Clarissa Radcliffe* was within less than one hour's steaming of the convoy but she never rejoined. What happened to her will be discussed later.

Three days after SC.122 departed from New York, HX.229 sailed with forty ships, and one day later the extra convoy, HX.229A, left with a further twenty-seven ships. This great clear-out left New York harbour with more room than it had seen for many weeks. These two convoys left New York, formed up and set course in much the same way as had SC.122. They were escorted by mixed groups of Canadian, American and British escorts; the smallest of the three convoys, HX.229A, actually had the largest escort of seven ships. It is not known whether this was due to a better availability

---

* Chapter 15, verse 6 of St Luke's Gospel: And when he cometh home, he calleth together his friends and neighbours, saying unto them, 'Rejoice with me, for I have found my sheep that was lost.'

of escorts when that convoy sailed or whether the greater value of its ships warranted this increased protection.

Both convoys missed the gale that had dispersed SC.122 but were troubled instead by fog and later by heavy snow. HX.229A was in thick fog for two and a half days and when the fog lifted only eight ships were found to be still with the Commodore, but most caught up again and only three ships fell out for good from the two convoys – the American tanker *Southern Sun* put back to New York almost as soon as HX.229A sailed; another, the *Sunoil*, had to put into Halifax with engine trouble, and the *Clan Matheson*, which was Vice-Commodore ship of HX.229, could not maintain the convoy speed and had to drop out and wait at Halifax for a slow convoy. Again, these sundry departures were all quite routine.

Once all three convoys had got well clear of the land they turned north-eastwards for Newfoundland, passing 150 miles off Cape Cod and later clearing Cape Sable, Nova Scotia, by the same distance. SC.122 was steaming at a steady 7 knots and the two HX. convoys, making 9 to 10 knots, were gradually overtaking. There was no zig-zagging; this had recently been abandoned by Admiral Horton as a waste of time for convoys of less than 12 knots. The convoys kept the same steady heading for several days, only altering course at the principal route-turning points.

For the merchant ship officers the main task was station keeping.

On the first two days, adjusting and varying the ship's speed to keep exact station from the next ahead occupied the watch-keeping hours. The difficulty of this can be understood when you think of four or five ships all of different size, power and propeller pitch, which by trial and error had to arrive at a mutual speed and keep some 400 yards apart, especially in pitch black nights with perhaps bad weather conditions. During this time we studied and became acquainted with our new neighbours and they became part of one's surroundings. (Third Officer R. McRae, S.S. *Coracero*)

Coal-burning ships were particularly difficult to keep to the convoy's speed with their furnaces requiring clinker removing every watch with a consequent loss of steam pressure. This was followed by a stoking up with new coal which usually sent a dense cloud of smoke pouring out of the funnel. The pall of smoke that perpetually hung over every convoy could be seen from up to fifty miles away and any master of a ship

that caused unnecessary offence with smoke was sharply reprimanded either by the Commodore or by one of the escorts.

Navigation was the responsibility of the second mates
(second officers); they had to send their estimated noon
positions each day by flag signal to the Commodore, who
then averaged them out for the convoy's position. The Commodore often sent signals back to his merchant ships informing them of future course changes, re-arranging the position
of ships, or dealing with any one of the numerous domestic
arrangements that needed attention during a convoy's passage. The Commodore was in touch with the naval escorts by
radio telephone and his staff could always decode the incoming naval signals so that the crew of the Commodore's ship
knew more about what was happening than anyone else in
the convoy.

The ordinary sailors went about their routine duties, keeping their traditional watches of four hours on and eight off.
There were always plenty of men on look-out duty searching
for any sign of the enemy, but a keen eye was also kept on the
Commodore's ship so that his signals should not be missed.
Only in good weather could the lookouts see the entire convoy, and in heavy seas a small ship might be in a trough of the
waves with not another ship to be seen until it breasted the
next wave and then thirty or forty ships would appear again.

The following description by an American seaman is of his
first trip to sea:

Finally, we were on our way – to England, we all knew, for the
code markings on the cargo told this. When we left New York, the
men refused the new rubber life-saving suits which the Chief Mate
tried to distribute. They said they weren't much good. I would have
liked one because we had training with them in boot camp but I
went along with my watch partners whom I considered experienced.

I remember my first watch at sea. There were three men per
watch and it was divided into one hour and twenty minute shifts –
one man was on the bow look-out with the Navy gunners; one man
on the bridge and one man stood-by in the mess hall on call. After
one hour and twenty minutes we would rotate until our four hours
were up. My first time relieving the man on the bridge, I approached Mr Heyme the Chief Mate, I saluted him snappily and
said, 'Relieving the watch, Sir.' The Chief Mate stood there a
moment, dumbfounded, and then he replied, 'Oh my, how lovely!'
Our training had taught us to relieve this way, formally. I soon
learned there were no formalities on merchant ships. (Able Seaman
P. Civitillo, S.S. *Mathew Luckenbach*)

Off-duty hours were spent playing cards or other games for hours on end, and reading. Drink was available on British ships but not over-indulged in while at sea; American ships were 'dry'. A great trial was the order that all private radios had to be handed in as soon as a voyage commenced; it was thought that the amount of 're-radiation' from these might enable U-boats to take bearings on a convoy's position. Only one radio was allowed for each ship. The early evening was the favourite time for relaxation but, soon after eight bells (8 p.m.), ships became quieter as men started to turn in to get as much sleep as possible while conditions in the convoy were still peaceful. Most men slept fully clothed with a life-jacket close by. The night was not a happy time, for it was then that danger was more likely to strike, and every sailor was pleased to see another dawn.

The voyage along the American and Canadian coasts was even more a matter of routine for the local escort vessels, this being their usual 'beat'. The captain of the senior escort vessel was always in overall command, even though the rank of the Convoy Commodore was usually far greater than his own. The waters through which the convoys were sailing had been the scene of the U-boats' second Happy Time a year earlier, just after the American entry into the war, but the intro-duction of convoys and the establishment of air patrols had persuaded Dönitz to move to easier hunting grounds, although he still sent one or two U-boats into the area some-times to force the Allies to retain the convoy system here and to tie down escort vessels.

Most of the local escorts were Canadians. Because of the huge wartime expansion of the Royal Canadian Navy these ships were manned almost entirely by Reservists and war-time sailors. Most of the junior officers were inexperienced R.C.N.V.R.s but the feeling of one of them that 'we were all badly trained, scared stiff and most of the time wished to God we had joined the Air Force' is maybe an extreme view. The ratings were mostly wartime men and, except for a few fishermen, were without any sea experience; indeed, of the many who came from the inland provinces – the 'prairiemen' – few had seen either the Atlantic or Pacific before joining the Navy.

An example of the inexperience of some of the Canadians

can be seen in this incident, which occurred while SC.122 was between New York and Halifax.

On midnight watch, Signalman Stapleton yelled at me 'Ship in trouble astern'. I looked and saw one of the rear ships silhouetted against an ominous red glow. 'It's on fire,' I thought, 'It's been torpedoed.' I called down the tube to the chart room where our salty Merchant Marine navigator was. Joe had a wealth of knowledge and experience. He quickly reached the bridge, looked through his binoculars and said, 'Damn it – Haven't you ever seen the moon come up before?' (Lieutenant T. MacDonald, H.M.C.S. *Blairmore*)

HX.229 and HX.229A each had one American destroyer among their local escorts. These were the *Kendrick* and the *Cowie*. Both were modern ships that had recently taken part in the North Africa landings and were now a temporary contribution from the United States Navy to the Western Local Escort Force. An officer on one of the Royal Navy ships present remembers the *Kendrick*.

It was unusual to have a U.S. destroyer with us. I remember we remarked on her enclosed bridge which gave no means to the officers to sight aircraft and this would have been an impossible position had there been any enemy aircraft about. No doubt it was pleasantly warm in the wheelhouse/bridge but we had learnt the hard way that one had to be able to withstand the elements on an open bridge. (Sub-Lieutenant A. D. Powell, H.M.S. *Chelsea*)*

During their fourth day out from New York each convoy reached a position off Nova Scotia called HOMP (Halifax Ocean Meeting Point) and several changes in the composition of the convoys occurred here. The exact position of HOMP varied with each convoy and was selected so that any changes necessary could be made in daylight. Fourteen ships were ready to sail from Halifax on the morning of 8 March to join SC.122 – thirteen merchant ships and the small steamer *Zamalek* which was to be SC.122's rescue ship.

* *Kendrick* and *Cowie* both served in the Sicily landings later in 1943 and in many further wartime engagements. It is probably only a coincidence that Sub-Lieutenant Powell should have commented upon *Kendrick*'s covered bridge and the possibility of her being vulnerable to air attack, but she was caught in a surprise attack by a German torpedo bomber one night in September 1943 off Oran. A torpedo struck *Kendrick*'s stern and exploded but she did not sink and there were no casualties. She did, however, shoot down the bomber. Many of the Canadian local escorts also 'graduated' from their dull work on the Atlantic coast and went to more active theatres of war; several were present at the invasion of Normandy. A full list of the SC.122, HX.229 and HX.229A local escorts is given in Appendix 2, p. 340. All survived the war.

There was some trouble over the provision of convoy rescue ships at this time. Ideally every convoy should have had one while it was crossing the main U-boat area of the North Atlantic, and it had been planned to provide twenty rescue ships for this purpose. But this figure had never been reached; six rescue ships had already been lost in convoy battles and only ten were available at that time – enough for only one convoy in four. The Americans were being pressed to provide rescue ships to fill the gap as their merchant ships were also being sunk in the convoys but, so far, none had appeared. Merchant sailors were always pleased when a rescue ship was with their convoy and morale went down if one was not present. Western Approaches had recently issued an order that a trawler or an ocean-going tug should accompany every convoy which did not have a proper rescue ship, but the necessary ships were just not available.

Rescue ships like *Zamalek* were manned by British Merchant Navy officers and men, but with a Royal Navy doctor and also naval signallers for the High Frequency Direction Finder with which she was equipped. Such ships, therefore, carried out a dual role – rescuing the crews of merchant ships that had been sunk and helping the escorts of a convoy to plot the signals of nearby U-boats. The rescue ships' small size and shallow draught made them a difficult target for a U-boat torpedo attack but, even so, it was a highly dangerous and nerve-wracking practice to stop near a torpedoed ship picking up survivors. *Zamalek* was an efficient and experienced rescue ship, although one of her firemen deserted at Halifax just before this voyage. The SC.122 convoy was lucky to have her. No rescue ship of any kind was available to sail with HX.229 or with HX.229A.

The procedure laid down for convoys without a rescue ship was that, if a merchant ship was torpedoed, the last merchant ship in the column of the ship that had been hit was to stop and act as rescue ship. This extremely unpopular order had been given to masters at all the convoy conferences at New York. The escort group commander could also detach one of his ships for rescue work but *only if he thought it justifiable to do so*; his main task was the 'safe and timely arrival of the convoy'. It will be seen later how different officers reacted to this situation.

Coming back to the fourteen ships due to join SC.122,

these left Halifax with three escorts – the destroyer H.M.S. *Leamington* and two Canadian corvettes, *Cowichan* and *Dunvegan*. *Leamington*'s report shows that, after she had had 'a brisk brush with a water boat who had evidently never learned the Rule of the Road', she picked up the merchant ships outside Halifax harbour for the short voyage to meet SC.122. *Leamington* and *Zamalek* had one thing in common; they had both been with the ill-fated PQ.17 convoy to Russia eight months earlier.

One of the merchant ships that sailed from Halifax was a small Icelandic cargo ship bound for Reykjavik with timber and food. The comments of one of her officers, when asked about this voyage, reveal the attitude of this man to wartime convoys.

You name the convoy with letters and numbers; I never knew that such a system existed. The Captain went to a meeting ashore, which was conducted by some directors of operations of this farce, and received there some orders to be followed on the voyage. These orders were in a sealed envelope which was not to be opened until we had got far out to sea and nobody knew anything even then save the Captain. (Second Engineer Björn Jonsson, S.S. *Fjallfoss*)*

Just over one day later *Leamington* met the main convoy with the help of a patrolling Digby aircraft from the Canadian air base at Dartmouth, Nova Scotia. The Digby found SC.122 and guided *Leamington* and her merchant ships to a meeting. Captain White slowed down his convoy slightly and the new merchant ships took their places in the port columns of the convoy. The three new escorts also stayed with the convoy, relieving *The Pas* and *Blairmore* which departed with two merchant ships that had only joined SC.122 in order to reach Halifax. The convoy picked up speed and steamed on, fifty-one ships strong with five escorts. Responsibility for the safety of this collection of shipping was now held by a young R.N. lieutenant – Lieutenant A. D. B. Campbell, captain of H.M.S. *Leamington*.

HX.229 had no changes to make off Halifax but HX.229A had both escort and merchant ship changes which were carried out in similar manner to those of SC.122 already described. Everything was efficiently carried out and there were no incidents. Two of the ships joining from Halifax now

* The *Fjallfoss* survived the war and, although built in 1919, is one of the few ships in the convoys covered by this book to be still in working service. She is now the Saudi-Arabian owned *Star of Taif*, operating from Jeddah.

became the largest and smallest in the three convoys. The 15,130-ton British refrigerated cargo ship *Akaroa*, in HX.229A, was the largest and the tiny 775-ton Icelander *Selfoss*, in SC.122, was the smallest.

The convoys were now in the Canadian Coastal Zone and under the tactical control of COAC (Commanding Officer Atlantic Coast), who at that time was Rear-Admiral L. W. ('Rastus') Murray, with headquarters at Halifax. The Canadians twice ordered small diversions to HX.229's route before the convoy reached Newfoundland. The first of these was because it was found that HX.229 had no need to go to the Halifax Ocean Meeting Point and the second probably diverted HX.229 from the path of an incoming convoy. These two diversion signals were to have unfortunate consequences. Overall control of the convoys still remained with Convoy and Routing at Washington. The Admiralty in London was being informed of every move but there was still nearly a week to go before the Admiralty would take over from Convoy and Routing when the convoys reached the CHOP Line.

It would take the convoys another three days to reach the position due east of Newfoundland where the ocean escorts were due to take over. The minor events of these three days can quickly be described. SC.122 nearly ran down a strange ship one night; it turned out to be a corvette searching for an incoming convoy and according to H.M.S. *Leamington's* report the stranger was 'driven off'. The Canadian corvette *Rimouski* obtained an Asdic contact on what she thought was a U-boat near SC.122 and she made several depth-charge attacks but it must have been an unfortunate whale or a shoal of fish or a layer of unduly dense water, for it is known that there were no U-boats about. In fact, there had never been any danger to the convoys from U-boats since leaving New York harbour. The nearest U-boat to New York had been out in the Atlantic, 1,400 miles away to the east.

H.M.S. *Chelsea*, which had commanded HX.229's escort all the way from New York, was relieved by H.M.C.S. *Annapolis*, another ex-American destroyer, half-way between Halifax and Newfoundland, and several more merchant ships fell out for various reasons. One of these was the Liberty ship *Stephen C. Foster*. She was in the same column of HX.229 as another Liberty, the *William Eustis*. These two ships had been built together in the Todd Shipbuilding Corporation yard at Houston, Texas, and had consecutive yard numbers. They had both been launched in January and were now on their first convoy voyage, both loaded with cargoes of sugar. The *Stephen C. Foster*'s welded plates started to split and she had to put into St John's. The *William Eustis* carried on and was to be lost within the week. The *Stephen C. Foster*'s split hull plates may have saved her, for she survived the war and did not go to the scrapyard until 1961.

It will several times be convenient to leave the detailed description of the convoys' progress and examine events elsewhere that were to affect the fortunes of the convoys. This first review will look at the U-boat situation in the North Atlantic while the three convoys were approaching Newfoundland, and the position in St John's harbour where the ocean escort groups were preparing to sail and take over the protection of the convoys.

All through the winter Dönitz had persisted with his policy of concentrating his U-boats on the North Atlantic convoy routes. New U-boats coming into service continued to exceed losses and, with at least one supply U-boat always on station in the Air Gap, Dönitz was able to maintain a daily average of nearly fifty U-boats actually available for operations on the North Atlantic convoy routes. This was many more than had ever before been available and, also, more than he would ever manage again. These boats were organized into separate *Gruppen* (groups) which first formed a patrol line and then attacked any convoy found by a member of that group. There were now enough U-boats to keep at least three such groups in being at any time.

While SC.122 and the two HX. convoys were steaming peacefully up the Canadian coast, their predecessors out of New York, SC.121 and HX.228, had both been caught by U-boat groups. SC.121, taking the northern route, managed to pass through a patrol line unseen but was spotted soon afterwards by a lone U-boat. The Westmark group, with seventeen boats, pursued the convoy for four days and sank thirteen ships, many of which had straggled in the heavy weather. Among the ships lost was the Norwegian *Bonneville* with the Convoy Commodore, Commodore R. C. Birnie, R.N.R., who was drowned. No U-boats were sunk in

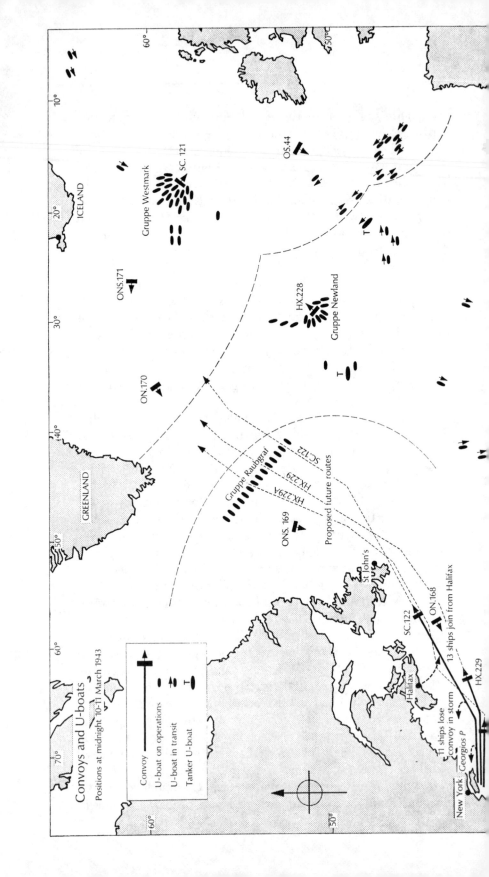

Convoys and U-boats

Positions at midnight 10–11 March 1943

Convoy

U-boat on operations

U-boat in transit

Tanker U-boat

GREENLAND

ICELAND

Gruppe Westmark

SC. 121

OS.44

ONS.171

ONS.170

HX.228

Gruppe Newland

ON.170

Gruppe Raubgraf

SC.122

HX.229

HX.229A

ONS. 169

ONS. 169

Proposed future routes

St John's

SC.122

ON.168

13 ships join from Halifax

Halifax

11 ships lose convoy in storm

Georgios P

New York

HX.229

this one-sided battle. HX.228 had taken a more direct route and was sighted by the most southerly boat of the Neuland group. During the night of 10/11 March a desperate battle took place in which four merchant ships, one destroyer and two U-boats were sunk. The destroyer was H.M.S. *Harvester* whose captain, Commander A. A. Tait, was leader of the escort group. Commander Tait also drowned.

It should not be thought that the North Atlantic was the only place where convoys were under attack. In the first week of March one convoy had been attacked off Durban, two on the Gibraltar run and another on the Russian run. Forty-one merchant ships were sunk by U-boats in all areas during the first ten days of the month.

A look at the map on p. 116 will show that there was an area east and north-east of Newfoundland through which all North Atlantic convoys had to pass. The convoys were forced through this bottle-neck because any major diversion to the south would have left the convoys for long periods without air cover and the smaller merchant ships would have risked running out of fuel before reaching England. Knowing this, Dönitz tried to maintain one of his U-boat groups permanently in the bottle-neck area to catch both incoming and outgoing convoys. This group was often within the radius of operations of Canadian and American aircraft stationed in Newfoundland and these aircraft tried to keep the U-boats as far out to sea as possible, but the group was a thorough nuisance to Allied planners attempting to route convoys into or out of the North Atlantic. It was Gruppe Raubgraf with fourteen boats that was now on station. It was well to the north at this time, searching for an incoming convoy, ONS.169, but this convoy evaded the U-boats partly due to weather so severe that fourteen out of the thirty-seven ships in the convoy became stragglers.

U-boat Headquarters knew all about the convoy cycles and watched very carefully for any variations to the regular pattern. Recent changes had left the Germans uncertain as to whether there was now an eight-day or a ten-day interval between sailings from New York of both SC. and HX. convoys. (The Germans called these the Sydney Convoys and Halifax Convoys despite the move, known to the Germans, to New York.) They also believed that at this time the convoys were being sent by routes north and south of the Great Circle to a pattern – four convoys by the northern route, four by the southern. The Germans were mistaken in this; there was no such pattern. But Dönitz's staff officers were certainly waiting for SC.122 and HX.229 to reach the open ocean past Newfoundland.

Their great helpers, the codebreakers of Kapitän Bonatz's B-Dienst, were also at work. In a previous chapter Bonatz told how his department never picked up every relevant signal and did not manage to decode all those that were picked up. Bonatz's personal diary and the weekly reports of his department now lodged at the Bundesarchiv in Koblenz reveal the exact extent of the B-Dienst's achievements with the convoys whose fortunes are being followed here.

U-boat Headquarters already had a B-Dienst message which contained the sailing date (9 March) of HX.229A, that convoy's future route and stragglers' route and the positions of its Halifax and St John's meeting points. The details were absolutely correct except for one fact – the Germans believed that the convoy involved was HX.229. (When U-boat Headquarters later found that HX.229 had sailed on the 8th, the keeper of its War Diary complained that it had sailed a day early!) It was most unusual for the B-Dienst to get a complete route in this way and the means by which the B-Dienst picked up this particular signal should be examined.

A signal became potentially insecure and its code liable to breaking the moment it was transmitted by wireless. For this reason, land lines or undersea cables should always have been used for signals from one land-based headquarters to other land-based headquarters. This particular signal about HX. 229A originated at COMEASTSEAFRON (Commander Eastern Sea Frontier) in New York at 22.10 on 4 March. The original signal sheet shows that it was sent to Convoy and Routing at Washington, the Admiralty in London, and to various headquarters at Ottawa, Halifax, St John's, Sydney and Iceland and to the Port Director at New York. A note on the signal says that the Port Director's copy was to be sent in plain language by an officer messenger.* All other addressees could have been reached by land line or cable. How did the Germans pick up this message?

* Copies of the convoy signals were provided by the Naval Historical Center, Washington.

The officer messenger to the Port Director can be eliminated at once; the B-Dienst report says that their information came *'nach einem Funkspruch'* (from a wireless message). There are two possibilities. It could be that the Atlantic cables were overloaded with messages at this time and someone, to save time, routed this signal by wireless although it was against standard procedure to do so. Another possibility is that this signal was later repeated to an addressee who could not be reached by cable. Later signals about the convoys included an addressee in addition to those quoted above – 'SOPA Greenland'. This was the Senior Officer Present Afloat of the U.S. Coast Guard's Greenland Patrol. The route in this signal did pass near Greenland while the routes for SC.122 and HX.229 were further south and may not have been transmitted to 'SOPA Greenland'. Whatever had gone wrong, U-boat Headquarters were in possession of a valuable piece of information. The only other complete route the Germans picked up at this time was that of an all-American convoy, UGS.6, a slow military convoy from New York to Casablanca. It was later attacked along its route.

Worse was to come. Once a convoy had put to sea, any signal to or from it had to be wireless transmitted and the security of that signal would depend upon the impregnability of its codes and ciphers. The escort commanders of both SC.122 and HX.229 exchanged several signals with New York and Halifax. These signals concerned various events such as HX.229's two small diversions, the need to provide rendezvous points for new escorts joining along the route, and SC.122's bad gale – the B-Dienst papers neatly list the names of all the stragglers after the gale. Eight accurate reports about SC.122 and HX.229 are recorded in the B-Dienst papers but none for HX.229A. Perhaps its escort managed without sending signals; perhaps the methodical Germans, expecting only two convoys, only looked for signals about two convoys.

So, as the convoys approached Newfoundland, the staff officers at U-boat Headquarters knew that two convoys were on the move; they had the complete future route of one convoy; they had an accurate record of the progress of SC.122 and HX.229 between New York and Newfoundland. But they did not realize that an extra convoy had sailed and they had no future routes for two of the convoys. Besides the thirteen U-boats of the Raubgraf group off Newfoundland, thirty-three more boats would shortly be released from their battles with SC.121 and HX.228; many of these were still fresh enough or could be refuelled for further operations, and at least as many more boats were on their way out from their bases as would have to be released for home.

The ships and the men who would shortly be facing these U-boats were mostly at St John's, the Newfoundland port which was the most easterly point of North America. A minor crisis was brewing here that would affect the future fortunes of the three convoys.

The provision of escort groups for the ocean crossing of convoys was the responsibility of the Headquarters of Western Approaches in Liverpool; a time-table was maintained for many months ahead, providing an escort group for each regular convoy. At the St John's end of the ocean route each group was, in theory, supposed to have at least two clear days for rest, repair and a little refresher training before sailing to join an England-bound convoy. The Western Approaches programme listed B5 Group to take over SC.122 on 12 March and B4 Group to take over HX.229 on 14 March. The sailing from New York of an extra convoy was not the cause of any difficulty; an extra group, the 40th Escort Group, had been sent specially from England and was ready to take over HX. 229A on 15 March. Three groups, therefore, had to be ready to sail from St John's in a four-day period. It is important that the composition and state of readiness of these groups be examined.

B5 Group was one of the veteran British escort groups that had been operating in the North Atlantic since late 1940, although it had spent the last nine months of 1942 on loan to the Americans and had been working off the American East Coast and in the Caribbean.* The senior ship was H.M.S. *Havelock* – a typical group-leader ship of the Havant Class – with Commander Richard C. Boyle, R.N., as captain and

---

* In January 1943 part of B5 Group had escorted a convoy of nine tankers on a new convoy route from Trinidad to Gibraltar. German U-boats found this convoy and attacked it, sinking seven of the nine tankers. The four-ship escort was so small that a U-boat trailed the convoy quite openly by day and, at one stage, an officer on the destroyer *Havelock* ordered a light signal to be flashed to the U-boat, 'Why don't you go away?' Back came the reply, 'Sorry, we have our orders.'

commander of the group. Boyle had not arrived in Western Approaches until late in 1942, but had already established a reputation in the group of being a careful but sound leader who followed established procedures at all times and demanded absolute efficiency from his own ship and the rest of the group. There were two other destroyers in the group, both of the First World War V and W Class – *Volunteer* which had just completed her first convoy crossing after a major refit, and *Warwick* which was temporarily absent, its place being taken by a new frigate, *Swale*. There were five corvettes in B5 Group at this time – *Godetia* and *Buttercup*, which were manned by Belgians, and *Pimpernel*, *Lavender* and *Saxifrage*.

B5 Group had recently brought Convoy ON.168 across from England, a tough voyage by the northern route with seventeen days of steaming into continuous gales, but Commander Boyle and his group were able to spend three restful days in St John's after coming in from this tiring convoy. The only ship that would not be ready to sail on time was *Volunteer*, which was in dry dock with leaking plates being repaired, but she would be ready to sail and overtake one day after the remainder of the group sailed.

B4 Group, due to take over Convoy HX.229, was in many ways similar to B5. The commander of the group was Commander E. C. L. ('Happy') Day in H.M.S. *Highlander*, another Havant Class destroyer. Day was a descendant of a naval Crimean War Victoria Cross winner, Lieutenant George Fiott Day, and he had served on North Atlantic escorts since January 1941 and commanded B4 Group for the past year and a half. Commander Day had been described as 'sound, blunt, brainy and very experienced'. Day's group at this time was sadly depleted. Of his other destroyers, the *Winchelsea* was absent refitting and its replacement, *Vimy*, was at Iceland under repair, but the old four-stacker *Beverley* was present. Of the six corvettes normally in the group only four – *Abelia*, *Anemone*, *Pennywort* and the Canadian *Sherbrooke* – were still with the group after the winter storms. B4 had experienced much trouble with its last convoy, ONS.169, and by the time the group was due to sail from St John's, *Highlander* would be in dry dock with serious leaks to be repaired and a new Asdic dome to be fitted, *Vimy* would be still in Iceland, *Pennywort* would also be delayed with engine trouble and *Abelia* and *Sherbrooke* would still be at sea escorting the remnants of ONS.169 which had almost dispersed in bad weather. *Beverley* and *Anemone* would be the only ships in the group ready to sail but, in *Anemone*'s case, after only one day's rest in port.

The ships of the 40th Escort Group, due to take over the third convoy HX.229A, were strangers to the North Atlantic. The 40th was one of Western Approaches' long-range escort groups, based at Londonderry, whose normal duty was the escorting of convoys on the Sierra Leone run. But the sailing of these convoys had been suspended for several months and the escorts thus freed had been employed in a variety of tasks since then. It is ironical that the need for the extra convoy was partly due to the re-routing of merchant ships after the suspension of the Sierra Leone convoys and that its escorts were also available for the same reason.

The composition of this group was quite unlike that of the normal North Atlantic group. It contained three 900-ton sloops – *Londonderry*, *Aberdeen* and *Hastings*, two large 1,546-ton ex-U.S. Coast Guard cutters on loan to the Royal Navy and now named *Landguard* and *Lulworth*, and two 1,045-ton frigates – *Moyola* and *Waveney*. Such a group of modern, fast, large escorts was a complete contrast to the mainly old destroyers and slow corvettes of B4 and B5 Groups. Commander John S. Dalison was the group commander – 'a showman who liked to talk of his Fighting Fortieth'. All the group ships were present at St John's except the group leader's ship, *Londonderry*, which was refitting; Commander Dalison had transferred to *Aberdeen*. The remaining six ships were all fit to sail on time and would be a suitable escort for the fast and valuable merchant ships of Convoy HX.229A.

The big problem facing the authorities at St John's was the weakness of Commander Day's B4 Group, due to take over the thirty-eight merchant ships of HX.229 on 14 March. The group would only have one destroyer and one corvette ready to sail on time, with the group commander's destroyer and three corvettes possibly able to sail later and overtake. There were a few spare escorts available and the obvious solution was to add as many of these as possible to the depleted group, but the danger of this was that the convoy's escort would finish up with a hastily assembled group of ships not used to each other's methods. It was always recognized that a medium-sized group that had trained and operated together was more effective than a large group whose ships were unused to each other.

Commander Day had another solution in mind. He hoped that his own ship, *Highlander*, would be ready only one day later than the pre-arranged sailing date and that the rest of his group would also be ready by that time. This would provide HX.229 with an escort of two destroyers and four corvettes. Day proposed to the Canadian naval staff at St John's that HX.229 should be ordered to sail into the sheltered waters of nearby Placentia Bay and wait there for two days while his group was made ready.

Such a delay, even though of a modest forty-eight hours, was a major change in the convoy timetable and the Canadians at St John's were either unable or unwilling to order the delay on their own. The proposal was certainly forwarded to Rear-Admiral Murray at Halifax and may even have gone as far as Convoy and Routing at Washington and the Admiralty. There are no records to show how far the request travelled but the answer was 'No'. Commander Day's own feeling is that the Canadians were unwilling to trouble the Americans with it and that Halifax turned it down.

Hurried steps were now taken to reinforce B4 Group. The destroyer *Volunteer*, which would complete its repairs one day after B5 Group sailed to pick up SC.122, was ordered to sail instead with B4. Commander Boyle was not asked to give up any more ships; he had the largest and slowest of the three convoys to protect and statistics showed that, if attacked, an SC. convoy could expect to suffer 30 per cent more losses than an HX. convoy. Indeed *Volunteer* was replaced by a spare American escort, the old destroyer U.S.S. *Upshur*, and Commander Boyle was also given H.M.S. *Campobello*, a new Canadian-built anti-submarine trawler that was crossing to England. There were two more spare destroyers available. These were the V and W Class *Witherington* and the four-stacker *Mansfield*. These Royal Navy ships on loan to the Canadians had only a limited fuel capacity; they did not normally make the full convoy crossing, but were used to reinforce the escorts of convoys within 600 miles of St John's. *Witherington* and *Mansfield* were added to B4 Group with orders to operate to the limit of their fuel endurance and then return to St John's. The escort for HX.229 now had four destroyers and one corvette ready to sail and a fifth destroyer, *Highlander*, and three corvettes that, with luck, should be able to sail in time to catch up the convoy before the main U-boat area was reached.

Commander Day had one more dilemma to face. Should he transfer to another ship or should he allow another escort captain to lead the group until *Highlander* could catch up? *Beverley* had an R.N. captain, Lieutenant-Commander Rodney Price, but so had the destroyer *Volunteer* lent by B5 Group; her captain was Lieutenant-Commander G. J. Luther. But *Volunteer* was a stranger to the group, was newly recommissioned, and Luther had made only one North Atlantic convoy crossing during which no U-boats had been met, although he was a pre-war qualified Anti-Submarine specialist who had until recently been a Staff Anti-Submarine officer with the Home Fleet. Also, *Volunteer*'s communications capacity was superior to that of *Beverley*.

On the morning of 13 March a fateful conference took place in Commander Day's cabin. His group was due to sail in a few hours to meet HX.229 the next day. Present were Day, Luther, probably Price, and also Captain J. M. Rowland, R.N., the newly-appointed Captain (D) at St John's who dealt with immediate escort problems.* The decisions taken at the conference are known. Commander Day had tried a second time to persuade the Canadians to delay HX.229 but with no success. He decided not to transfer to another ship but hoped to be able to catch up quickly as soon as *Highlander* was repaired. Until this happened, Lieutenant-Commander Luther, the Anti-Submarine specialist, would command the escorts. Before the meeting broke up, Day and Luther spoke about the tactics to be adopted in the event of a U-boat attack. Day, the old campaigner, stressed the need to keep a tight defence. Luther, the eager newcomer, revealed that he hoped to have the opportunity to act offensively and 'kill' U-boats. Day was not happy that ships in his group would be under the command of an officer who did not appear to see the problems in the same way as he did, and he hoped that the delay in his taking over again would not be a long one.

---

* While in command of H.M.S. *Vanoc* Captain Rowland had helped sink U.100 with the U-boat ace Joachim Schepke in March 1941.

Luther returned to *Volunteer* a little apprehensive but pleased at the prospect of having command, if only temporarily, of a group.

The three convoys would be escorted in all by seven destroyers, nine corvettes, three frigates, two sloops, two ex-Coast Guard cutters and one trawler, although not all these ships were yet ready to sail and the most experienced group commander present would be left behind when his group sailed.* One of these ships was American, one was Canadian, two were manned by Belgians and there were a few Free Frenchmen on the destroyer *Beverley*. The remainder were all quite ordinary Royal Navy escort vessels. Some of these twenty-four ships would soon be facing a greater concentration of U-boats than had ever before threatened the North Atlantic convoy routes.

* Appendix 3, p. 342, gives the composition of the three escort groups in detail.

6

The Approach to Danger

'Being in all respects fit for sea and ready to engage the enemy, H.M.S. *Havelock* will proceed . . .' In his cabin on board *Havelock* Commander R. C. Boyle probably did not spend long reading the time-honoured naval phrases with which his Sailing Orders commenced. He was probably more interested in the type of convoy he was to meet, and where and when he was to take over from the local escorts. Commander Boyle knew of the large concentrations of U-boats out in the Atlantic from the latest intelligence reports, but maybe not yet that in the early hours of that day *Havelock*'s sister ship *Harvester* had been sunk with heavy loss of life in the fight around HX.228 and that one of his fellow group commanders had been drowned.

At 5.0 p.m. local time *Havelock* cast off without any ceremony and made for the impressive entrance to St John's harbour with the sheer cliffs on either side descending into the crystal clear water. The frigate *Swale* and the corvettes *Buttercup*, *Godetia*, *Lavender*, *Pimpernel* and *Saxifrage* followed *Havelock* out to sea. The new trawler H.M.S. *Campobello* had already sailed with a local merchant ship to join the convoy and the American destroyer *Upshur* sailed from Argentia. Within hours, Admiral Dönitz's staff in Berlin knew that Commander Boyle and B5 Group had sailed. The B-Dienst picked up and decoded a signal which gave the time of departure and, from the group's signal codesign 'T.U. (Task Unit) 24.1.19', correctly identified it as the escort group that had brought Convoy ON.168 from England a few days earlier.

Once clear of St John's Commander Boyle turned south-east for the ninety-mile voyage to W O M P – the Western Ocean Meeting Point – at which he would meet SC.122 next morning. When darkness fell the escorts kept station by radar and with their Asdics working were actually carrying out an anti-submarine sweep; this was purely a routine measure and they did not really expect to meet any U-boats at this early stage. The nearest Germans were 480 miles away.

It is not always an easy matter to find a convoy at sea, but B5 Group made contact without difficulty early the next morning and soon after dawn *Havelock* approached the commander of the local escort, Lieutenant A. D. B. Campbell in H.M.S. *Leamington*, to take over the convoy papers. *Havelock* closed *Leamington* on a parallel course; *Leamington's* quartermaster fired a thin line across to *Havelock's* foredeck; a heavier line was then passed and with it the waterproofed packet of convoy papers. These would tell Commander Boyle the name and destination of every ship in the convoy, and the passing of the papers also signified the handing over to Commander Boyle of the responsibility for the convoy's protection. The new escort vessels took their places in the screen around the convoy while *Leamington* and the four Canadian corvettes of the local escort pulled away and made for St John's.

Two days later this procedure was repeated when Lieutenant-Commander G. J. Luther in H.M.S. *Volunteer* with a somewhat depleted B4 Group took over HX.229 and, after a further day, Commander J. S. Dalison in H.M.S. *Aberdeen* with his 'Fighting Fortieth' Escort Group met HX.229A. Several merchant ships were only going as far as St John's and left the convoys at this point while others had joined from St John's. The composition of the three convoys was now as follows:

SC.122 – 50 merchant ships, 1 rescue ship, 9 escorts;
HX.229 – 38 merchant ships, 5 escorts;
HX.229A – 37 merchant ships, 6 escorts.

These ships steamed steadily on in good weather. The faster HX. convoys had considerably closed the gap between themselves and SC.122 and the three convoys passed Newfoundland at little more than one-day intervals. The local escorts had performed the valuable duty of bringing the merchant ships safely across more than 1,000 miles of ocean, roughly one third of the distance from New York to England. Although several ships had fallen out for various reasons, not a ship, not a man, not a ton of cargo had so far been lost. But the convoys still had a distance to travel as great as that between London and the Ural Mountains or between New

York and Los Angeles. The best speed of the fastest of the convoys was little more than that of a man riding an ordinary bicycle at a steady speed.

There is no doubt that the escort sailors were pleased to be at sea again. No matter how short their rest period had been at St John's nor how dangerous the coming convoy might be, they were on their way home with the possibility of a few days leave at the end of the voyage. Part of an earlier chapter gave a short description of how the crews of the merchant ships settled down to the routine of the convoy passage; here a similar look will be taken at the men of the ocean escorts.

The normal routine at sea was that of cruising stations with the crew divided into three watches each doing two four-hour spells of duty. Officers and petty officers had their own accommodation but the ratings slept, lived and ate in their respective messes, all the communications men together, etc. Ordinary Telegraphist R. T. Brown describes the daily routine in one of *Volunteer*'s messes.

I do not recall any pipe for reveille but there was a stir about 07.15 when the first person climbed from his hammock followed by others at odd intervals. There was no need to dress as we slept in our clothes. The first one to rise made the tea which was the only good thing about breakfast. The bread, biscuit and jam was a help-yourself arrangement. The bread had to be vigorously shaken to rid it of cockroaches. The 'Forenoon Watch to Muster' was piped at 07.50 and those going on duty ascended the iron ladder suitably clad for the elements above and relieved the early-morning watch.

In the forenoon, all ratings worked ship. The forenoon watch (08.00–12.00) were on duty, the morning watch (04.00–08.00) were detailed to clean the mess and prepare the midday meal for the galley and the remainder made themselves generally useful in their respective departments or were given jobs by the First Lieutenant. The midday meal was then prepared and usually took the form of what was called 'pot mess'. Into a large pot were put tinned stewing steak, peas, beans and fresh potatoes – when available – with suffi-cient water to cover this concoction. Most days we made a pastry case with flour and tinned dripping and filled it with dried fruit or jam. It was quite an achievement to transport these up the iron ladder and through the hatch to the galley where they were pre-sented to the chef to cook. We always took it for granted that he would complain to us about cooking the 'clacker' (the pastry case) when the weather was bad but it was always cooked for us.

At 11.00 'Up Spirits' was piped and the senior mess ratings mustered in the galley flat to draw the mess ration of rum in bulk. It was his job to share the ration once back in the mess and always to remember the ration for those on watch. (The daily rum ration was $\frac{1}{2}$ gill per man.) At the same time, the stewpot and pastry was collected from the galley and all ratings off duty had their meal. The meal for the forenoon watch was left in the pot and hung up on what we called a sky-hook in the deckhead. At 13.00 'Hands to Make and Mend' was piped and this was the traditional way of informing off-duty ratings that they had the afternoon off. Some turned in to catch up on lost sleep as we very seldom had more than four hours at a stretch. Others sat around talking in undertones until asked to pipe down by a voice from one of the hammocks. If the weather was fine, it was a time to get some fresh air on the upper deck. This was also the time for washing clothes and personal washing; there were no baths or showers. We dried our washing in the boiler-room where there were a couple of lines.

Tea was brewed about 15.30 and at 16.00 the watches changed again. Supper was taken at 18.00; this meal usually comprised of 'herrings in' or baked beans, and bread whilst the latter lasted. 'First Watch to Muster' was piped at 20.00 after which a hush settled over the mess decks as, one by one, ratings turned in.

This routine at sea was the same every day apart from Sunday when the occasional church service was held in the forenoon. This was conducted by the First Lieutenant and hymns were sung to the accompaniment of a mouth organ played by one of the seamen.

And for an officer's view:

Sleep was the prize most of us looked for when we came off watch but of course it wasn't quite so easy as that. No matter who you were there were always jobs to be done. Someone wanted to see you about something or another, perhaps a personal problem or maybe to tell you there was something wrong with some equipment, Asdic or Radar – we called it RDF in those days – for example. So off you went to look into it and if you were lucky it took twenty minutes and if you weren't it took you an hour and a half.

Each of the executive officers had certain 'departments' for which he was responsible. In a destroyer one chap would have the torpedo armament and, say, searchlight and signals under his wing. Someone else would be in charge of the depth-charge crews and a keen chap, if he wasn't satisfied with the speed his crews reloaded the throwers during practice runs, would have them out during his watch below – and theirs! – hauling and heaving the heavy brutes about the confined space of the quarter-deck until he and they reckoned they had knocked a second or two off the reloading time. Then there was the navigator. He was always called 'the pilot' and when he wasn't on watch or asleep or having a meal he'd be up in the chart-house busy with his charts. Each day he'd try to get a morning and evening fix of the ship's position from the stars at dawn and dusk if there were any to be seen and if the sun was out at noon – which it usually wasn't! – he'd try for a noon fix too. The job of correspondence officer usually fell to the lot of the most junior officer and the writing of letters and the filling in of forms was a job to be done whenever it could be fitted in at sea. Ciphering and the numerous confidential books and documents was another task some-one had on his plate. Ratings coded and decoded routine signals but the high security messages were sent out in cipher and officers only were allowed to deal with these. If there was a flap on, the air would be full of them and the doctor, if you had one, would be roped in with anyone else off watch to give a hand at the wardroom table as speed was sometimes vital.

The 'Chief', the name by which the Engineer Officer always went, would spend a great deal of his time down in the engine and boiler rooms as the ships during the war were all overworked – like the men – and there was always something giving trouble, a pump or a condenser or a valve or some other piece of machinery. (Lieutenant D. G. M. Gardner, H.M.S. *Highlander*)

The most lonely man on an escort was the captain, spending nearly all his time either on the bridge or in his sea cabin just below the bridge. He bore the full responsibility for everything that happened anywhere on his ship at any time, had to bear the anxiety of entrusting his ship to new officers-of-the-watch 'just out of the hatbox' and probably worried how well his officers and crew would perform in a crisis. There was no shortage of officers wanting commands in escorts, but it was a lonely and anxious existence.

Not already mentioned was the constant motion of the ship which 'became very tiring due to the continuous tensing of the muscles the whole time', the misery of keeping watch in severe weather and the danger of moving along the upper deck even when secured to the lines rigged for this purpose, the clamping down of portholes which resulted in 'all spaces below quickly becoming fuggy and smelly and the condensation on the cold steel sides of the ship resulting in the whole of the ship's interior being continually wet'. Above all was the fact that 'our whole mode of life was one of expectation of Action Stations at any time'.

But the escort was not just a single fighting ship, it was part of a group with each ship having its part to play in the overall defence of the convoy. Until an attack developed, the group commander could do little more than dispose his available escorts around the convoy to prevent a U-boat getting inside the escort screen, although he rarely had enough escorts to maintain a completely secure screen, with individual escorts suffering Asdic or radar breakdown, others having to be detached to look after a merchant ship in trouble, or just not enough escorts available. The following diagrams show how the escort commanders of SC.122 and HX.229 arranged their screens soon after taking over their respective convoys. Commander Boyle in *Havelock* had a reasonably strong escort for SC.122; Lieutenant-Commander Luther in *Volunteer* had a weak escort for HX.229. The close screen was normally maintained by corvettes while the faster ships, if they could be spared, were sent to sweep further out on the possible approach paths of U-boats. The group commander tried to stay fairly central and the position ahead of the convoy was an ideal one for his destroyers. In the diagrams Commander Boyle is keeping a much tighter screen but most of his ships are slow corvettes. Lieutenant-Commander Luther, although he was short of escorts, had the destroyers *Witherington* and *Mansfield* operating well away from the convoy, but these ships could be brought into the close screen at night or at any other time that Luther wished.

The majority of the ships in all three escort groups had worked together before, but three ships were complete strangers to their groups and, by coincidence, all three were shortly to be important factors affecting the fortunes of their respective convoys, although in widely differing ways.

The First World War destroyer H.M.S. *Volunteer* was the first of these ships. She had only just been recommissioned after a long refit, and had made only one very stormy eastbound convoy crossing before being detached from her own group and given temporary command of HX.229's escort. The transfer of *Volunteer* to this group had been made so late that none of the other captains in the group had ever met Lieutenant-Commander Luther, their temporary escort commander, or even knew his name! Only *Volunteer*'s first lieutenant and perhaps a quarter of her crew had been on Atlantic convoys before; the remainder were fresh, keen and well-trained, but inexperienced. Luther – 'Gordon John' to the officers, 'Long John' to the ratings – owned a white Sealyham terrier called Tiddler which spent much of its time on the bridge, from whence there were 'periodic calls for a broom and pan'; Tiddler's name was included on the Action Stations and Abandon Ship lists.

The other two 'strangers' were in SC.122's escort; they were the British trawler *Campobello* and the American destroyer *Upshur*. *Campobello* was the new anti-submarine and minesweeping trawler built in a shipyard on the Canadian

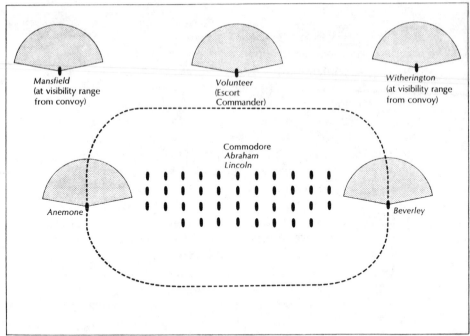

Diagram 2. HX.229's escort and convoy disposition

The intention is to prevent a U-boat from getting within effective torpedo range of the merchant ships. The shaded cone ahead of each escort is the area covered by its Asdic. Most escorts also had radar with a range of 4,000 yards, sufficient to provide a complete radar-covered zone around a well escorted convoy, but a surfaced U-boat at night was not easy to detect with radar. Escort screens could be altered at any time and, at night, the escort commander often took up a new position behind the convoy.
in action. (Wireman R. A. White, H.M.S. *Pennywort*)

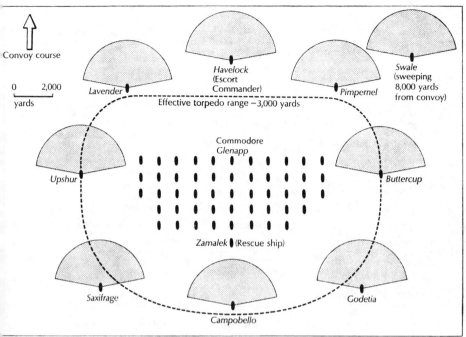

Diagram 1. SC.122's escort and convoy disposition

Lakes and now being taken to England by her Royal Navy crew. *Campobello* had already suffered several setbacks when she had been trapped in ice at Quebec and again in the Gulf of St Lawrence. Her voyage from there to St John's had been tracked by the Germans through signals decoded by the B-Dienst. The Germans could not identify this new name, *Campobello*, and called her *'unbekannt'* – an 'unknown'. This tiny ship was now in position S in SC.122's screen, that is bringing up the rear and guarding the convoy from the least likely approach of a U-boat attack.

The U.S.S. *Upshur* was an old American four-stacker destroyer which had served on the Atlantic convoys from before America's formal entry into the war until February 1942, but since then had been mostly in the Caribbean. She had just undergone a three-week overhaul in Boston Navy Yard during which a new radar and a new HF/DF set had been fitted. The radar was hand-cranked and the Officer of the Deck, merely by glancing up at the aerial, could tell whether the operator was flagging. This voyage with SC.122 was *Upshur*'s first in the North Atlantic for over a year and because of very cold weather at Boston after many months in the Caribbean several of her crew had been left behind in hospital. Signalmen were in particularly short supply and two of *Upshur*'s junior officers, Lieutenant Herb Gravely and Ensign Dunbar Lawson, whose more recent Naval Academy training included the sending of Morse by lamp, were detailed to stand watch together and do their own signalling. Both officers have memories of their efforts to communicate with the British escorts. Ensign Lawson first:

Royal Navy signalmen are noted for their speed and excellence and I must have completely exasperated them with my requests for repeats. It is common practise to acknowledge receipt of a word with a dash when signalling by light. A really good signalman will often hold a steady light towards the sender, thereby bragging in effect that he can receive as fast as the sender can transmit. When I came on watch, the R.N. signalman sometimes made me acknowledge the receipt of every *letter* in order to ensure that I received it correctly!

Lieutenant Gravely:

I remember clearly the *patience* and *understanding* we received from other escorts. For the first few days they were so gently slow and so patient with the two of us. As we polished our skill, they almost cheered with our progress. Please thank them in your book from the two of us.

While the three convoys steamed on along their respective routes there was much activity at the intelligence centres and operations rooms in Berlin, London, Washington and Halifax. More signals were being intercepted and deciphered; fresh plans were being made and then sometimes altered again within hours; move was followed by countermove. Although both sides had some knowledge of the other's intentions, neither side had complete knowledge of those intentions.

The War Diary of U-boat Headquarters contains the following entry for 12 March 1943:

> *Auf Grund der eingegangen BX-Meldungen entschliesst sich die Führung zur Operation auf den dadurch erfassten HX.229.*
> On the basis of decoded messages received from the B-Dienst, the leadership decided to commence the operation against HX.229 which had been detected.*

There is no doubt that the 'leadership' mentioned here involved the personal decision of Admiral Dönitz and that this decision was taken in the evening of 12 March. Orders were sent out for all U-boats still involved in the dying battles for SC.121 and HX.228 to break off their operations. Those with enough fuel and torpedoes were ordered to turn back and form two new patrol lines sweeping westwards; they were to be reinforced by fresh boats coming out from Germany and France. The two new groups were named Stürmer and Dränger. Stürmer's lines were to be established by the evening of the 14th, two days hence, and Dränger would then form up to the south of Stürmer. The boats of Gruppe Raubgraf were ordered to maintain their present position in the convoy-route bottleneck area north-east of Newfoundland.

Dönitz was in effect harnessing the entire force of U-boats in the North Atlantic for one massive operation against Convoy HX.229. His reason for doing this was that he had received on that day excellent intelligence about that convoy.

---

* From Befehlshaber der U-Boote, Kriegstagebuch 1 January to 30 June 1943, p. 93. The U-boat Headquarters War Diaries were captured by the British in 1945, but Admiral Dönitz managed to get hold of a copy several years after the war and this is now in the Bundesarchiv at Koblenz where it may be consulted only with the written permission of Admiral Dönitz. I am indebted to him for this permission and to the Bundesarchiv for supplying photocopies.

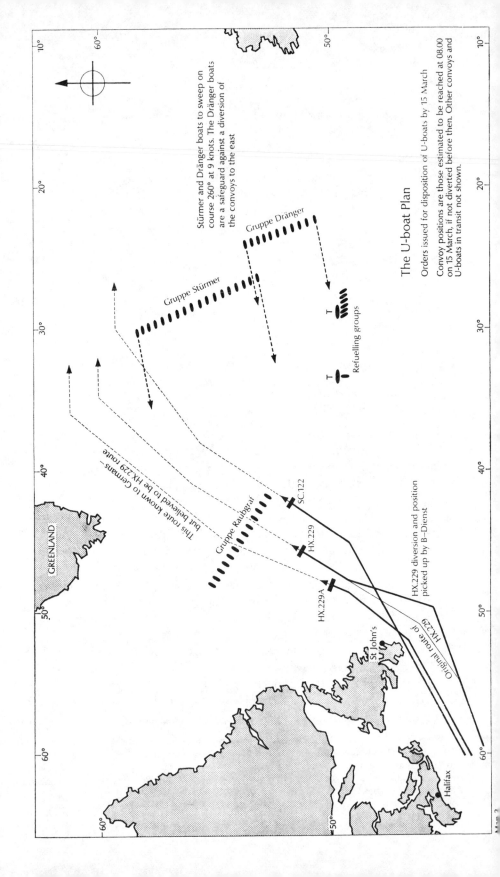

Stürmer and Dränger boats to sweep on course 260° at 9 knots. The Dränger boats are a safeguard against a diversion of the convoys to the east

Gruppe Dränger

Gruppe Stürmer

T

Refuelling groups

T

The U-boat Plan

Orders issued for disposition of U-boats by 15 March

Convoy positions are those estimated to be reached at 08.00 on 15 March, if not diverted before then. Other convoys and U-boats in transit not shown.

This route known to Germans — but believed to be HX.229 route

Gruppe Raubgraf

SC.122

HX.229

HX.229A

HX.229 diversion and position picked up by B–Dienst

Original route of HX.229

St John's

GREENLAND

Halifax

60°

50°

60°

50°

50°

60°

N

This was the result of a diversion ordered by the Canadians at Halifax before HX.229 reached Newfoundland. This diversion was only of a minor nature, probably to avoid an incoming convoy, but it had required a signal which Halifax sent out at 16.16 G.M.T. on the 12th. Within no more than two hours the B-Dienst had deciphered this signal and the text of it was in Dönitz's hands. The signal contained the three pieces of information that were needed to intercept any convoy – a position, the course being steered, and the speed of the convoy. This vital intelligence, together with the future route that Dönitz believed he had of HX.229, should be enough to result in the convoy's interception.

The intention was that the Raubgraf patrol line should catch HX.229 on its north-north-easterly course from Newfoundland and then the Stürmer and Dränger boats could dash across and join the battle. If HX.229 was diverted to the east before reaching the Raubgraf line or slipped through it, then the 600-mile long line of Stürmer and Dränger boats would still form a barrier right down the centre of the North Atlantic.

It was an ambitious plan based on information that was almost entirely accurate. The only faulty part of Dönitz's intelligence was that the future route in his possession was not that of HX.229 but of HX.229A, but this did not matter much because HX.229's own route was near enough to the one Dönitz possessed. The flamboyant choice of code names for the three groups shows how U-boat Headquarters hoped to inspire their men out in the Atlantic to vigorous action. *Raubgraf*, the name of the existing group, means 'Robber Baron'; the new code names *Stürmer* and *Dränger* mean 'Daredevil' and 'Harrier'.

It would take two days for most of the U-boats to get into position for this grand plan. During this time a number of minor events occurred; some of these turned out to be of little importance, but others would alter the plans of both sides.

On the same day, the 12th, that Dönitz was preparing to trap HX.229, the American officers of Convoy and Routing at Washington were altering the plans of SC.122. Each morning the Americans held a conference at which the current convoy positions and the estimated U-boat positions were discussed. Captain Knowles's tracking room's forecast for this day included:

'20 (U-boats) estimated patrolling approx. between 50N, 38W and 56N, 46W from recent DFs.'

This was Gruppe Raubgraf, and when plotted on the American map this long line of U-boats appeared to stretch right across the position which SC.122 was due to reach in two days time. Convoy and Routing decided to reroute SC.122 and sent the following signal: SC.122 REPEAT SC.122 AMEND ROUTE FROM NEW POSITION (A) 48.22 46.50 TO NEW (B) 57.10 46.50 THENCE (L) OMITTING (J) (K).*

This was a very unfortunate signal, although it was sent in perfectly good faith by the Americans. In fact the Raubgraf patrol line was only half the length estimated by the Submarine Tracking Room, but it is easy to see how the mistake was made. Four U-boats had recently left Raubgraf for home and had probably signalled, perhaps deliberately to cause confusion, from positions well to the south-west of the Raubgraf line. If left alone, SC.122 would have passed clear of Raubgraf. Instead, SC.122 was given this long diversion to the north that would drastically extend the voyage of this slow convoy and also place it more within reach of the real position of Raubgraf. SC.122 obeyed the diversion order on the following day and took a course due north but it was not destined to stay on this course long before further developments altered the plan yet again. But the signal had been picked up by the B-Dienst and the 'new position A' with its latitude and longitude would help the Germans interpret a later message.

Soon after midday on the 13th, U-boat Headquarters received two further pieces of information. The first was a report by a Luftwaffe radio listening unit that a new westbound convoy, ON.172, had been located in a position only two days out from England and quite close to the U-boats moving west to take up position in Gruppe Stürmer. Although several U-boats were only 240 miles from ON.172, the report was merely noted and no action was taken on it. Instead the U-boats were allowed to proceed with their movement towards the hoped-for interception with HX.229 which was still 1,200 miles away! The reason for this decision was that the message concerning ON.172 gave only a position with no

* Details of the daily estimated U-boat positions and the text of the diversion signals provided by Naval Historical Center, Washington.

course or speed. This convoy was not, therefore, as good a prospect for attack as HX.229.

Less than thirty minutes after receiving the Luftwaffe report about ON.172, U-boat Headquarters received more news – one of the Raubgraf boats had unexpectedly sighted another westbound convoy, ON.170, approaching Newfoundland. The Raubgraf U-boats closed in on this and what became a confused skirmish rather than a battle followed. The weather was very poor with gales and blizzards; the escorts were alert and the action was taking place well within the range of the Newfoundland-based American and Canadian aircraft. Contact with the convoy was soon lost after one merchant ship had been sunk and several U-boats shaken up by depth-charges. The main significance here of this brief encounter was that the exact location of the Raubgraf boats now became known to the Americans. All three of the eastbound convoys were heading directly into the area where the Raubgraf boats had just been in contact with ON.170, the leading convoy SC.122 on its new northerly course being only 150 miles from the scene of the action. Convoy and Routing moved quickly and signals were sent to SC.122 and to HX.229 diverting them well to the east. HX.229A was still some distance away from the danger area but a signal was sent to this convoy on the following day, the 14th, ordering a diversion not to the east but even further north around the other side of the area in which the Raubgraf U-boats were known to be operating.

Within four hours the two diversion signals had been decoded by the B-Dienst and the texts were on the table at U-boat Headquarters. These two messages each contained new positions and when taken with the previously decoded signals gave Dönitz's staff an almost completely accurate idea of where the two leading convoys were at that moment and what their future movements would be, but the Germans did not act at once. The Raubgraf boats were left trying to regain contact with ON.170 for a further twelve hours, but without success, and it was not until the afternoon of the 14th that they were ordered to break off and to move at top speed south-east in an attempt to form a new patrol line astride SC.122's route. The intention to concentrate against the one convoy, HX.229, had now been extended to an attempt to catch both SC.122 and HX.229.

The encounter between the Raubgraf boats and ON.170 had a further result. The American destroyer *Upshur* had been detached from SC.122 on the 13th in order to reinforce ON.170 but was sent back to SC.122 on the 14th. The two moves both required signals, and the second of these was decoded by the B-Dienst whose records show that:

An unknown single unit or escort group should, if the situation allows, leave ON.170 on the morning of the 14th and, at economical speed, go to Convoy SC.122 which at 13.00 would be at 49.47N, 43.57W, course 79 degrees, speed 7 knots.*

This message, which was almost entirely accurate – SC.122 was actually two hours ahead of time at that position – does not appear to have been passed to U-boat Headquarters. Perhaps it was decoded too late to be of any use.

In fact the Raubgraf boats never had a chance of catching SC.122 and their new patrol line did not form until nearly twenty-four hours after SC.122 had passed. Many sources say that this failure was due to heavy weather delaying the formation of the new line, but U-boat Headquarters, despite all their information from the B-Dienst, had badly miscalculated this convoy's progress and, by leaving the Raubgraf boats for so long trying to regain contact with ON.170, had lost all chance of this group getting in front of SC.122, although that convoy still had to negotiate the great line of the Stürmer and Dränger boats further to the east. The new line to be taken up by the Raubgraf boats would, however, be a menace to HX.229, whose route would pass very close to the southern end of the patrol line.

This story of move and countermove will be left at this point. By dawn of 15 March, SC.122 and HX.229 had settled down on courses that were slightly north of due east but both were intended to resume a more northerly course later. HX.229A was just passing Newfoundland on its original north-easterly course. The U-boats of Gruppe Raubgraf were dashing south-east to their new patrol line and the Stürmer and Dränger boats had formed and started their steady sweep to the west. The convoys appeared to have avoided the Raubgraf boats for the present but the 600-mile Stürmer/ Dränger line represented a formidable future obstacle.

It might be of interest to look at some of the U-boats in the

* X-B Bericht Nr. 12/43, Appendix 1, p. 7, from Bundesarchiv.

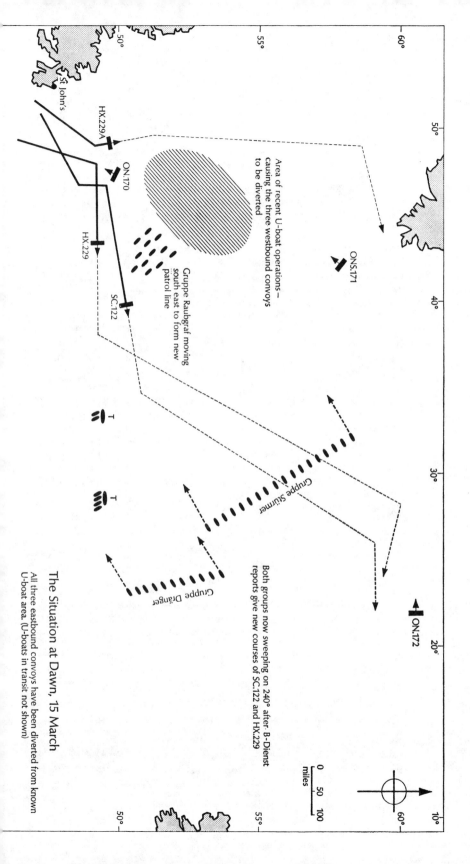

St. John's

HX.229A

ON.170

HX.229

SC.122

Area of recent U-boat operations –
causing the three westbound convoys
to be diverted

ONS.171

Gruppe Raubgraf moving
south east to form new
patrol line

Gruppe Stürmer

Gruppe Dränger

Both groups now sweeping on 240° after B-Dienst
reports give new courses of SC.122 and HX.229

ON.172

0   50   100
miles

The Situation at Dawn, 15 March

All three eastbound convoys have been diverted from known
U-boat area. (U-boats in transit not shown)

patrol lines. One of these boats was Oberleutnant Erwin
Christopherson's U.228, and the following entries from the
private diary of her First Watchofficer, Oberleutnant Carl
Schauroth, describe the recent movements of a typical U-boat:

10 March: Evening – operation broken off, set course to the
west. Night – Gruppen Burggraf, Westmark, Ostmark dissolved.
11 March: Steaming south-west. Orders – hurry to refuel from
U-Zech in Bruno Dora. [Signals referring to a U-boat always gave
the captain's name not the U-boat's number; U-Zech was U.119,
Bruno Dora was a position on the U-boat chart.] Also another convoy is already approaching but very slowly. We must be careful of
fuel consumption. Broadcast from B.d.U.: Another 100,000 tons 'in
the basket'. [The diary then listed details of nineteen ships claimed
sunk in SC.121 and HX.228.]
12 March: Today we are exactly half a year in service. It's
strawberries and cream and this evening pancakes. Still 205 degrees
on diesel-electric. Shall we catch the convoy? Broadcast from B.d.U.:
'We remain a decent firm.' (*Wir bleiben eine anständige Firma.*)
13 March: Still on the way to U-Zech, our 'milch-cow'. First-
rate weather. Written a letter to parents, hopefully the U-tanker
can take it. Towards evening the weather becomes worse.
14 March: A real day of rain. Terrible watch duty. Large liferafts to starboard. Boxes, straw, shattered timber, beams and oil,
oil, oil. There must have been a real drama here a short time ago.
The wind is getting up.
15 March: We are not refuelling from U-Zech but from U-
Wolfbauer [U.463]. Course 235 degrees. Always to the west. Top
speed. If only we could have some better weather.

Kapitänleutnant Hansjürgen Zetsche of U.591 had been
one of the successful captains against SC.121. Nearly three
weeks earlier he had come across a lifeboat from the Dutch
ship *Madoera* which had been torpedoed by another U-boat.
Zetsche had taken aboard the *Madoera*'s Dutch first officer,
leaving behind the other occupants of the lifeboat, all of whom
are believed to have perished. This prisoner had been aboard
the U-boat right through the SC.121 convoy action and had
witnessed the sinking by U.591 of two merchant ships.
Zetsche claims that the Dutch officer 'was very cooperative;
we had friendly discussions and he helped calculate U.591's
position excellently', and he asked me to put him in touch
with the Dutchman. Unfortunately the Dutch officer's
memory of the U-boat captain is that 'he hated the English
and the English language very much. I am not interested in
meeting him again.' Kapitänleutnant Zetsche was to take
his prisoner into the coming convoy battle also, but he was to

have no success this time. He eventually handed the Dutch officer over to another U-boat which took him back to France. The commander of this boat, Kapitänleutnant Helmut Manseck of U.758, was found by the Dutchman to be 'rather more strict'.

Many of the boats that had been on patrol for some time were in need of refuelling. It was quite normal for a U-boat to operate well beyond its own fuel capacity; this was a great act of faith in the U-boat Headquarters staff officers who controlled the tanker U-boats, or 'milch-cows' as they were known. There were two of these on station at this time, U.119 which was a Type XB, really a minelayer but converted for tanker work, and U.463, a Type XIV purpose-built tanker. These two had taken up positions in the Air Gap just south of the normal convoy routes. Operational U-boats always added a small coded message to every signal to U-boat Headquarters indicating their fuel stocks in cubic metres. The condition of every boat was watched carefully and they were ordered to a rendezvous with a tanker U-boat whenever fuel ran low. Both of the tanker U-boats were busy now in this interval between convoy operations. It was a great test of navigation and seamanship to find the tanker, float across fuel pipes supported by life jackets, take on fuel, provisions and essential stores and then get out of the way before the next U-boat was due. The Allies suspected the presence of these tanker U-boats but were having great difficulty in preventing their operations. They were undoubtedly a great contribution to the maintenance at sea of large numbers of U-boats and thus of the successful convoy battles of this period.

Some of the U-boats were fresh out from Germany. These had recently come through the 'Iceland Passage' where they had suffered from severe storms. This man had been one of the midshipmen hurriedly transferred from the Naval Air Service School to a U-boat.

I only had my best uniform that was more suitable for dancing than U-boat life; they gave me four-hour look-out watches and without any special clothes I was always wet. The first three days I was so sick that I just wanted to dive into the sea and get away from it all. The older members of the crew made fun of us two new midshipmen and gave us fat bacon on pieces of string. (Fähnrich Wolfgang Jacobsen, U.305)

Another boat to arrive recently was U.523, a new Type IXC

boat which had met such severe weather that it was later found that the outer hull plates under the bow had been buckled by the constant smashing down onto huge seas. The boat's entire stock of binoculars had also been affected by the bad weather, their lenses becoming coated with irremovable mist. This was reported to U-boat Headquarters and a few days later another U-boat appeared alongside and handed over six new binoculars which had been picked up for U.523 from one of the tanker U-boats – a good example of the attention to detail and service provided by U-boat Head-quarters.

The older boats that had come out from Biscay had not experienced the severe weather of the Iceland Passage, but they had been forced to run the gauntlet of the heavy Allied air patrols in the Bay of Biscay. In the early morning of 4 March a Wellington of 172 Squadron based at Chivenor in Devon had spotted U.333 and attacked with depth-charges that were placed so accurately that one struck the U-boat's deck and disintegrated without exploding. But the German gunners were also alert and they shot the Wellington down in flames. The crew of the Wellington, captained by Flying Officer Gordon Lundon, had only two weeks earlier sunk a U-boat, U.268; now Lundon and his mainly Canadian crew were all dead at the hands of another U-boat – a typical blow and counterblow of the U-boat war. Several days later the Germans found parts of the depth-charge casing and the firing mechanism wedged into their boat's upper deck and these were later handed over for inspection on U.333's return to base.

Forty-two U-boats were to become involved in the coming convoy operation, although not all would succeed in making contact.* Thirty-seven of these boats were the standard Type VIIC North Atlantic boat and five were the larger IXCs making their first patrols before reporting to their bases in Biscay for eventual more distant operations. It is a measure of the recent rapid expansion of the U-boat arm that fifteen of the forty-two were brand-new boats and on their first patrols, and of the remainder most were less than one year old. The U-boats and crews hoping to do battle with the approaching convoys were just typically average represen-

---

* Appendix 4, p. 344, gives details of the U-boats taking part in this operation.

tatives of Dönitz's command at this stage of the war – good boats with well-trained crews under sound captains but with no great depth of experience behind them. There were the usual mixture of bold and careful commanders and there will later be examples of the behaviour of both types. There was not one established ace among the commanders and none of them would ever achieve that status.

The patrol line was by this time a well-tried feature of U-boat operations. The whole system depended on centralized control by U-boat Headquarters and on the navigational skill and seamanship of the U-boat crews. A patrol line was born when Dönitz's staff sent a signal addressed to a number of U-boats informing them that they were to become members of a Gruppe with a new code name. The same signal would give each boat's own position in the initial line-up, the positions of the two ends of the new patrol line, the time it was to assemble and the speed and direction of its sweep. Further orders addressed to the Gruppe as a whole could amend the direction and speed of the sweep and it was up to individual boats to plot their own part in this. At no time were the boats in touch with each other and, contrary to what several Allied naval officers still believe, no single U-boat captain was appointed to control the Gruppe either in the patrol line or in any subsequent pack attack on a convoy.

Signals to U-boats were, of course, always in code, and at this time would normally be sent out by the transmitting station at Nauen near Berlin which was reserved only for U-boat signals. There was another station at Kalbe, near Magdeburg, with a particularly powerful, very-low-frequency transmitter which could reach a U-boat submerged to a depth of fifteen metres. As for signals back to U-boat Headquarters, there was a golden rule – once a patrol line had been established, the boats in it were to keep absolute silence unless a convoy was sighted.

The distance between U-boats in a patrol line was about twenty sea miles. A constant watch was kept for the faint smudge of smoke on the horizon that would be the first indication of a convoy's presence, and the U-boat would dive every so often to listen on her hydrophones which in good conditions had a greater range than the human eye. A Gruppe that was sweeping forward to meet an expected convoy was an *Aufklärungsstreifen* – a 'reconnaissance' or 'patrol line' – while a Gruppe that was ordered to remain static on a given line became a *Vorpostenstreifen* – an 'outpost' or 'picket line'.

U-boat men much preferred to be in one of these lines than on the 'free manoeuvre' type of operation in which a boat was given an area of ocean and left to search for targets in it, but the actual search for the convoy could be a trying time.

When a U-boat had been at sea for days or even weeks without seeing a ship, morale on board was not of the best especially if we kept hearing on the wireless of other boat's successes. We were always very pleased to get the order from B.d.U. to form a patrol line; we knew from experience that this gave the best opportunity of finding a convoy. We put the best men on look-out duty; very often there was a bottle of champagne or cognac on return to harbour for the man who made the first sighting. Morale reached the depths when storms or rain blotted out visibility and went even lower if the sweep went on too long without a sighting. We felt then that the convoy had slipped through or round the line. (Kapitänleutnant Kurt Baberg, U.618)

It was a time of ordinary routine work with men made angry by the motion of the ship, wet clothes, a tense atmosphere between technical crew and the seamen 'who did not spot the convoy quickly enough' in order to get to the attack. They had longed to get rid of the first torpedoes lying on the floor of the bow compartment which were blocking the 'traffic' and making the work of the torpedo personnel difficult. The Chief Engineer made it very clear that the seamen could be eliminated from the boat altogether because the technicians could easily take over their jobs. (Oberleutnant Herbert Zeissler, U.338)

So the U-boats of the Raubgraf, Stürmer and Dränger groups formed their lines and swept on looking for the convoys. The weather was mostly very rough and conditions for the men on look-out were grim. Despite their oilskins they were soon wet through and, if heading into the weather as the Stürmer and Dränger boats were, freezing spray or even green water broke over the conning tower with every sea that the U-boats met. Binoculars soon became misted up and the look-outs desperately tired and cold. In the Stürmer line, U.338, a new boat on its first patrol, was surprised to meet another boat, the veteran U.439 on its seventh patrol. The bad weather had caused navigational difficulties. The two captains, Manfred Kinzel and Helmut von Tippelskirch, compared positions and then parted.

*

March 15th was a day during which the fortunes of the two leading convoys were affected more by the weather, with a bad storm blowing up, than by the Germans. The slower of the two convoys, SC.122, was still leading but was now only a little over half a day's steaming ahead of HX.229.

SC.122's escorts all had sufficient fuel to complete the crossing, but on this day Commander Boyle attempted to top up the tanks of his own ship *Havelock* before the weather got too bad. If the convoy did come into contact with U-boats a good reserve of fuel in this high-speed destroyer would be a great asset. Two of the tankers in the convoy had been designated as 'escort oilers' and were equipped to pass fuel to the escorts, but one of these, the *Benedick*, only had equipment for the 'trough and metal hose' method of oiling which required the two ships to steam close alongside each other while fuel was passed. This required very calm conditions and *Benedick* was not used on this occasion. The second oiler, the *Christian Holm*, was equipped for the 'floating line' method in which the tanker streamed a lightweight hose astern which was picked up and attached to a fuel inlet on the escort's forecastle.

Havelock had first tried to oil the previous day but the hose used, an American fire-fighting hose, would not fit properly onto *Havelock*'s fuel inlet and after three and a half hours only one eighth of a ton of fuel had come aboard. That night, *Havelock*'s engineers had made a proper adaptor and now Commander Boyle tried again. Despite the worsening weather the line was picked up and this time connected successfully. But after only thirty minutes the merchant ship *Empire Galahad* from the next column to port suddenly veered off with steering gear trouble and approached the *Christian Holm* on a collision course. The tanker had no option but to get out of the way and herself veered away to starboard. *Havelock*'s speed was too slow for a similar rapid manoeuvre and the hose pulled taut and parted; only four tons of oil had come aboard. Commander Boyle gave up the attempt and returned to his place in the escort screen. Oiling at sea was still in its early days and the experiences of *Havelock* and *Christian Holm* were probably typical of the difficulties encountered at this time.

All day and into the night the gale blew. An officer on the corvette *Lavender* remembers that 'there was a considerable gale which brought home to me a sense of the Power of Almighty God – also the sea-keeping qualities of a Flower Class corvette'. The storm resulted in two ships becoming separated from SC.122 – the 775-ton Icelandic merchant ship *Selfoss* and the 545-ton trawler H.M.S. *Campobello*. They were the two smallest ships with the convoy. The captain and first mate of the *Selfoss* decided not to attempt to catch up the convoy or to take the stragglers' route that they had been given but to make straight for their destination, which was Iceland. This brave little ship sailed right through one of the biggest U-boat concentrations of the war without being spotted. Six days later she arrived safely in Reykjavik after sailing alone for almost 1,000 miles.

H.M.S. *Campobello* was not so lucky. She had struggled to maintain her place in the escort screen but had gradually dropped behind. One of her officers, Lieutenant G. B. Rogers, describes subsequent events.

Campobello may have been strained by the ice that she had been trapped in in Canada more than was thought or we might have hit something without realizing it. We found that we were leaking badly somewhere under the coal in the bunkers. The coal itself prevented us from pinpointing the actual leak though we tried trimming from the sides. We could not apply a collision mat properly as we just did not know where to put it. The inflow of water was greater than our pumping and baling ability and after a time the water reached the bottom of the boilers. As coal fires and hot boilers do not mix with cold sea water in quantity we finally had to draw fires and let the steam pressure down. This put paid to any more mechanical pumping and cut off all power throughout the ship.

The bunkers, boiler room and engine room were all interconnected and occupied a good half of the ship's hull. It was obvious that only large salvage pumps and a tug could save the ship and these were just not available. As we would have several hours left before having to abandon ship we were able to let the convoy get well away before breaking wireless silence. In the meantime all hands changed into their best clothes and were able to get a good meal.

When he heard that *Campobello* was in trouble, Commander Boyle sent the corvette *Godetia* back to give help and also tried to get a tug out from St John's but none was available. Lieutenant M. A. F. Larose, *Godetia*'s Belgian captain, carries on the story.

Campobello's captain said he would have to abandon ship; I informed *Havelock* who replied 'Remove ship's company and sink her'. Lieutenant Delforge then took one of our boats across and *Campobello's* boat was also used. It was a fairly simple job to transfer her crew. We then tried to sink her with armour-piercing shells but we only made small holes in the hull. This was taking too long so I decided to use a depth-charge. We adjusted the setting to 50 feet, passed *Campobello* at full speed about 70 yards off and fired the depth-charge. It fell on *Campobello's* gun platform, rolled over the side and exploded practically under the ship which disappeared in a few seconds. We left at full speed to rejoin the convoy.

The experienced Belgian seamen of *Godetia* had done a typically sound job in rescuing the entire crew of *Campobello* but a valuable, brand-new ship had been lost – the first casualty of the three convoys. The wider implications of this incident were that Commander Boyle had lost one member of his escort group and *Godetia* would be several hours before rejoining. Boyle had also been forced to break wireless silence and the B-Dienst did pick up a signal giving yet another accurate fix on SC.122's position, course and speed at this time. This may have been a signal sent to *Godetia* to help her rejoin.

No more than 100 miles to the south-west of SC.122 was HX.229, with Lieutenant-Commander Luther still in temporary command of the escorts. Luther's strength was increased to six ships on the afternoon of the 15th when the corvette *Pennywort*, which had been delayed at St John's, caught up and joined. But the gale which was affecting SC.122 had actually struck HX.229 a few hours earlier and was more severe at that convoy's position, and the first effect of this was that the strength of the escort immediately fell back again to five ships. The captain of H.M.S. *Witherington*, one of the two St John's-based destroyers ordered to stay with HX.229 'to the limit of their endurance', was forced to heave to in the heavy weather with damaged deck plating, so, while the convoy pushed on to the east, the destroyer turned west into the seas to ride out the storm. *Witherington* never managed to find the convoy again and she returned to St John's.

That night the ships of HX. 229 had a bad time. The storm was coming from right behind the convoy with huge seas overtaking the ships and sometimes causing the unpleasant experience known as 'pooping', when a particularly large wave broke over the stern of a ship and ran along the deck. One sea came right through the saloon of the *Canadian Star* and two little girls who were sleeping there were washed right out on to the deck near the rails, but they were not hurt. Another heavy sea struck the *Walter Q. Gresham*, a Liberty ship on its first voyage, and swept away a lifeboat, tearing the davits clean out of the deck and leaving two gaping holes.

On the night of the 15th the gale reached the peak of its ferocity and, in fact, shortly before midnight we passed through the centre of the storm when for approximately half an hour there was very little wind and only heavy seas running; then at the end of the half hour the wind resumed its full force from an opposite direction. This is normal and expected when traversing the vortex of a storm. Shortly after midnight we hit and shipped a mountainous sea which swept down our starboard side and broke amidships, smashed No. 3 lifeboat and reduced it to matchwood which the following sea smartly removed and nothing was left. As was common practice our boats were always swung outboard, hanging from their davits ready to be lowered and, when in this position, hung some forty feet above sea level. This will give you an indication of the height of the seas running that night. (Third Officer R. McRae, S.S. *Coracero*)

By dawn on the 16th the worst of the storm would be over. Three ships fell out of the HX.229 convoy; two of these were still in sight a few miles in the rear, the third, the American Liberty ship *Hugh Williamson*, must have taken the stragglers' route and sailed right through the U-boat area, for she is known to have reached England safely.

The U-boats of Gruppe Raubgraf had also been in this storm. Five boats had recently been released by U-boat Headquarters from the original group with shortage of fuel or mechanical trouble, leaving only eight boats. At the southern end of this shortened patrol line was U.91. In the late evening of the 15th her lookouts caught a glimpse of a destroyer estimated to be on a north-easterly course but visibility was so bad, only about 500 yards, that the destroyer was soon lost in the gloom. Shortly afterwards the U-boat dived and the noise of several ships was heard on its hydrophones. The U-boat surfaced again and sent off this report to U-boat Headquarters who immediately ordered U.91 and three more boats in the patrol line to search the area thoroughly, but they never managed to regain contact. It is almost certain that the escort seen by U.91 was the destroyer *Mansfield* on the port side of HX.229's screen and that the noises heard on the hydrophones were those of the merchant ships in this convoy.

The storm had actually been pushing the ships of HX.229 along well and, at 15.00 on the 15th, Lieutenant-Commander Luther calculated that a speed of 10½ knots was being made in a direction that was almost a direct route to England. He begrudged the turn to the north-east that he was due to take in a few hours time, which would add much sailing time to the voyage, and ordered a signal to be sent to Convoy and Routing requesting that the route be altered 'in view of high speed of advance and strength of prevailing weather', and asking to be routed instead through 53 degrees North, 25 degrees West. Such a route would represent a direct run home for the convoy. It was unusual for an escort commander to break radio silence in this way but there is no record of this signal being picked up and made use of by the B-Dienst.

The sending of this signal caused quite a stir in *Volunteer*'s wireless room, but for another reason described by Ordinary Telegraphist D. Greenhouse.

When it came to transmitting the signal, every transmitter we tried failed to raise a response. As a last resort our Petty Officer Telegraphist Anderson decided to try the veteran transmitter which was installed when *Volunteer* was built in 1917! This was contained in a large wire cage which took up half the available space in the W/T office. The valves in the 'thing' were like goldfish bowls. After much fiddling about, the monster showed signs of life and the P.O. Tel, amidst crackling, blue flashes and sparks, began tapping out the call sign. The sparkers not actually manning the listening sets looked on in amazement. To our even greater astonishment, when the smoke and steam had subsided, he received a 'dah-de-dah' (K) 'carry on'. Carry on he did, in good steady Morse code.

All credit to the quality of workmanship and materials which must have gone into the manufacture and installation of that beautiful old transmitter. I think the efforts of Petty Officer Anderson are worthy of mention also; we didn't think he was in the race to get the thing working.

Although this signal was addressed to Convoy and Routing, the Admiralty also picked it up. Both HX.229 and SC.122 were only a few hours away from the CHOP Line and the Admiralty, who would soon be taking over control, signalled to Convoy and Routing urging that both convoys should now be ordered to take direct routes for England. The Admiralty, by means which will be described later, had a good idea of the presence of the Stürmer and Dränger patrol lines. The routes proposed by the Admiralty would pass right through the Stürmer and Dränger lines. In other words, the Admiralty was now prepared to give up any further attempts to route SC.122 and HX.229 *around* this huge line of U-boats but was recommending to the Americans that the two convoys should attempt to fight their way *through* the U-boat line to the nearest air cover and on the shortest route to England. Convoy and Routing received this bold suggestion at 18.00 on the 15th but did not order any further alterations for another nineteen hours by which time the situation had changed yet again.

The position at midnight on the 15th was as follows. The slow convoy SC.122 was steering east-north-east in calmer weather now and well ahead of the Raubgraf U-boats, but still facing the 600-mile long line of the Stürmer and Dränger boats which were sweeping towards the convoy but were 300 miles away to the east. The faster HX.229, still in stormy weather, had for the second time been fortunate in evading the Raubgraf U-boats and was steaming on an easterly course. Lieutenant-Commander Luther in *Volunteer* was still in command of a weak escort of only five ships; Commander Day in *Highlander* was due to sail from St John's, now 550 miles away, in two hours time. As a result of various diversions from their original northern routes, both of these convoys had been forced into a position deep in the Air Gap and were seemingly trapped between the Raubgraf U-boats behind them and the Stürmer and Dränger boats ahead of them. The more valuable ships of H.X.229A and also three westbound convoys – ONS.171, ON.172 and ON.173 – were all travelling the northern route in apparent safety.

One of the U-boats that had been released from Gruppe Raubgraf was Kapitänleutnant Gerhard Feiler's U.653. Feiler had only one torpedo left and this was defective, his boat was low on fuel, his starboard diesel was giving trouble, one of his officers and four lookouts had been lost when they had been washed out of the conning tower in a storm the previous month and now one of the petty officers was ill. In this sorry state, U.653 was making its way on the surface to a rendez-vous with one of the tanker U-boats before returning to its base at Brest. Between 03.00 and 04.00 on the 16th Ober-steuermann Heinz Theen was in command of the bridge watch.

The wind was very strong and it was very dark. I saw a light directly ahead, only for about two seconds; I think it was a sailor on the deck of a steamer lighting a cigarette. I sent a message to the captain and by the time he had come up on the bridge we could see ships all around us. There must have been about twenty, the nearest was on the port side between 500 metres and half a sea mile away. We did an alarm dive. As the ships of the convoy went over the top of us we could hear quite clearly the noises of the different engines – the diesels with fast revs, the steamers with slow revs and the turbines of the escorts made a singing noise. After about two hours we surfaced behind the convoy and sent off a sighting report. Then we took up a shadowing position at a distance from which we could see the masts of the ships and when we were taken up by a high wave we could see the bridges and funnels.

U.653 had sailed right into the middle of the ships of HX.229, had got off its short sighting report of just three code letters in morse, and was now in the morning of the 16th calmly shadowing the convoy – all without being spotted.

At 08.25 Central European Time, U-boat Headquarters received the report placing a convoy in position BC 1491 of the U-boat chart on a course of 70 degrees.

7

The Shadowing

There can be no doubt that everyone at U-boat Headquarters was delighted to receive U.653's sighting report. It had required a certain amount of nerve and determination to keep every U-boat available in the North Atlantic tied to this one operation in the hope of trapping the two eastbound convoys instead of breaking the U-boats up into smaller group attempts to catch the three westbound convoys also known to be at sea. Now, after being twice so close to obtaining a firm contact on HX.229 but losing that contact, here was U.653 comfortably shadowing one of the eastbound convoys in just the right place – the Air Gap – and at the ideal time of day – dawn – that would enable a reasonable number of U-boats to join before dark and make a pack attack that night. U.653's signal almost certainly brought Admiral Dönitz hurrying to the Operations Room.

The War Diary of U-boat Headquarters records only the bare facts of the decisions Dönitz took but it is not difficult to deduce the reasoning and intentions behind those decisions. To start with it was believed that the convoy that U.653 had sighted was SC.122 and that HX.229 was behind that convoy – a mistaken belief based upon the miscalculations about the rate of progress the two convoys had been making. In fact SC.122 was 150 miles ahead and somewhat to the north. It is quite clear from the decisions taken that morning that Dönitz was now determined to get both convoys. From past experience he could expect to have forty-eight hours clear of any major aircraft intervention and another forty-eight hours after that before the air cover of the convoys became so intense that the operation would have to be broken off. With four days and nights available there was no need to give up hope of finding the second convoy and of striking a double blow.

It did not take long for the first part of this plan to be put into operation. It was decided that all eight U-boats in the

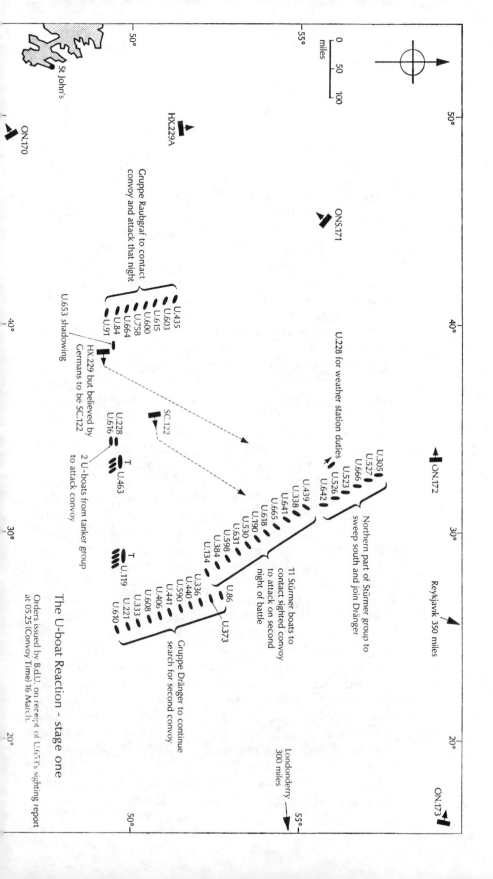

The U-boat Reaction - stage one

Raubgraf patrol line, which was only eighty miles away, should proceed at top speed towards the convoy. Two more boats had recently taken on fuel from one of the nearby tanker U-boats, and these were also ordered towards the convoy. These ten boats – U.84, 91, 228, 435, 600, 603, 615, 616, 664 and 758 – should all be able to make contact in time to operate against the convoy that night. In addition to this, Dönitz ordered that eleven boats from the centre of the long Stürmer–Dränger line, although 420 miles ahead of the sighted convoy, should break out of the line and make their top speed towards the convoy. These boats – U.134, 190, 338, 384, 439, 530, 598, 618, 631, 641 and 665 – should come into contact with the enemy on the following morning, the 17th, for submerged attacks during the day or to continue the pack attack on the second night.

At the same time six of the remaining seven Stürmer boats at the north end of the patrol line were ordered to turn south and close the gap left by the sending into action of the remaining Stürmer boats. This shortened line would have a dual purpose – to act as a longstop in case contact with the sighted convoy was lost, and also to continue the search for the second westbound convoy. The seventh boat in the northern part of the old Stürmer line, U.229, had only three torpedoes remaining and Dönitz had other work for her. Oberleutnant Robert Schetelig was ordered to go to an area off the southeast corner of Greenland. Here he was to send twice-daily weather reports, have a free hand against any Allied ships he found in that remote area and also keep an eye open for the *Regensburg*, one of the two German blockade runners that were overdue on their voyages from Japan. What happened to U.229 and to the blockade runner will be described later.

These orders went out and, within an hour or so of U.653's sighting signal being sent off, twenty-one U-boats were making their way towards HX.229. Convoys were sometimes codenamed after the U-boat captain who had first spotted the convoy and the signals that went out at this time refer to the convoy as being that of '*Funkspruch Feiler*' – 'the Feiler signal' – but back at U-boat Headquarters it was entered up in the War Diary more prosaically as *Geleitzug Nummer 19* – 'Convoy Number 19'.

Within only a few hours of these dispositions being made another message arrived at U-boat Headquarters which

Merchant Ship – British. The *Nariva*, an
8,714-ton refrigerated cargo ship of the
Royal Mail Lines, sailed with Convoy
HX.229 carrying 5,600 tons of frozen meat
from the Argentine but became one of the
victims of the convoy battle.

Merchant Ship – American. A good shot of
one of the famous U.S. Liberty merchant
ships. This is the S.S. *John Fiske* which
sailed in Convoy HX.229A. The Libertys
carried a crew of forty-one merchant seamen
and twenty-six U.S. Navy Armed Guard
gunners. The eleven gun positions can be
clearly seen, as can the stack of 'doughnut'
type floats in front of the bridge and the
four liferafts secured to the mast stays ready
for swift launching in an emergency. The
John Fiske survived the war.

Escort Vessel – Havant Class Destroyer.
These modern destroyers were being built
for the Brazilian Navy but were taken over
by the Royal Navy and used as Escort Group
Leaders in the North Atlantic. H.M.S.
Highlander of B4 Group was intended to be
senior escort with Convoy HX.229 but she
was delayed for repairs in St John's and
missed part of the convoy action.

Escort Vessel – Flower Class Corvette.
H.M.S. *Pimpernel*, typical of the corvettes
that provided the majority of the close
escort vessels of the North Atlantic convoys
during the war. *Pimpernel* took part in SC.122's
battle as a member of B5 Group whose
identification mark is painted on her funnel.

A captain and his ship. Lieutenant-Commander Gordon John Luther, the young officer who finished up commanding the weak escort of Convoy HX.229 with the V and W Class destroyer, H.M.S. *Volunteer*.

U-boat captain ashore. Kapitänleutnant
Rudolf Bahr, twenty-six-year-old captain of
U.305. He had previously served on the
cruiser *Prinz Eugen*. Bahr was killed with
all his crew on 17 January 1944.

Christmas at sea. The engine-room of U.618
celebrate Christmas 1943. This U-boat
operated for nearly two years, sinking five
merchant ships and shooting down two
Sunderlands and one Wellington of Coastal
Command. She operated against Convoy
HX.229 but without success.

U-boat captain at sea. Oberleutnant Hans-Joachim Bertelsmann of U.603 photographed while on patrol in March 1943. His boat had just sunk the first merchant ship in Convoy HX.229. He and his crew were all lost a year later.

Inside one of the concrete U-boat pens in one of the Biscay bases. Three U-boats are being serviced in this one bay. Each of the four main ports could have up to six such bays.

The fuselages of American-built Boston
bombers being loaded as deck cargo at an
unnamed East Coast port.

A Canadian Wren watches another escort
vessel leave St John's harbour. This ship,
the Canadian corvette *Bowmanville*, survived
the war and later passed through a variety
of hands to finish up in the navy of
Communist China.

Oiling at sea. This shot shows H.M.S.
Volunteer taking oil by the trailing hose
method from one of the tankers in a convoy.
The operation was not always successful.

Storm. The U.S. Coast Guard Cutter *Ingham*.
Ingham and her sister ships were the largest
escorts working in the North Atlantic
during the winter of 1942–3; she came as a
reinforcement to the escort of SC.122.

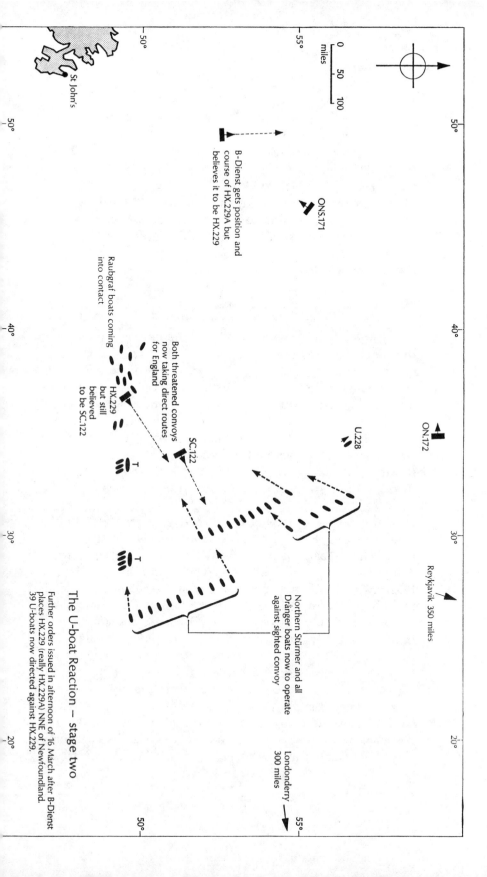

St John's

50°
55°

0 50 100
miles

50°

B-Dienst gets position and
course of HX.229A but
believes it to be HX.229

ONS.171

Raubgraf boats coming
into contact

HX.229
but still
believed
to be SC.122

Both threatened convoys
now taking direct routes
for England

SC.122

40°

ON.172

U.228

Reykjavik 350 miles

Northern Stürmer and all
Dränger boats now to operate
against sighted convoy

Londonderry
300 miles

30°

20°

55°

50°

The U-boat Reaction – stage two

Further orders issued in afternoon of 16 March after B-Dienst
places HX.229 (really HX.229A) NNE of Newfoundland.
39 U-boats now directed against HX.229.

caused some consternation. This was the text of a two-day-old American signal to HX.229A ordering the convoy onto a diversion to the *west* of the old Raubgraf area off Newfoundland. Unfortunately the B-Dienst failed to appreciate that the signal referred to HX.229A and the copy which appeared at Dönitz's headquarters on the afternoon of 16 March reported that convoy HX.229 was now north-east of Newfoundland on a course slightly west of due north, although several previous decodes had placed HX.229 on an easterly course and well to the east of Newfoundland.

This development confused the U-boat staff officers. It is obvious that they did not consider the possibility of three eastbound convoys being at sea and they accepted the signal as being a genuine one for HX.229, the inference being that the convoy spotted that morning by U.653 was now the only one in that area. If this was so, one of the purposes of the six remaining Stürmer boats and the eleven Dränger boats maintaining their patrol line – the continuing search for the second convoy – was no longer valid. The decision was immediately taken to break up this line completely and these seventeen U-boats were all ordered to make their best speed towards the *Funkspruch Feiler* convoy. They were expected to reach the convoy in time for action on the evening of the 17th.

In his post-war memoirs Admiral Dönitz refers to the receipt of this decoded signal and his conclusion that either the B-Dienst had made a mistake or it was a fake signal deliberately sent by the Allies to mislead him. The last of these two possibilities must have troubled Dönitz a little; if true, it meant that the Allies had realized that their naval codes were being broken by the B-Dienst and might well have sent other false information in their signals. The Admiral need not have worried; although the Allies were beginning at that time to suspect that some of their naval signals were being decoded, it is unlikely that false information was being fed to the Germans in this way. HX.229A was later to have its own difficulties for other reasons, but the Germans never did know of its existence and it certainly played its small part in causing confusion at U-boat Headquarters on the afternoon of 16 March.

By nightfall of that day, thirty-eight U-boats had been ordered to operate against the convoy sighted that morning,

and thirteen more were with the supply U-boats or were
coming out from Germany or France, although not all of
these would be allowed by Dönitz to take part in the coming
operation. It is probable that at no other time in the war had
such numbers of U-boats been directed onto one convoy.

The U-boats thrashed their way at top speed on the
surface towards the convoy.

Great relief. At last. Rapid calculation of the course, even more
vigilance by the bridge lookouts, very careful checking of all
important instruments, perhaps a trial dive. (Kapitänleutnant
Herbert Uhlig, U.527)

On the receipt of that first message the mood in the boat changed
completely. Everybody was waiting tensely for the next signal giving a further course of the convoy. We went flat out so as to reach
the convoy before dark regardless of the state of the sea and waves
often broke over the conning tower. All the tiredness and dis-
appointments of preceding days vanished when you at last had the
chance to see the target with your own eyes and you strained to spot
the first spikes of the masts. (Kapitänleutnant Kurt Baberg, U.618)

While HX.229 was being shadowed during the daylight
hours of 16 March, a number of small but important events
occurred affecting the progress and wellbeing of the convoy.
On just this one occasion these events might best be
described if they are set out in the form of a log, each event
being explained as fully as necessary. The information here
comes from many sources and, in order to give a better idea
of the state of the light, all times both here and throughout
the action will be in Convoy Time which on that day was two
hours behind Greenwich Mean Time and three hours behind
Central European Time.

05.00. Convoy HX.229 is steering a course of 089 degrees,
speed 9 knots. Thirty-five merchant ships in convoy, two
stragglers in sight astern, escort of three destroyers and two
corvettes. Lieutenant-Commander G. J. Luther in H.M.S.
Volunteer in command of escort.

05.25. U.653 sends the short sighting signal to U-boat
Headquarters and continues to shadow the convoy. The
Admiralty picks up the signal and recognizes that it is a
sighting signal. Shore-based HF/DFs advise that it originates
from a position near HX.229. U.653 will send further reports
every two hours, every time the convoy changes course and
whenever an approaching U-boat asks for a bearing.

07.17. The Admiralty signals HX.229 advising that the convoy is probably being shadowed.

08.37. U.758 (Kapitänleutnant Helmut Manseck) sights the convoy, signals U-boat Headquarters that he is in contact and settles down to shadow until dark. Two U-boats now in contact.

09.00. Convoy reaches Position V of its route, 49.10 North, 38.01 West, and alters course by a slow wheel to 028 degrees in accordance with Convoy and Routing's signal of three days earlier. This turn surprises the shadowing U.653 who finds a destroyer from the port screen of the convoy bearing down on her. The destroyer is H.M.S. *Mansfield* but its radar is unserviceable and U.653 is not seen by *Mansfield*'s lookouts. The U-boat dives, allows the convoy to pass and then comes back to the surface to resume shadowing.

09.40. U.664 (Oberleutnant Adolf Graef) sights convoy. Three U-boats now in contact.

10.20 (approximately). U.615 (Kapitänleutnant Ralph Kapitzky) sights convoy and thinks about making a submerged attack on one of the stragglers but its periscope is unserviceable. Four U-boats now in contact.

11.10. U.91 (Kapitänleutnant Heinz Walkerling) sights the stragglers but not the convoy. He later dives with the intention of making an underwater attack on one of the stragglers but apparently no attack is made.

11.42. *Volunteer*'s HF/DF operator picks up a strong signal of long content and obtains a bearing of 353 degrees on it. Unfortunately *Volunteer* is the only HF/DF equipped ship in the convoy's escort, so no fix could be obtained by a cross-bearing with another HF/DF ship. This long signal was sent by U.653 to U-boat Headquarters after Kapitänleutnant Feiler had more carefully plotted the convoy's course and speed since his initial sighting. Lieutenant-Commander Luther decides he can spare only one escort to put down the shadower.

11.52. H.M.S. *Mansfield* (Lieutenant-Commander L. C. Hill) is ordered to run out on the bearing of the U-boat signal for 15 miles and, if no contact is made, to remain there searching for the U-boat until 15.00. *Mansfield* does not manage to

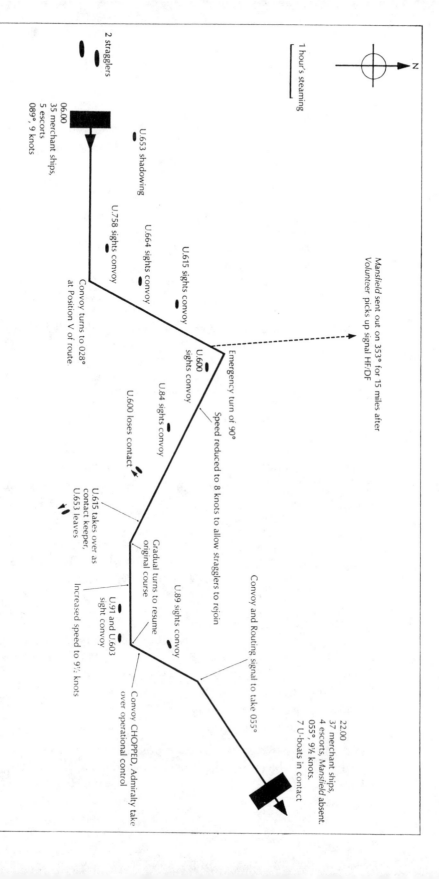

Z

1 hour's steaming

2 stragglers

06.00
35 merchant ships,
5 escorts
089°, 9 knots

U.653 shadowing

U.758 sights convoy

U.664 sights convoy

U.615 sights convoy

Convoy turns to 028°
at Position V of route.

Mansfield sent out on 353° for 15 miles after
Volunteer picks up signal HF/DF

U.600
sights convoy

Emergency turn of 90°

Speed reduced to 8 knots to allow stragglers to rejoin

U.84 sights convoy

U.600 loses contact

U.615 takes over as
contact keeper,
U.653 leaves

Increased speed to 9½ knots

Gradual turns to resume
original course

U.89 sights convoy

U.91 and U.603
sight convoy

Convoy and Routing signal to take 055°

Convoy CHOPPED, Admiralty take
over operational control

22.00
37 merchant ships,
4 escorts, *Mansfield* absent.
055°, 9½ knots.
7 U-boats in contact

contact U.653, which remains shadowing. As a further pre-caution, Lieutenant-Commander Luther recommends to Commodore Mayall that the convoy makes an emergency turn of 90 degrees to starboard. An escort commander is allowed to divert his convoy 20 miles either side of its ordered course but if he goes further than this he must inform the relevant headquarters as soon as possible. Luther signals the diversion to Convoy and Routing two hours later; from the many U-boat signals being picked up around the convoy at that time there is no further cause for keeping radio silence. Commodore Mayall orders the emergency wheel and the convoy settles to a new course of 118 degrees.

12.06. U.600 (Kapitänleutnant Bernhard Zurmühlen) sights the convoy and a corvette of the escort screen which appears to follow the U-boat without making an attack. But no U-boats are sighted by corvettes at this time and U.600 is quite safe. Five U-boats now in contact.

12.35. The Admiralty orders that the destroyer H.M.S. *Vimy* be sailed from Iceland at once to reinforce the escort of HX.229. At this time the convoy is 960 miles from Iceland! *Vimy* is still under repair with defects to her Asdic and although the work is rushed through as quickly as possible *Vimy* will not sail for another thirty-six hours.

12.37. Lieutenant-Commander Luther asks Commodore Mayall to slow the convoy down to 8 knots so that the stragglers astern can catch up. Luther is obviously prepared to sacrifice a little speed so that he can remove from his mind the worry of these two stragglers. They will catch up four hours later and speed will then be increased to $9\frac{1}{2}$ knots.

14.00 (approximately). U.84 (Kapitänleutnant Horst Uphoff) sights the convoy and tries to make an underwater attack on one of the stragglers, but the heavy swell and the tendency of the U-boat to 'cut under' every time he gets into position forces Uphoff to give up the attempt. Six U-boats now in contact.

15.00. The destroyer *Mansfield* gives up her search for the shadowing U-boat at the time ordered by Lieutenant-Commander Luther and sets course to rejoin the convoy, which is expected to be just over one hour's steaming away.

But Luther has omitted to inform *Mansfield* of the emergency course change, a mistake that a regular group commander with a full staff would not have made. When *Mansfield* reaches the expected rendezvous point and finds no convoy in sight, Lieutenant-Commander Hill hesitates before using his radio telephone so as not to force Luther to reveal the convoy's position, but eventually *Mansfield* does call up for a bearing. *Mansfield* will not reach the convoy until after midnight.

15.06. U.600 has been trying to get past an awkward corvette which appears to have been trailing her for three hours, but each time U.600 increases speed she emits too much diesel exhaust smoke. The troublesome corvette eventually causes this U-boat to lose contact. Five U-boats now in contact.

16.00. Lieutenant-Commander Luther asks Commodore Mayall to bring the convoy by easy stages back to its intended course of 028 degrees. This is completed over the next two hours.

16.10. It is obvious that action with U-boats will soon take place, so Luther decides to top up the fuel tanks of his own ship from one of the escort oilers in the convoy. Apart from *Mansfield*, which is still absent, *Volunteer* is the most likely escort to run short of fuel. *Volunteer* goes into the convoy and takes station behind the *Gulf Disc* but the sea is still so rough that after one hour's effort *Volunteer* has not even managed to pick up the hose floating astern of the tanker and Luther orders the attempt to be abandoned.

At an unknown time during the afternoon, U-boat Head-quarters orders U.615 to take over as contact keeper and reporter and U.653 is released in order to make its way to the tanker U-boat and take on fuel for its voyage home. Four U-boats now in contact.*

17.00 (approximately). U.91, which has earlier been trying

---

* Almost exactly one year later, on 15 March 1944, U.653 under a fresh captain was sunk by Swordfish aircraft from the escort carrier H.M.S. *Vindex* and by the British sloops *Starling* and *Wild Goose* in mid-Atlantic. This was the thirteenth 'kill' of the famous Captain Frederick Walker who, at that time, was commanding the 2nd Escort Group. There were no survivors from U.653. Kapitänleutnant Feiler survived the war.

to get into a firing position against a straggler, sights the main convoy. Five U-boats now in contact.

18.00 (approximately). U.603 (Oberleutnant Hans-Joachim Bertelsmann) sights the convoy. Six U-boats now in contact.

18.00. The CHOP Line is reached and the Admiralty takes over control of the convoy from Convoy and Routing.

18.46. U.89 (Kapitänleutnant Dietrich Lohmann) sights the convoy. This U-boat is very low on fuel and was on her way to refuel from a supply U-boat before returning to France. Despite this, Lohmann decides to stay with the convoy and join in the action for at least one night. Seven U-boats now in contact.

19.05. *Volunteer* receives a signal, actually from Convoy and Routing but in response to a previous request by the Admiralty, to take a course of 055 degrees. This is a direct route for England from the convoy's present position and represents the giving up by the Admiralty on the previous day of any hope of routing the convoy round the Stürmer–Dränger line of U-boats. Although the situation has now changed and the U-boats are already in contact, this new course is an ideal one in the present circumstances.

19.25. The convoy wheels to 055 degrees and from now on will make what is in effect a straight run for England with only a few local diversions ordered by the escort commander for tactical reasons.

19.30. Lieutenant-Commander Luther sends a signal to the Admiralty giving his position, course and speed, a weather report, and the information that the destroyer *Witherington* dropped out on the previous day. This is the first indication the Admiralty has that the escort group is down to five ships.

Now about 130 miles away to the north-east, the slower convoy SC.122 had had a much quieter day. Commander Boyle in *Havelock* had been forced to break radio silence to request a tug for the trawler *Campobello*, and to report that he had left behind the corvette *Godetia* to look after the *Campobello*, so no harm was done when he added a further request to that signal. Boyle knew that two American escorts, the destroyer *Babbitt* and the Coast Guard cutter *Ingham*, were at Reykjavik and had been designated to come out later to meet SC.122 and take back to Iceland the six merchant ships destined for Reykjavik. Boyle asked that *Babbitt* and *Ingham* be sailed from Iceland at once to replace *Campobello* and *Godetia* in his screen. The Admiralty responded to this and the two American ships sailed from Reykjavik that evening, although SC.122 was 900 miles away and *Godetia* was then actually rejoining fast after sinking *Campobello* and picking up her crew.

When the eleven U-boats of Gruppe Stürmer had been ordered to break out of their patrol line and make at speed for HX.229, their courses would pass very close to SC.122. Although the nearest Stürmer boat had been 210 miles from SC.122 when ordered to set off for HX.229, and could not reach SC.122's position before late evening at the earliest, Commander Boyle's HF/DF escorts started to pick up U-boat signals during the afternoon and at 16.43 (Convoy Time) the Admiralty signalled the convoy advising that it was probably being shadowed. But a study of German records reveals that no U-boats were in contact with SC.122 at this time and it is probable that the signals picked up were some distance from this convoy and were from U-boats making for HX.229. The Admiralty signal may have been intended as no more than a warning, but the ships of SC.122 believed they were in the same danger as were the HX.229 ships and in both convoys every possible preparation was now being made to meet a U-boat attack.

In Convoy HX.229, which had no rescue ship, Commodore Mayall had two days earlier drawn the attention of the rear merchant ships of every column to their instructions that they were to stop and act as rescue ship to any ship torpedoed further up their column; it is not recorded what provision was made for picking up the crew of the last ship in the column if it was torpedoed. On the *Tekoa*, a cabin boy had threatened to desert at Panama and had been locked in his cabin until the ship was at sea again. After leaving New York the boy had refused to work and Captain Hocken had ordered him to be confined to a locker on bread and water and his pay for the whole voyage to be stopped. The boy was now released because of the U-boat danger but he still refused to carry out any of his duties. The Dutch tanker *Magdala* suffered a sad loss during this day when her only wireless operator died as a result of a heart attack, perhaps brought on by the storm of the previous night. Radio Officer J. F. J. van Dongen, from Rotterdam, was the first man to die in Convoy HX.229.

While we had been in St John's we had been issued with safety lights which could be clipped on to your uniform and which could be switched on if you found yourself in the water to indicate your position. We rather laughed at these when we got them as we didn't exactly think they were morale boosters to an escort about to set out on a convoy. Shortly after we got the signals about the U-boats, however, the First Lieutenant said in his usual cheerful way, 'You know, I don't think it would be a bad idea if we clipped on those lights we got at St John's.' We were all probably thinking of this but I don't think anyone would have wanted to take the first step, but the First Lieutenant's remarks, made with just the right degree of humour, enabled everyone to get their lights without any feeling of doom or despondency. In fact, from that time on, we wore the lights as a matter of course whenever we were at sea. (Sub-Lieutenant R. G. Goudy, H.M.S. *Volunteer*)

On the evening of 16 March, I remained in the Saloon for a few minutes after dinner talking to the Commodore and he told me that he had received a message, from London I presume, informing him they had reason to believe we were being shadowed but, he said, there is another convoy near us and it is possible it could be them. Later in the evening all engine-room personnel who were not already on watch proceeded to the engine room. All the Chinese greasers arrived, each with a suitcase packed. I did not tell them they would not be allowed to take suitcases to the boats – time for that if the necessity arose. (Chief Engineer R. W. Douglas, M.V. *Glenapp*)

Our captain was told by the Senior Officer of the Escort that U-boats were about and he told us everything he knew over the loudspeakers; after all, there wasn't anyone we could tell about it. In fact, I think on this occasion, we were told that two packs were after us which was unusual. We couldn't do much more than hope for the best. (Stoker R. V. Procter, H.M.S. *Anemone*)

Myself and some of the catering staff were off duty having a cup of tea on deck when I think it was a destroyer sped through the convoy flying the flags T over S and I asked them if they knew what that signal meant. None of them did know and I told them that from my previous experiences I understood it to mean that enemy submarines were in our vicinity. (Able Seaman H. J. Brinkworth, S.S. *Nariva*)

And in SC.122:

Just at sunset a two-flag signal crept up to the masthead of the Commodore Ship and you did not need to look at it to know it was W.C. – Enemy Submarines in the Vicinity. You can imagine the play that was made by us on the choice of W.C. for such a signal.

Immediately one had a horrible feeling of apprehension and of nervous tension which increased as the darkness set in – which of us will be afloat in the morning and, if we 'get it', who will be with us in the morning? (Second Officer J. D. Sharp, S.S. *Zouave*)

Everyone became keyed up – this was our job – to protect the convoy and get one of those sneaking Jerry b . . .s who preyed on non-combatants and sank the food ships for the folks back home. Grumbles ceased and a general air of purposefulness pervaded, driving out the 'it's only a job of work' feeling. (Lieutenant M. C. C. F. Shaw, H.M.S. *Saxifrage*)

Practice Action Stations were held on the escorts with more intensity than normal that evening. First lieutenants would see that a hot meal was available. The captain of each ship would make his own decision as to when to order full Action Stations; when this did happen, watertight doors would be closed all through the ship, men would have to remain permanently on duty for many hours or possibly days. Each captain left his decision as late as possible but it is probable that by nightfall the crews of many escorts were at Action Stations.

Let the reader spare a thought for Lieutenant-Commander Gordon John Luther. This young professional naval officer had several years of theoretical anti-submarine experience but only one, incident-free, North Atlantic convoy crossing. He must have dreamed many times before then of leading his own destroyer into action against an enemy submarine. As the last of the daylight of 16 March slowly faded, Luther stood on the bridge of *Volunteer* commanding not just one destroyer in action against one U-boat but a weak collection of completely unknown escort vessels – two slow corvettes and two tired old ex-American destroyers in addition to his own ship. The stream of U-boat signals now coming from out of the darkness on the horizon left Luther in no doubt as to the presence of a strong U-boat pack. The laden merchant ships of the convoy columns were the visible evidence of his huge responsibility.

One patrolling aircraft that evening could have put down every one of these U-boats but, alas, this was the Air Gap and the powers that be had yet to allocate a single V.L.R. aircraft to this part of the North Atlantic.

Gordon John Luther prepared to face the crisis of his naval career. He had only four escorts available to dispose around the convoy because *Mansfield* was still away. There was no rescue ship. Gone were any hopes of acting offensively and of 'killing' U-boats; there could be no thought now of anything but tight defence. Luther placed his own ship in position M, ahead of the port column of the convoy, and from the evidence of the U-boat signals the most likely direction of attack.

22.00. Almost fully dark. Convoy HX.229 with thirty-seven merchant ships and four escorts on course 055 degrees, speed 9½ knots. H.M.S. *Mansfield* absent. Seven U-boats in contact.

8

The Battle of
St Patrick's Day

After the convoy had been reported in our area, our commander, Bertelsmann, went onto the bridge himself and the further reports that came in were sent up there to him. So he was in the perfect position to keep an eye on the weather and the search for the convoy. We went at top speed to find it; the sea was 6 to 8 and the seas kept breaking over us. 'Flags' of smoke seen! Sighting report sent off! Orders for U.603 – 'Keep contact until enough U-boats present'. I took over the sea-watch of a sick comrade because they had talked of snow and I wanted to see snow in the North Atlantic. After an hour's watch I received a heavy sea in my back which almost swept me away. With an iron grip I hung onto the railings of the *Wintergarten* (the platform behind the conning tower); after that we were ordered to strap ourselves to the railings. I survived the rest of the three hours with the thought, 'hard as Krupps steel', 'tough as leather', and so on. This contact-keeping lasted until dusk and then came the order to attack. (Mechanikerobergefreiter Max Zweigle, U.603)

Here were the classic preliminaries to the '*Rudeltaktik*' U-boat pack attack – the initial sighting and the succession of reports from the shadower, the joining of more boats who, in turn, reported their presence to U-boat Headquarters, the determined watch in foul weather in order to keep the convoy in sight. While the drenched and freezing Zweigle was strapped to the *Wintergarten* railings, his captain had been working hard to manœuvre the U-boat into the ideal attack position well ahead of the convoy and slightly to one flank.

There had been eight U-boats in the Raubgraf group and another nearby at the time of U.653's first sighting of what Dönitz's staff believed to be SC.122 but was actually HX.229. Seven of these boats were in contact with the convoy by dusk; this was easily sufficient for the pack attack and Dönitz's headquarters sent out a further signal – U-boats in contact were to attack the convoy.

The sea was still rough after the storm, but the night was

quite clear and an almost full moon produced visibility estimated to be ten miles. This would normally be a hazard to both sides but, with such a weakly escorted convoy, the clear conditions now helped the U-boats more than the escorts. Still upon the conning tower of U.603 with his lookouts, Oberleutnant Bertelsmann carefully guided his boat on the surface through a huge gap between the destroyer *Beverley* and the corvette *Pennywort* until he was well inside the screen and ahead of the starboard flank of the convoy. Lieutenant-Commander Luther had been expecting the U-boats to come in from the port and this starboard side of the convoy had the least protection. U.603 was not spotted either by the escorts' lookouts or by their radar.

U.603 had been on patrol for seven weeks and only had four ready-use torpedoes left after sinking two tankers, both Norwegians, in previous actions. Bertelsmann decided to close in and fire the four torpedoes in two salvos at the only two ships in the starboard column of the convoy. Three of the four torpedoes were the sophisticated new FAT type which could run right through a convoy, then turn a half circle and run back, possibly making several runs until either hitting a ship or the power in the torpedo motor ran out; the fourth torpedo was an older G7e.*

While U.603's captain conned the boat from up top the First Watchofficer, Oberleutnant Rudolf Baltz, was at his post in the interior of the conning tower at the UZO (*U-Boots-Ziel-Optik*), the special sighting glass built into the front of the conning tower and connected to the torpedo calculator which was used for torpedo attack by a surfaced U-boat. The often-filmed U-boat captain, dramatically positioned at his periscope, was not a feature of the surfaced night attack. The last few minutes before firing were a vital and tense time for the crew of any U-boat. This was the culmination of all their training and the fruit of all their long, tiring weeks at sea; it was also their moment of greatest danger, exposed as they were on the surface between the escort screen and the merchant ships. U.603's radio operator sent out the obligatory signal warning other U-boats in the

---

* I have been given two meanings for FAT: *Flächenabsuchendertorpedo* (shallow searching torpedo) and *Federapparatetorpedo* (coiled-spring torpedo); the 'coiled spring' obviously described the path of the FAT torpedo. The G7e was 'Mark G, seven metres long, electric motor'.

immediate vicinity that FAT torpedoes were about to be fired. These boats would have to keep out of the way for at least twenty minutes. Oberleutnant Baltz called out the last few directions and these were fed into the calculator. In the torpedo compartment Mechanikermaat Kurt Schubert turned the handles of the cranks that gave the final setting to the torpedoes loaded in the tubes. At exactly 20.00 Bertelsmann judged that he was close enough to his targets and called out '*Feuereraubnis!*' – 'Permission to fire!' – and Baltz pulled the firing trigger linked to the torpedo tubes calling out as he did so '*Fächer eins . . . LOS! Fächer zwei . . . LOS!*' – 'First salvo . . . FIRE! Second salvo . . . FIRE!'

The motors of the four torpedoes started up, a rush of compressed air forced them out of their tubes, and with a series of soft 'whooshes' they were on their way, causing the bow of the U-boat to rise slightly with the loss of weight. If the batteries were in good condition, the torpedoes soon worked up to a speed of 28 knots and the automatic steering gear guided each torpedo onto its pre-set course.

The convoy had actually just been completing its last turn before dark when U.603 attacked and this is perhaps the reason why only one of Bertelsmann's four torpedoes found a target. The two ships in the exposed starboard column – *Harry Luckenbach* and *Daniel Webster*, both American – were missed but the *Elin K*, a modern Norwegian motor ship at the head of the next column, was hit. The torpedo struck *Elin K* in the aft hold and exploded and, although the bulk cargo of wheat and manganese deadened somewhat the sound of the explosion, the ship made water very fast and there was no hope for her. Chief Officer Berge was on watch on the bridge; he immediately sounded four blasts on the siren and ordered the red masthead light to be lit and two rockets to be fired, all according to the standing instructions for a ship that had been torpedoed. By the time Captain Robert Johannessen reached the bridge, the *Elin K* was clearly sinking and he gave the order to abandon ship.

The *Elin K* had only forty men on board – Norwegian seamen, a Royal Norwegian Navy guncrew and a few Australian seamen.

I was on duty with the gun crew but I don't remember too much about it because I was hit on the head by something. I went below with another sailor and we helped a friend who was in sick bay with malaria fever. We then went back for the friend's kitten but it was too late. When we were ordered to abandon I remember feeling very angry at not being able to fight back and I was also frightened in case the Germans would open fire at the lifeboats. (Seaman First Class Leiv Engebretsen)

It was very dark but somebody had a torch and within a minute everybody was busy getting the lifeboats launched. I remember I was trying to loosen a knot on one of the ropes that held the boat into the ship's side but I couldn't get it untied. All of a sudden one of the officers came and cut the rope with an axe. We jumped on board and the lifeboat was lowered in no time. I remember the boat was banged against the ship's side two or three times before we managed to get clear; by that time, the aft deck was awash. The ship sank very fast and, when we found the other lifeboat and heard that everybody had been saved, we all shouted a hurra. (Motorman John Johannessen)

The *Elin K* went down so quickly that, by the time the third ship in the same column steamed past, the torpedoed ship was already standing vertically on end and the bulkheads could be heard breaking loose. The *Elin K* then made a rapid corkscrew motion and disappeared, but thanks to the seamanship and discipline of this experienced crew, the only life lost was that of the sick man's kitten. The fourth and last ship in *Elin K*'s column was the Dutch cargo ship *Terkoelei* but she sailed straight past the lifeboats without stopping, despite the order that the last ship in any column was to act as rescue ship. Several of the *Elin K*'s crew had been torpedoed before and realized that the escort commander might not be able to send an escort back to rescue them. One officer, shipwrecked twice before, suggested that they row south a little to get out of the convoy's vicinity and then west to pick up the prevailing winds. Quite stoically, the Norwegians settled down for a long voyage in their two lifeboats.

From *Volunteer*'s bridge, Lieutenant-Commander Luther had seen the second of *Elin K*'s distress rockets but he was not even sure that a ship had been torpedoed at all; he had seen no explosion and the *Elin K* had sunk so fast that no ship had been seen dropping astern of the convoy. Luther ordered the standard move for moonlit conditions known as Half Raspberry in which all escorts turned outwards and swept their own sector with radar and Asdic but without using starshell. The four escorts carried out the manœuvre but without any U-boat being detected. The survivors of the *Elin K* were not destined to be in their lifeboats for long. The corvette *Pennywort* sighted the two lifeboats as she was completing her Half Raspberry sweep and, without waiting for orders from Luther, stopped to pick the Norwegians up.

I was most impressed by the calmness and efficiency of the Norwegians. A lifeboat full was brought alongside by their captain, under great difficulty due to the shocking weather conditions. He saluted us, oars were tossed and laid down the centre of the boat. He ordered the crew aboard our ship, stepped aboard himself and the lifeboat was then cut adrift. Their seamanship and general conduct was much to be admired. I'm afraid I cannot say the same of some of our later survivors but I would prefer not to enlarge on that. (Sub-Lieutenant L. M. Maude-Roxby, H.M.S. *Pennywort*)

This first U-boat attack and the sinking of this one merchant ship has been described in some detail but it had really been an undramatic and low-key incident. Only one of Bertelsmann's four torpedoes had scored, the *Elin K* had sunk quietly rather than exploding or bursting into flames and its crew, thanks partly to their own skill and partly to their luck in being spotted by the *Pennywort*, were all safe. A good ship together with 7,500 tons of wheat and manganese had been lost and 339 sacks of mail from Australia for England would never be delivered.

Oberleutnant Bertelsmann had carried out the standard escape tactic of diving under the convoy. He was disappointed when only one explosion was heard but later he heard what he thought was a second explosion – it was probably the *Elin K* breaking up. When U.603 eventually surfaced behind the convoy and sent off the attack report to U-boat Headquarters, Bertelsmann claimed one ship hit and one probably hit but no definite sinkings. With no more torpedoes available U.603 was ordered to take over as contact keeper and she performed this duty, with some breaks, for the next two days. It is an unusual coincidence that after sinking two Norwegian ships earlier in this patrol the *Elin K* should also be Norwegian and the first ship sunk by U.603 in the next patrol would be of the same nationality.*

After the quiet sinking of the *Elin K* nothing happened for the next hour. Although there were several U-boats about, the convoy's last turn just before dark might have caught the

---

* The three other Norwegian ships sunk by U.603 were *Stigstad*, *Glittre*, and *Brandt*. Oberleutnant Bertelsmann suffered from rheumatism and did not sail with U.603 for the next two patrols but he was back in command and was lost with his entire crew when U.603 was sunk by the U.S. destroyer *Bronstein* on 1 March 1944.

U-boats by surprise; this turn, together with U.603's warning signal that she was firing FAT torpedoes, probably caused other captains to delay their attacks. The men in the convoy ships hoped that the U-boat that had torpedoed the *Elin K* had been a 'loner' and when nothing further happened in the next hour they began to hope that this might be so. The convoy steamed on with only three escorts now – *Volunteer, Beverley* and *Anemone*. *Mansfield* had still not returned from her daylight sortie and *Pennywort* was astern picking up the *Elin K*'s crew.

U.758 with Kapitänleutnant Helmut Manseck in command had been in contact for over twelve hours.

I had shadowed the convoy all day keeping at extreme range on the starboard side, just keeping the smoke and the tips of the masts in sight. I remember that the ships were doing well and not making much smoke. We had been about twelve miles out during the day and came in to four to five miles at dusk.

When I came in to make my attack I found that I had misjudged the speed of the convoy and that we were almost level with it but I decided to attack from there rather than try to get ahead again; we came in from just ahead of 90 degrees. I could see six, eight, or ten ships and selected a solid, overlapping target of the third ships in the starboard columns. We fired our four torpedoes, then turned sharply away to port and ran out.

Manseck's two FAT and two G7e torpedoes were fired at 21.25. There had been no question of him having to penetrate the escort screen; the only escort posted on the starboard side had been *Pennywort*. The difficulties encountered at St John's in making up an escort group for this convoy, the absence of a rescue ship and the failure of the *Terkoelei* to stop and pick up the crew of the *Elin K*, were now snowballing to cause great difficulties for the escorts. Kapitänleutnant Manseck later claimed victims for each of his four torpedoes – two cargo ships sunk outright and a tanker and a cargo ship hit. In fact Manseck had overclaimed and had hit only two ships – the Dutch cargo ship *Zaanland* and the American Liberty ship *James Oglethorpe*, third ships in the second and third columns from starboard.

Some of the men who were on the *Zaanland* describe their experiences, Chief Officer P. G. van Altveer first.

I was just passing number 4 hatch, when it suddenly happened. I felt the iron deckplating rattle and shake under my feet. I saw a flash of light and I felt a torrent of water all over my body and then

I was blasted away. At first I thought that I was swept into the sea but then I lost consciousness. After about ten minutes I regained consciousness and found out that I was lying on my back between the winches of Number 4 hatch and the mainmast. I saw the stars and clouds right above me and realized my situation. I got up very carefully and walked towards the boat-deck being well aware of the danger that the deckplating might have been torn open by the explosion of the torpedo but everything appeared to be safe to proceed. I later found that I had two broken ribs.

G. H. Baird-Jones was an English cadet on the Dutch ship.

I was standing in the starboard wing of the bridge, looking over the side towards the stern, when there was a tremendous explosion and a huge flash and a strong smell of burning. The ship had been torpedoed on the starboard side abreast of the mainmast which collapsed, falling onto the after end of the boat-deck and causing injury to some of the crew. To my amazement I felt extremely calm; I had often wondered how I should feel if this ever happened to me. Although I had been at sea for two years prior to this, it was the first time I had been involved in a submarine attack. I found afterwards that the whole of one side of my uniform was full of small holes; I wondered whether the wheat we were carrying had peppered me.

Zaanland's master, Captain Gerardus Franken, could hear water rushing into the engine-room 'with great force' and fuel oil was gushing up from the tanks below, flooding the deck and running overboard where it did at least calm the surface of the water around the sinking ship. Captain Franken ordered his crew to abandon ship. Second Officer J. Waasenaar was in charge of one of the boats.

The boats were wildly tossing up and down. I don't remember climbing into my boat but I do remember vividly struggling with the lower davit blocks to disengage them from the lifeboat. One moment there was slack in the boat falls, the next moment they were dangerously tight; we were buck-jumping all the time. One moment our eyes were level with the ship's railing, the next moment we saw the boot-top flashing past us. When we got the afterblock free, it started to swing dangerously over our heads. It knocked my cap off and that cap seemed so damned important to me that, for a moment, I forgot that far more important things were at hand. I desperately searched for the cap without finding it. Disengaging the forward block was impossible and, in the end, we cut the falls.

We saw the other lifeboats also waiting for what was going to happen to our ship. We did not have long to wait; she sank very rapidly, stern first she went down. With her bows high in the air, we heard a rumble like thunder – probably the boilers crashing through the bulkheads – we saw sparks on the forecastle probably caused by the anchor chains running away and then it was all over.

Once again, a well-trained crew of mainly pre-war sailors had been able to abandon their ship in good order.

There was a far different scene on the *James Oglethorpe* in the next column. This was the Liberty ship that had been the first one built in Savannah. She had signed on her crew at that port and this was her first voyage. Ever since the *Elin K* had been hit, *James Oglethorpe* had been steaming a zig-zag course but without getting too far from her convoy position. Shortly before she was torpedoed, a lookout sighted a surfaced U-boat astern; this was reported to the master but, for some reason which can only be guessed, Captain Long ordered the naval guncrew not to open fire and he did not have time to report the sighting to the Commodore or to the escorts. This U-boat was not the one which subsequently torpedoed the *James Oglethorpe*. The remainder of an unhappy story can be told quickly. The torpedo had started a fire in one of the holds loaded with cotton and the vessel started to list to starboard. Captain Long did not order Abandon Ship and must have decided to try and save his ship but part of his crew did not stand by him and a panic-stricken rush to launch the boats took place. It will never be known who started this but two of the three mates were soon ordering boats and rafts away and many men assumed that the captain had ordered the ship to be abandoned. One lifeboat had one of its falls cut prematurely and the occupants were thrown into the sea 'like peas out of a pod'. Thirteen men were drowned following this accident or because they dived in and were swept away before they could reach a boat. Other lifeboats got away safely with thirty of the crew, including two of the mates.

Back on the *James Oglethorpe*, Captain Long was left with his only remaining mate – Second Officer Joseph Duke, a Savannah man – and thirty-one other men. This depleted crew put out the fire; the engines were still running and the ship showed no signs of sinking, probably because the large tonnage of cotton in her cargo had swollen and prevented too much water entering. But either the steering had been damaged or the helm had been abandoned because, for some time, the ship steamed a succession of huge circles in the rear of the rapidly disappearing convoy.

Once more no merchant ship had stopped to rescue survivors, the Dutch *Terkoelei* again steaming straight past the sinking *Zaanland* while *James Oglethorpe* had been the last ship in her own column. Four lifeboats from *Zaanland* and two from the American ship were all in the same area, where their greatest danger was from the crippled *James Oglethorpe* whose turning circle kept bringing the Liberty ship through the group of lifeboats and there were several nerve-racking moments. Chief Officer van Altveer was in one of the boats.

Suddenly, I saw a ship heading straight for us and I feared that she would overrun us; the accommodation ladder was hanging over the side and I thought we should get jammed underneath the spur (lower end) of the ladder. Most of my lifeboat's occupants rose from their seats to be ready for jumping overboard but I shouted, 'Sit down. Do sit down!' – I feared the boat might capsize. It was a very exciting moment. She passed us on her port side, at no more than five yards.

Then, we saw a man floating on the sea, apparently he had left that ship by way of the accommodation ladder but it was too risky for us to pick him up. Firstly it was necessary to get clear of the turning circle of that ship and secondly my boat was very deep in the water, even overloaded. Furthermore, the condition of the sea and the weather did not allow me to take an extra risk in turning round. I would have had to take the bow away from the sea and high swell and incur the risk of being turned over or capsized by a cross-sea. I decided to stay on my course although it was a very painful decision. I felt very sorry for that man but I could not risk the lives of all the men in my boat.

And so an unknown American seaman had to be left to drown.

Second Officer Waasenaar's lifeboat was also nearly run down by the *James Oglethorpe* but, in this case, it led to a life being saved.

While I was trying to get out of the path of that American ship I saw, for just a second, the flicker of a red light. I kept the bearing in mind and, as soon as I was able to, I tried to get to the spot. After a while we came upon a black mass in the water; it was a body. The bosun thought it to be a negro and still alive. With difficulty the sailors got him aboard, he was covered by fuel oil, heavy and slippery. He uttered a few words and the bosun shouted, 'He talks Dutch. Damn it, he is our captain.'

Captain Franken had been the last man to leave the *Zaanland* but had fallen into the sea and been carried away while trying to get into the last lifeboat. He became so fouled up with oil and so sick that he had tried to finish things off quickly by cutting the strings of his lifejacket but was rescued just in time.

There is no need to spend much more time with the *Zaanland* men. The corvette *Pennywort* came up from picking up the crew of the *Elin K* and again started rescue work being joined for a short time by the destroyer *Beverley*, but *Beverley* was ordered by Lieutenant-Commander Luther to return to the convoy. The entire crew of the *Zaanland* was saved and also the men from the *James Oglethorpe*'s boats. Lieutenant O. G. Stuart, the Canadian captain of *Pennywort*, then closed the *James Oglethorpe*, which was by then under control, and talked with her master. Captain Long told Stuart that he had decided to remain on board with what was left of his crew and would turn back and try to make St John's independently. Stuart tried to persuade his *James Oglethorpe* survivors to return to their ship but there were no takers for this offer and *Pennywort* was soon ordered back to the convoy. *James Oglethorpe* with her faithful master and what was left of his crew must be left at this stage; their fortunes will be described in a later chapter.

Lieutenant-Commander Luther had immediately ordered another Half Raspberry when *Zaanland* and *James Oglethorpe* had been torpedoed, although there were only three escorts with the convoy to carry out this counter move. U.758, which had made the attack, was not affected by the Half Raspberry as there was no escort on its side of the convoy. The U-boat quietly ran out on the surface away from the convoy and, for this crew, the battle was over. U.758 had no more torpedoes and Manseck set course for a rendezvous with a tanker U-boat to refuel before the long journey back to his base at St Nazaire.

On the port quarter of the convoy, however, the corvette *Anemone* was just completing the second leg of her Half Raspberry sweep twenty minutes later when an alert lookout sighted a surfaced U-boat clearly visible exactly in the path of the moon 3,000 yards away and between *Anemone* and the convoy. *Anemone*'s captain, Lieutenant-Commander P. G. A. King, R.N.R., immediately turned his ship towards the U-boat. For twelve minutes the corvette pounded down towards the U-boat, which took no evasive action and had obviously not seen the corvette. For this reason, King ordered the corvette's forward gun to hold its fire.

Anemone was only 300 yards away when the U-boat sud-

denly dived. An escort could rarely ask for a better chance
than this for a successful depth-charge attack, the only drawback being that if the full pattern of ten charges – two each
from the throwers on either side and six from traps which
rolled the depth charges over the stern – were set to explode
at shallow settings the resulting explosions might well
damage the corvette. Lieutenant-Commander King quickly
ordered that five depth charges only were to be dropped, set
to explode at 50 feet. There was neither time nor need for an
elaborate Asdic attack; the depth charges were released when
Anemone passed through the swirl left by the diving U-boat.
Even this reduced pattern shook *Anemone* up so much that her
radar and her radio-telephone were both put out of action
temporarily. The Asdic set continued to function well but
contact with the U-boat was lost in the turmoil of the depth-
charge explosions.

The men on *Anemone* were convinced that her depth
charges had been near enough to cripple the U-boat but they
looked in vain for the expected air bubbles, oil and debris. So
Anemone stood off a little to allow Asdic conditions to settle
down and sixteen minutes later a firm underwater contact was
picked up some distance away. Again, *Anemone* ran in to
attack but almost at once the U-boat surfaced 1,500 yards
ahead. A surface chase followed with *Anemone* gradually
gaining until the U-boat dived again twelve minutes later.
Anemone's Asdic picked it up at once and a new underwater
attack was started. Unfortunately, when the corvette was 200
yards short of the attack position, an electrical fault caused
some of the firing bells to ring prematurely and the men
manning the four side throwers promptly fired their depth
charges. The resulting explosions then upset the Asdic set
and the six charges in the stern traps were never released.

The men on the *Anemone* waited for the Asdic to recover
and then started searching yet again. Their patience was
rewarded twenty minutes later by another contact and three
more attacks were made on this, two with full depth-charge
patterns and one with the Hedgehog – a device mounted on
the corvette's forecastle which could fire a salvo of twenty-
four small charges well ahead of the ship. This recently
introduced weapon was fine in principle, and allowed an
attack to be made before an escort reached the U-boat's
position, but on this occasion the gale of the previous day

had saturated the firing mechanism; only four of the charges fired and these failed to hit the U-boat. After one of the conventional depth-charge attacks, however, those aboard *Anemone* thought they saw 'a further muffled explosion in the wake throwing up a large dome of water which was quite different to the characteristic depth-charge explosion and occurred thirty seconds after the last depth-charge had fired'.*

Asdic contact was not regained after *Anemone's* fifth attack and, as he had been absent from the convoy for almost two hours, Lieutenant-Commander King tried one more short search which produced nothing, then gave up and set course for the convoy. *Anemone* had dropped thirty-five depth charges and fired four Hedgehog missiles. In his report King stated that his first attack, just after the U-boat first dived, 'looked a certain kill' and he could not understand 'how the U-boat got away with it'.

The U-boat that had been attacked by *Anemone* was U.89. This boat had been released from the Raubgraf group two days earlier with engine trouble and a shortage of fuel but Kapitänleutnant Dietrich Lohmann had joined in the attack on the convoy on his own initiative. It was probably U.89's engine trouble that prevented Lohmann from outrunning *Anemone* on the surface and forced him to dive and face the depth-charge attack. The contest had been an interesting one because of the particular ships involved. *Anemone* was one of the early wartime corvettes and had sunk an Italian sub-marine, the *Nani*, back in January 1941, one of the first corvette successes of the war. The low numbered U.89, launched at Lubeck in 1941, was also a vessel with long experience of the Atlantic war. U.89 had suffered only very minor damage through *Anemone's* attacks and the encounter illustrates just how difficult it was for a single escort to sink a U-boat with depth charges, even in the most promising circumstances. It is not known what caused the 'muffled explosion' and the 'large dome of water' after one of the attacks; perhaps U.89 had vented its tanks suddenly to go deeper. Kapitänleutnant Lohmann and his crew cannot be

---

* From *Anemone's* Report of Proceedings in Public Record Office ADM 199/575. Subsequent quotations from HX.229 escorts will be from the same reference while SC.122 escorts will be ADM 199/579; quotations from the Convoy Commodore's Report for HX.229 will be ADM 199/576 and SC.122's will be ADM 199/580.

asked about this; they were all lost on their next patrol. The commander of the Londonderry Escort Force later criticized Lieutenant-Commander King for not dropping the full ten-charge pattern in the first attack, but King replied that he understood that corvettes were not permitted to drop a full pattern with shallow setting. After this convoy *Anemone's* engines 'were never the same again', with frequent big-end trouble, and her crew believed that these shallow depth-charge explosions were responsible.

Anemone's effort had by no means been wasted. U.89 now gave up any hope of attacking the convoy and set course for the tanker U-boat, and another U-boat in the area, Oberleutnant Adolf Graef's U.664, had been near enough to be shaken up by the depth charges. Graef also dropped out of the operation for the time being. Unfortunately two more U-boats had caught up with the convoy during the period in which U.758 left for home after her attack and U.89 and U.664 were forced to lose contact by *Anemone*.

Lieutenant-Commander Luther had been kept in touch about the progress of *Anemone's* depth-charge attack but there had been no question of sending another escort to join in the action because for some time *Volunteer* had been the only escort remaining with the convoy. After finishing his own Half Raspberry manœuvre, Luther decided to sweep across the rear of the convoy. To his intense dismay, Luther found yet another merchant ship listing heavily to starboard and blowing off steam. Its crew were taking to their lifeboats. This ship was another of the convoy's American Liberty ships, the *William Eustis*, also on its first voyage. The U-boat responsible for her condition was U.435, which had only come into contact with the convoy twenty minutes earlier; Kapitänleutnant Siegfried Strelow had come straight in for a snap attack and fired two torpedoes. He later claimed two hits on a 7,000-ton tanker; the *William Eustis* was 7,196 tons but no tanker. There was no point in Luther ordering any more Half Raspberries for the non-existent escort and he was now left with the difficult choice of hunting for the attacking U-boat himself, staying to protect the convoy from further attacks, or stopping to pick up the American crew – the last ship in the column had again failed to stop. Lieutenant G. C. Leslie, Luther's first lieutenant, describes the dilemma.

The Captain and I had discussed the possibility on several occasions and our conclusion was that rescue was very important at a time when the worst disaster in the Atlantic battle would be a failure of morale in the merchant ships. In the absence of a rescue ship and with the failure of the last ship in the column to stop and pick up survivors, the Escort Group Commander had an almost impossible decision to make.

Volunteer's captain did not take long to reach a decision. The destroyer made a wide Asdic sweep around the wrecked Liberty ship in case a U-boat was nearby, then slowed down to start picking up the American seamen.

Two of the *William Eustis*'s boats had been smashed by the torpedo explosion but the entire crew had got away in the remaining four boats and on the liferafts.

They were all wearing the type of safety lights I mentioned earlier. These were red and, in the darkness, they looked like fairy lights bobbing up and down above the water as you could not see the boats or the men at any distance in the dark. We turned the ship towards the lights and, as we got near, we stopped and put our scrambling nets over the side and this enabled men to get from the boats into the ship without too much difficulty. The sea was not rough at the time and I doubt very much whether more than a few even got their feet wet. (Sub-Lieutenant R. G. Goudy)

I remember being on deck when the survivors came aboard and they removed their frogman-type rubber suits. They had on what looked to me like their complete Sunday suits; I helped one person remove his rubber suit, he had highly polished shoes, etc., all ready for stepping off ashore. He presented me with the rubber suit, which I kept for several months. (Petty Officer R. V. Jackson)

Several of *Volunteer*'s crew were somewhat annoyed to find that the Americans had brought full suitcases and felt that the *William Eustis* had been abandoned too soon. Lieutenant-Commander Luther was not pleased when the American captain admitted that he had even failed to dump his code books and other confidential papers and when the American chief engineer suggested that, if *Volunteer* could wait until daylight, he might be able to get the *William Eustis* going again. This was clearly out of the question, but Luther could not leave the Liberty ship afloat with its code books on board.

We went very fast past her and fired four depth charges from the throwers as we went by. They all exploded underneath her and lifted her about five feet out of the water – it was quite a dramatic sight. We went whizzing off back to the convoy. The survivors told us that she had been carrying 7,000 tons of sugar. Later we worked out how many cups of tea that sugar would have made and were horrified to think we had just sunk about three weeks' sugar ration for Britain. (Lieutenant G. C. Leslie)

The torpedoing of the *William Eustis* marked the end of the first phase of what had so far been a somewhat one-sided battle. Three U-boats had made attacks on HX.229 in three hours and four merchant ships had been torpedoed. If one analyses the position at 23.00 Convoy Time it would probably be concluded that this was a moment of crisis for the convoy. Three escorts, including the ship of the escort commander, were absent rescuing survivors, *Anemone* was also behind the convoy depth-charging its U-boat contact and the fifth, *Mansfield*, had still not rejoined from its daylight search for the shadowing U-boat. The thirty-three merchant ships of the convoy steamed steadily on their course of 053 degrees; there was a wind of force 3 to 4 from the north, a swell that was now moderating, and the bright moon gave a visibility of between five and eight miles. The convoy was completely without escort at this time and would remain so for a further one and a half hours. Six U-boats were in contact in what were perfect weather conditions for surface torpedo attacks. The merchant ships of convoy HX.229 were in truly desperate straits.

Not one torpedo was fired at the convoy in that hour and a half. There are good reasons why two of the U-boats failed to take advantage of the defenceless targets – U.603, which had torpedoed the *Elin K*, had no ready-use torpedoes and had been appointed as the contact-keeper, and U.435, which had just torpedoed the *William Eustis*, was busy reloading her tubes. But the remaining four boats, three of which had been in contact for many hours, simply did not press home their attacks. These U-boats had been operating off Newfoundland in the Raubgraf group for several weeks. Their crews were physically tired and several of the boats had developed mechanical defects. But bold, aggressive captains would have overcome these difficulties and come in to slaughter the merchant ships. The First Watchofficer of one of these reluctant U-boats describes the actions of his boat.

All day we had kept in touch with the convoy in the way we had been taught at school, keeping at extreme range, gradually overtaking, establishing the convoy's general course and radioing this to B.d.U. During the twilight we reduced the gap and when it was very dark we closed in from the south. We could see the merchant ships clearly; they seemed to be lifting themselves like turtles' backs on the horizon. The captain decided, a decision unclear to me, to allow the convoy to pass and then break into it from astern. While we were on a course to the west to get into the rear of the convoy, we met a whole stream of U-boats travelling east to get at the convoy from the flank.

When we at last turned north to come back onto the convoy's course we came upon an escort which was guarding a ship left behind; it is possible that the escort was taking on fuel but we couldn't tell for certain. [This may have been *James Oglethorpe* with *Pennywort* standing by.] When, at long last, the captain decided to follow the convoy we didn't find anything but we could later make out, far to the east, the faint gleams of exploding ships which were being attacked by other boats. We followed at top speed but we were too late that night. (Leutnant Claus von Egan-Krieger, U.615)

Midnight came, the many seamen of Irish extraction perhaps wondering what St Patrick's Day would bring them, and in fact at that time there was a little relief when *Beverley* rejoined and thirty minutes later further help arrived in the shape of *Mansfield* which had been absent since the previous afternoon. Lieutenant-Commander Rodney Price, R.N., in *Beverley*, took charge of this escort of two ex-American destroyers.

But the U-boats were also reinforced about this time with the arrival of U.228 and U.616, two boats that had just refuelled from the tanker U-boat, U.463, which was only sixty miles to the south of HX.229's route. Their arrival brought the effective number of U-boats now in contact to eight.

The old American freighter *Harry Luckenbach* was the leading ship of the starboard column and enough torpedoes had already come past her to hit three ships in neighbouring columns. So exposed was her position that the *Harry Luckenbach* had gone out in front of the convoy and steamed a nervous zig-zag course until ordered to return to her column by one of the escorts. Three quarters of an hour after midnight the convoy's spell of immunity came to an inevitable end when one, or possibly two, torpedoes struck the *Harry Luckenbach*'s starboard side. There was a big explosion, a huge column of smoke and flame and this poor old freighter went down in four minutes.

No less than ten torpedoes were actually fired at the convoy

at this time. U.435, having reloaded her tubes after torpedoing the *William Eustis*, had fired a full salvo from her four
forward tubes and a fifth torpedo from her stern tube at ships
on the convoy's port side and a few minutes later U.91 had
done exactly the same from the starboard side. It was almost
certainly U.91's torpedo or torpedoes that had hit the *Harry
Luckenbach* in her exposed position on the starboard corner of
the convoy. U.91's captain was Kapitänleutnant Heinz
Walkerling, who had almost caught HX.229 two days
earlier when he had spotted one of HX.229's escorts but had
lost sight of it again in the storm. Now Walkerling had his
revenge in the destruction of the *Harry Luckenbach*, although
he claimed the sinking of two ships. Kapitänleutnant Siegfried
Strelow, of U.435, was even more enthusiastic in his elaborate success signal and gave details of the five ships that he
claimed to have hit although his torpedoes had hit nothing
at all.

Lieutenant-Commander Price in *Beverley* had seen the
torpedo explosion and then *Harry Luckenbach's* distress
rockets. He took his own ship to search to starboard of the
convoy and sent *Mansfield*, whose radar was out of action, to
port. *Beverley* must have swept very close to U.91, which had
swung away on the surface after firing its torpedoes, but the
U-boat was not spotted. On the other side of the convoy
Mansfield picked up an echo on her Asdic, but Lieutenant-
Commander Hill was not too sure of the quality of this
contact and dropped only one pattern of ten depth charges
over it before returning to the convoy. His assessment was
quite correct; no U-boat was depth-charged at that time.

It is not known how many of the *Harry Luckenbach's* crew
were killed by the effects of the torpedo explosions but,
although she went down so quickly, enough men managed to
get away to fill three lifeboats. Once again, small reports in
several documents can be pieced together to make one
account, although in this case an incomplete one. At the end
of her sweep down the starboard side of the convoy, *Beverley*
spotted the three boatloads of survivors, now three miles
behind the convoy. Lieutenant-Commander Price decided
that he could not leave the convoy with only the radarless
Mansfield, and reported the lifeboats to Lieutenant-Commander Luther in *Volunteer* so that one of the escorts then
coming up from astern could deal with the survivors. Forty

minutes later *Volunteer* herself sighted the lifeboats, but Luther had already left the convoy once to pick up survivors and he decided that he must this time leave the lifeboats to another escort; a message was sent to *Anemone* ordering her to find *Harry Luckenbach*'s lifeboats and pick up the survivors after she had finished the depth-charge attacks on her U-boat contact. Unfortunately *Anemone* never found the lifeboats but some time later, while returning from her rescue work, *Pennywort* came upon them; however, this corvette already had 108 survivors on board and had been absent from the convoy for six hours. Lieutenant Stuart tried to call up Lieutenant-Commander Luther and consult him about the three lifeboats, but the radio telephone was too busy. Stuart had been listening to the radio messages of other escorts, and he believed that these lifeboats contained the crew of a salvageable merchant ship which had been abandoned too soon and that the men in the lifeboats could reboard their ship. In any case, he expected to contact Luther soon and would be allowed to come back; in the meantime *Pennywort* steamed on. But Luther was now realizing that he simply could not keep leaving the convoy without cover to rescue survivors and, when *Pennywort* eventually did speak to him, he ordered her to rejoin the convoy at once. It is probable that yet another escort vessel saw the *Harry Luckenbach*'s lifeboats. H.M.S. *Abelia*, steaming hard from St John's to catch up HX.229, is believed to have sighted these lifeboats, but was under strict orders to join the convoy without delay and had to ignore them.

The reader can only imagine the anguish of Captain Ralph McKinnon and his crew, most of them New York or New Jersey men, when they saw four possible rescue ships steam on and leave them. The harassed Lieutenant-Commander Luther never did realize that these survivors were not accounted for. Not one man of the *Harry Luckenbach* was ever seen again. The total death roll, either in the torpedoing or in the lifeboats, was fifty-four merchant seamen and twenty-six naval men of the Armed Guard. It will never be known how they perished. A violent storm later swept this area and perhaps they found a mercifully quick death in that.

There followed another lull for the merchant ships of the

convoy and their harassed escorts and there were no attacks for an hour and a half after the *Harry Luckenbach* went down. None of the absent escorts rejoined during this period and the protection of the convoy remained in the hands of *Beverley* and *Mansfield*. *Volunteer* was coming up fast from astern but the two corvettes were very slow in rejoining, having only a 6-knot speed advantage over the relatively fast convoy which, spurred on now by the obvious U-boat danger, was making nearly 10 knots.

The next moment of danger came at 02.18. U.616 had been ahead of the convoy trying to get into a good attack position for over an hour. Her captain, Oberleutnant Siegfried Koitschka, had seen the two destroyers of the escort and one of these, almost certainly *Beverley*, suddenly turned and came dead ahead of U.616. The frustrated Koitschka took rapid aim and fired off all four of his bow torpedoes at the destroyer, but *Beverley*'s high speed and erratic zig-zag saved her and no one aboard was ever aware of how close she had come to destruction.

Thirty minutes later *Volunteer* arrived back with the convoy and Lieutenant-Commander Luther must have been hoping that the worst might now be over; although five merchant ships had so far been torpedoed, as far as he knew there had been no torpedo attacks for the past two and a half hours; there were now three escorts with the convoy and every hope that two more would soon rejoin; first light would come in an hour and a half. But, even before *Volunteer* had time to take over command of the escort again, the hopes of Luther and of many others were dashed. The dull thuds of explosions followed by sirens, distress rockets and the red masthead lights of torpedoed ships, showed that another U-boat had struck.

This was the work of Kapitänleutnant Bernhard Zurmühlen's U.600, which had been in contact with the convoy for over four hours, carefully avoiding escorts and submerging each time another U-boat signalled that it was firing FAT torpedoes. Then, during the long spell of quiet and while the convoy had been almost without escort, Zurmühlen had come in on the surface ahead of the starboard columns and fired a full salvo of five torpedoes. His success signal claimed to have hit five ships. This claim was not as wild as that of some captains; three ships had been hit – the American freighter *Irénée du Pont*, the British refrigerated ship *Nariva* and the British tanker *Southern Princess*.

The *Irénée du Pont* had been hit first. Ensign Frank Pilling has earlier described how his party of American naval officers taking passage to England had been apprehensive at New York over the U-boat threat. Pilling makes a good eyewitness again.

I became fascinated by the artificial torpedo tracks of moonlight and white water. I was not surprised when we were hit, for suddenly I saw, deep under the surface, two streaks of greasy light, parallel, moving fast, coming in at an angle. There was no time to shout a warning; in one instant there were the tracks, in another a great shattering crash. Holding onto the rail I was not thrown down, but seeing what appeared to be flying metal in the air, I rushed towards the opposite side of the ship. The second torpedo shook the *Irénée* with amazing violence. She quivered and staggered so wildly for a moment that I lost my footing and lurched crazily before running forward to the port ladder. Passing the captain, I heard him say, 'Well, this is it!' Then I joined the mad rush of humanity on the ladder.

Events now followed in meaningless confusion. I made my way to my assigned liferaft. There was no one there, so I joined people trying to break loose the raft opposite. It would not come away – rusted or frozen perhaps. So we all ran forward to the rafts of the welldeck, which was awash and beginning to settle. After violent efforts we slid one of these into the ocean but the first man into it slipped its toggle, doubtless fearing the sinking ship would carry it under. Thus our slim hope of survival went drifting away. There was no choice but to jump overboard and swim for it. We were joined by the entire complement of the port lifeboat which had upset through mismanagement and panic. Cries for help were all over the ocean.

If an Allied merchant ship was torpedoed and any part of the crew survived, an official Survivors' Report was later compiled. It is possible to read between the lines of the *Irénée du Pont*'s report and find an unhappy tale of panic. Ensign Pilling has described the mad rush that followed immediately on the torpedoing. Only two of the ship's boats were properly launched, and many of the crew dived into the water without attempting to launch the remaining boats. Several current codebooks and seventeen diplomatic pouches were left on board; the above documents 'could not be found' although the ship remained afloat and on an even keel for at least the next five hours. The report goes on to say that the ship was eventually abandoned by order of the master nearly five hours after the torpedoes struck; all other sources say that the *Irénée du Pont* was empty within minutes of being torpedoed!*

The second ship to be hit was the *Nariva*. Second Officer G. D. Williams was on duty at the time.

Nariva was torpedoed with a ear-shattering roar and the deck bucked and heaved violently under my feet. A huge tower of black smoke, tons of water and debris was flung into the air just forward of the bridge. Captain Dodds came out on to the port wing of the bridge when I remembered the adage that what goes up must just as assuredly come down and, without ceremony, I pushed the 'old-man' back into the wheel-house and not a second too soon for tons of water and debris fell on the bridge with a crunch and clatter. I well remember right then being momentarily amused almost into a fit of giggles when the elderly seaman at the wheel shouted above the noise, 'All right, Sir, I've got her. She's alright now.'

The ship was making water fast and the forward welldeck was soon awash. The order was given to abandon ship and the boats launched. We pulled away from the ship but then we saw the life-raft which was secured to the port mainmast shrouds released with a splash into the water and several men jump after it where they clung desperately and shouted for help. As my boat crew tried clumsily to backwater their oars we saw the raft drift slowly forward along the ship's side and, to our horror, we watched helplessly as a great inrush of water sucked the raft and its occupants into the cavernous hole blasted in the ship's side by the torpedo and even now, as I write this account, I can still hear the screams of the men inside the hull. But then, thank God, the same rush of water that had drawn them into the hull of the ship, at the next roll, swept them out again by which time we were so much closer and could grab the raft lines and drag the men to safety in our boat, one of whom turned out to be our dear, elderly chief engineer who, as if in gratitude for his rescue, became violently sick all over me.

Nariva remained floating, probably buoyed up by the cork of her insulation, and Captain Dodds decided to keep his boats nearby ready to reboard his ship after dawn if she was still afloat then.

The third ship hit in this attack was the *Southern Princess*, a thirty-year-old tanker (previously named *San Patricio*) which had once been converted to become a whaling factory ship in the Antarctic but was now being used as a tanker again. At 12,156 tons she was easily the largest ship in the convoy, and was carrying 10,000 tons of fuel oil in her cargo tanks and two railway engines and several invasion barges on the flat whale-flensing deck above the main deck. One torpedo had

* Survivors' Report provided by U.S. Naval Historical Center.

hit on the starboard side under the bridge, the explosion breaking down the bulkheads between the cargo tanks and the crew accommodation, and oil was soon flooding into the quarters and also out of the torpedo hole onto the surface of the sea. The explosion ignited the gas in the top of the rear-most cargo tank and the oil in the holds soon caught fire. The resulting huge, towering blaze became so intense that paint-work on a ship in the next column was blistered and onlookers could hear agonizing screams from men trapped on the forepart of the *Southern Princess*. They feared there would be a huge loss of life – 'We watched in stupefying silence and fascinated horror, hardly believing that it was real.'

The situation on *Southern Princess* was not as serious as might be imagined. There were exactly 100 men aboard, twenty-nine of whom were survivors of previous sinkings and were returning to England as passengers. Only four of those aboard failed to get away safely – two of the passengers and two teenage boys. One of these was a messroom boy who, before the ship was hit, had told one of the officers that he could not keep awake any longer and was going below for a sleep, and the other was an eighteen-year-old American boy seaman.

The last ship in the *Irénée du Pont*'s column was the New Zealand Shipping Company's refrigerated ship *Tekoa* with Captain Albert Hocken in his first command as master. Earlier, after he had seen five ships torpedoed nearby and not one merchant ship at the tail of a column had stopped to rescue survivors, Hocken had signalled the Commodore, 'Am I to act as rescue ship without further signal from you?' and received the reply that the order still held. When the *Irénée du Pont* was hit, Captain Hocken immediately gave orders that *Tekoa* was to stop and commence rescuing survivors. The designated rescue ship for *Nariva* and *Southern Princess* did not take the same action.

Lit up by the blazing *Southern Princess*, *Tekoa*'s crew set to work. Rope ladders and cargo nets were lowered over the lee side and lines thrown to the lifeboats which soon came along-side. Many of the *Southern Princess* survivors were covered in oil, and *Tekoa*'s bosun rigged a hosepipe and unceremoniously hosed down each of these as they came aboard. *Mansfield* was detached from the convoy to screen the rescue work and she also took on board several of the more scattered survivors.

The blazing *Southern Princess* continued to light up the whole scene, and the ammunition for her guns started to explode 'like a Brock's firework display'. *Tekoa* passed within a few feet of *Irénée du Pont*, 'still afloat and like a ghost ship, low in the water forward and with a torch left burning on her deck'.

Two corvettes that were attempting to rejoin the convoy from astern also turned up. *Anemone* made for the *Nariva*, also still afloat but a little apart from the other ships with her crew in lifeboats nearby. *Anemone's* approach from out of the dark put the wind up one boatload of *Nariva* men, who thought the corvette was a U-boat coming in to machine-gun them. *Anemone* started to pick up these survivors but when *Pennywort* passed close by, trying to find the convoy, she could see that rescue work was well in hand so continued on her way.

Also attracted to the scene by the blazing tanker were no less than five U-boats, and there is no doubt that the crews of *Tekoa*, *Anemone* and *Mansfield* and the survivors they were taking aboard were in the greatest danger, but, by a fortune which might seem to be God-given, they escaped. U.228 fired three torpedoes at a 'destroyer', probably *Mansfield*, but missed. U.91 fired another three torpedoes and claimed to have hit a '6,000-ton ship', possibly the 6,125-ton *Irénée du Pont*, but if this ship was hit again it still did not sink. U.615 came along but her captain was too uncertain of himself to press home an attack. U.603 was also there but could do nothing; its torpedo tubes were empty, and as it was the official contact keeper it hurried on after the convoy. U.616's captain decided to wait until dawn.

Unfortunately there are no photographs or film of this dramatic but confused scene – the roaring, blazing tanker spitting out exploding shells like a giant fire-cracker, the two dead cargo ships wallowing low in the water, the gallant *Tekoa* drifting gently among the lifeboats and the oil-covered men in the water with survivors scrambling up her steep side like ants, the three small escort vessels arriving and stopping to help or hurrying off to another crisis elsewhere, the German eyes watching from the edge of the darkness. The Germans had often witnessed such a scene before; they had a name for a rescue ship – '*Knochensammler*', 'a bone-collector'.

Four nervous hours later, by which time it was daylight,

the rescue work was done. *Tekoa* had picked up 146 survivors, mostly from *Irénée du Pont* and *Southern Princess*; *Anemone* had ninety-four men from the *Nariva*, and *Mansfield* twenty more from the *Irénée du Pont*. Only four men had been lost from the *Southern Princess* and none at all from *Nariva*; these were the first British ships to be torpedoed in the convoy. The American *Irénée du Pont*'s losses were twelve men, and another who died later from his injuries. One of the deaths was that of Ensign Boyce Norris who, in New York, had tried so hard to find another way of crossing the Atlantic than by convoy merchant ship.

Tekoa and *Mansfield* set off in the bleak dawn to rejoin the convoy. All three of the torpedoed ships were still afloat, although *Southern Princess* soon turned turtle, shooting her deck cargo of locomotives and invasion barges into the ocean depths, but her fuel cargo continued to burn for some time. *Anemone* stayed with the *Nariva*, intending to put the crew back aboard, but this can be dealt with in another chapter.

After the last three ships had been torpedoed the convoy had continued on its old course of 053 degrees, pleased to get away from the glare of the blazing *Southern Princess*. *Beverley* had ordered another Half Raspberry, which was carried out without result, and later, when *Volunteer* had formally taken over command of the escort again, Lieutenant-Commander Luther ordered the three escorts with the convoy 'to act offensively by making frequent dashes outward at high speed, dropping occasional single depth charges in the hope that it might deter an impending attacker'. It was while returning from one of these high-speed forays that *Mansfield* was spotted by the nervous Liberty ship *Robert Howe*, on the port side of the convoy, who 'mistook us for a submarine and fired at us with all he could muster, luckily with no effects'. *Mansfield* had then been ordered back to help with the rescue work around the last three ships to be sunk.

Commodore Mayall's subsequent report reveals how much he had become frustrated by the night's events.

Up to this time I had not received any word from escorts and could not follow their movements; I knew some were astern looking after survivors. I considered the U-boats were either spread out on our line of advance or were relaying along it. Something had to be done to try and shake them off.

Fifteen minutes after the last three ships had been hit, the Commodore ordered two emergency turns to port on his own initiative – an unusual event. The orders went out by siren from the Commodore's ship *Abraham Lincoln*, but this plan was frustrated by an unknown merchant ship that had either not heard or had not understood the signal, and it nearly rammed the *Abraham Lincoln*. Commodore Mayall then decided to turn the convoy back onto its original course. His report goes on to say that 'these emergency turns had the desired effect of disorganizing the U-boat plan' – a good example of the belief that the Germans were acting according to a master plan or under one U-boat captain when they were actually taking part in a confused free-for-all.

It was at this stage that Convoy HX.229's fortunes took a turn for the better. *Beverley*, on the starboard side of the convoy, suddenly picked up a radar contact at 3,000 yards astern. Lieutenant-Commander Price immediately turned, reduced *Beverley*'s speed to 12 knots so that no bow wave would be produced, and headed towards the radar contact which was between *Beverley* and the glare of the *Southern Princess* on the horizon. By eliminating the white bow wave and approaching from out of the darkness, Price hoped to get close to the U-boat, if the contact should prove to be one, before being spotted.

The radar contact was indeed a U-boat. U.228 had recently taken a long-range shot with three torpedoes at *Mansfield* at the rescue scene and had then overhauled the main convoy again. It was getting into position for another attack, this time on a tanker it could see on the starboard side of the convoy; the intended victim was either the British tanker *Luculus* or the Dutch *Magdala*, two of only three ships left in the three starboard columns of the convoy. *Beverley* managed to get within 1,200 yards of the U-boat before being detected, but when the 'blip' disappeared from her radar screen it was replaced immediately by a firm underwater Asdic contact which was moving rapidly to *Beverley*'s left – range 800 yards. Price ordered the wheel to port but then the contact was found to be moving rapidly right – range 400 yards. Over went *Beverley*'s wheel to starboard, but everyone on her bridge realized that the U-boat, by her swift reversal of course, was well inside the old American destroyer's large turning circle.

As *Beverley* passed the U-boat's estimated position, seven depth charges were dropped at settings between 50 feet and 140 feet in order to achieve the 'sandwiching' effect of explosions. This was a good attack by *Beverley*; U.228 was severely shaken by the depth charges and a small leak developed. Oberleutnant Erwin Christopherson took his boat straight down 'into the cellar' – to 180 metres – and then went 'silent'. *Beverley* made her long turn and swept back but could not regain Asdic contact. As he knew the convoy only had one escort at that time, Lieutenant-Commander Price wasted no more time on the U-boat and returned to his screening position, but the Commodore (D) at Londonderry later observed that the destroyer should have remained longer over the U-boat – an easier decision to make in an office at Londonderry than on *Beverley*'s bridge.

Beverley was back with the convoy about one hour before dawn and nothing further was to happen during that hour. In theory eight U-boats were still in contact but seven of these were well behind the convoy reloading torpedo tubes, repairing mechanical faults and depth-charge damage, or recovering from various consequences of a hard night's work. It had been a good night for the Germans. All eight boats of Gruppe Raubgraf had found the convoy, plus another that had recently been released with mechanical trouble from the group, and two more boats had come from the nearby refuelling group. Six of these eleven U-boats had made eight attacks on the convoy and twenty-eight torpedoes had been fired hitting eight merchant ships – four American, two British, one Norwegian and one Dutch. Of the 590 men aboard the torpedoed ships, 447 were, or soon would be, safely aboard various rescue ships. Many of these survivors owed their lives to the humanitarian decision of the escort commander, taken in the most harassing circumstances, and to the devotion to duty of *Tekoa*'s young master.

It is a convenient time to leave the embattled Convoy HX.229.

9

The Wild Donkey

One of the U-boat captains destined to emerge from this convoy battle with success and credit was Kapitänleutnant Manfred Kinzel, commander of U.338. When this boat was being prepared for launching at the Nordsee Werke at Emden in April 1942, too many of the restraining ropes were cut, the U-boat launched itself prematurely, and its first 'success' had been the sinking of a small river craft that had been in the way. U.338's crew had decided to call their boat 'The Wild Donkey' because of this incident and an appropriate symbol was now painted on the conning tower. Kinzel had been on loan to the Luftwaffe from 1937 to 1941 and had only served one short spell of three months in another U-boat before taking a captain's course and then taking command of U.338. This U-boat had arrived in the Atlantic only three days earlier on its first patrol.

U.338 was one of the Stürmer boats that had the previous day been ordered towards HX.229's position with orders to reach that convoy during the day of the 17th and to be ready to attack on the second night of the convoy action. But, soon after midnight of the first night while HX.229 was being so heavily attacked, U.338's lookouts quite unexpectedly sighted merchant ships, obviously part of a convoy, dead ahead and no more than a mile away. Kinzel came quickly to the bridge but he could see no escorts protecting these ships. He gave orders that an immediate surface attack should take place and U.338's crew, many of whom had been asleep, were called to their stations. These merchant ships were in Convoy SC.122, 120 miles north-east of HX.229 and on a roughly parallel course. The Belgian corvette *Godetia* was still absent after picking up the *Campobello*'s crew, otherwise the full strength of Commander Boyle's group was with the convoy – two destroyers, a frigate and four corvettes. Boyle had disposed his escort in one of the standard Western Approaches screens – Night Escort 6 – with six ships spaced evenly around the convoy in the close screen and the fast frigate *Swale* ranging four miles out on the starboard bow. Night Escort 6 required one escort ahead of each of the convoy's two outer columns but nothing between, and U.338 had quite fortuitously passed unseen between these two leading escorts.

The U-boat men acted swiftly. A sighting report was got off to U-boat Headquarters and all torpedo tubes made ready. Oberleutnant Herbert Zeissler, First Watchofficer, was on the bridge with his captain.

Although it was our first action, we took the bearings and ranges quite calmly. The UZO was partly out of action – the connection to the torpedo calculator had been broken by a heavy sea – so we had to make the attack partly by eye and had to aim the torpedoes by turning the boat onto each target. We could only see four columns of ships. We fired the first two torpedoes at the right-hand ship we could see; we then had to turn to port to aim the second pair of torpedoes at the lead ship of the second column. By then we were very close indeed, about 150 metres, from another ship – I could see a man walking along its deck with a torch [this would be the tanker *Benedick*].

We heard two torpedo explosions and our Obersteuermann Trefflich, an enthusiastic Saxon, embraced me. Some of the ships fired at us with machine-guns but the fire fell short. We turned hard a'starboard and fired the stern torpedo at the ship at the head of the next column but we never heard whether it hit or not. We dived then and the convoy came over the top of us.

U.338's crew had only heard two explosions and Kinzel later claimed just these two hits, but this audacious attack resulted in four ships being torpedoed. Two of Kinzel's torpedoes had hit the targets they were intended for – the British cargo ship *Kingsbury* at the head of the convoy's third column from the port side and the Dutch cargo ship *Alderamin* at the head of the fourth column; there had been two more columns to the right-hand side that the Germans had not seen. Another torpedo had missed the *Kingsbury* but struck the ship immediately behind, the old British tramp steamer *King Gruffydd*. The stern torpedo had also scored, travelling some distance through the convoy before hitting the brand-new *Fort Cedar Lake*. It was this ship that should have sailed with her sister ship *Fort Anne* in HX.229 but had been relegated to the slow convoy.

The four stricken ships all lost way and their captains all ordered their crews to abandon ship. Fireman Patrick Murphy came on deck from the *Kingsbury*'s engine-room and found

that his lifeboat had slipped one of its falls and was hanging
vertically from the other.

It seemed a dead cert that there would be quite a delay before we
could depart. I remembered that I had purchased a wedding ring in
New York for my wife as she had lost her own whilst washing
clothes, also I was very cold indeed not having a jacket on. I made
my way carefully along the sloping deck to the foc's'le and
descended a few steps into the rooms. They were already awash and
there was no one else there. I put a leatherette windbreaker jacket
on and placed the wedding ring in my pocket. At the last moment I
picked a dry blanket off a bunk in case it came in handy later on.

When I returned along the deck, the deckhands were still
energetically trying to get the lifeboat ready to lower. They
needed room to work so we firemen had to stand clear and while the
time away passing comments on what we would be doing this St
Patrick's Day if we were in Cardiff. With the ship sinking under
our feet, one had to have a fine sense of humour to speculate on
conviviality ashore. One chap was hopping around almost naked,
having had to leave his bunk at short notice; the blanket I carried
came in of use to him.

Eventually all but three of the *Kingsbury* men were safely
away in boats. Left on the ship were Captain William Laidler,
Radio Officer King still at his wireless sending out distress
signals, and Able Seaman Sammy Ward who had been on the
helm and had stayed with the master since the torpedoing.
Ward released one of the liferafts and then jumped into the
water to stop it drifting away but was immediately caught
by a heavy swell and carried away himself; he was not seen
again. Captain Laidler and the radio officer managed to get
to the raft just as the *Kingsbury*'s bows reared high in the air
and then the ship slid backwards beneath the waves. The
liferaft was left spinning round like a top over the place
where *Kingsbury* had sunk. Able Seaman Ward, a Shetlander,
was posthumously awarded the King's Commendation for
Brave Conduct for standing by the master until all boats
were safely away and for jumping into the sea to secure the
liferaft.

In the *King Gruffydd* the torpedo explosion trapped five
men in the foc's'le and they could not be released; their
screams were pitiful. The ship, with its cargo of 5,000 tons of
steel, was sinking quickly, and most of the remainder of the
crew abandoned ship hurriedly, possibly because of the 500
tons of high explosive also in the cargo, but this did not
explode. Second Officer F. R. Hughes had attended to the

dumping of the codes and confidential papers but then found that his boat had already been lowered.

I could see that it was now too late to board the boat by means of the rope ladder as the boat was now well clear of the ship's side and drifting slowly towards the stern of the vessel. The ship had settled further by the head and it was obvious that she was sinking and we had to get clear as soon as possible. I told the others that our only chance was to jump into the water from the after welldeck just forward of the poopdeck and swim out to intercept the lifeboat as it drifted down. They followed me along the deck but didn't jump into the water after me. They both went down with the ship as did the captain, who remained on the bridge when I left.

I swam or rather struggled out to the lifeboat which was now about twenty yards from the stricken ship. When I reached the lifeboat I had to hang on to the gunwale while the survivors in the boat rowed it further away from the sinking vessel – they were not about to stop rowing in order to pull me aboard. I looked back and saw that the *King Gruffydd* was now lying at an angle of 45 degrees with the propellor and rudder high above the water. Suddenly, as I watched, she dived vertically under the waves.

The crew of *Fort Cedar Lake* had little trouble. Captain Collings describes what happened.

It was a very orderly affair; we stayed afloat for several hours. I even had time to bring the Chief Steward's book showing the credits given for canteen sales, much to the disgust of the crew who later had to settle up. The crew had given some trouble on the way round from the Canadian shipyard but, while we were still afloat, they were very steady and obeyed all orders and the eventual abandoning of the ship was entirely calm.

Captain Collings does not mention that his ship was burning fiercely throughout this 'orderly affair'.

The rescue ship *Zamalek* started picking up the crews of *Kingsbury* and *King Gruffydd*. One of *Kingsbury*'s boats capsized while alongside *Zamalek*, and the chief engineer and a young Welsh businessman who was a passenger were so numbed that they could not hold their grip on *Zamalek*'s scrambling net; they fell back into the sea and disappeared. One of *King Gruffydd*'s lifeboats also capsized in the heavy swell and more than half of this crew, mainly Liverpool seamen, were lost. There were no casualties in the *Fort Cedar Lake* and her crew were later picked up by *Zamalek*.

The Dutch ship *Alderamin* was in more difficulty and the corvette *Saxifrage* was ordered to rescue her crew. The *Alderamin* had gone down very quickly and there had been an unfortunate incident. Two of the boats were damaged by the torpedo explosion and a third capsized on being launched. The only remaining boat was in good condition and was lowered safely with the chief engineer and three men aboard. Instead of waiting for others to come down into the boat, this officer then released the falls to get clear of the sinking ship, leaving most of the *Alderamin* men to get away on liferafts which were poor substitutes in the heavy swell and bitterly cold conditions. The lifeboat had a motor on it which was heard to start but despite calls for help, it never came back to pick up the many men on rafts or in the water. *Saxifrage* picked up thirty-seven men and the crew of the corvette were very impressed by *Alderamin*'s master, Captain C. L. van Os, who swam around the lifeboats and then right around the corvette to make sure that all his men were safe. *Zamalek* also saved some of this crew but fifteen men were never found and three more who were rescued later died of exposure. Captain van Os was later given command of S.S. *Alpherat*, which was bombed and sunk near Malta on 21 December 1943. Captain van Os was saved for a second time by *Saxifrage* when this corvette picked up the *Alpherat*'s crew.

There was another, less happy, sequel to the sinking of *Alderamin*. In October of that year an Extraordinary Board of Shipping sat in London to investigate the loss of the ship and, in particular, the conduct of the chief engineer who had cast off the only sound lifeboat prematurely and then refused to come back to pick up men from rafts and in the sea. The Dutch officer freely admitted that he was completely overcome with fright, that he was not sure of his ability to manoeuvre the lifeboat in the heavy swell, and could not get two of the other three men in the lifeboat to help him. The Board debated whether the officer was guilty of a penal offence under Article 47 of the Shipping Act but decided that there were serious doubts about this and the matter was dropped.

So Kapitänleutnant Manfred Kinzel with his attack, from almost right under the bows of the leading ships of the convoy, had finished off four merchant ships of over 24,000 tons with nearly 30,000 tons of valuable war cargo.

We were pleased but not unduly excited. In some way, we had expected success and we thought it quite natural that we should do this. We had done our training properly and we had complete confidence in our captain, Kinzel; we were a very close crew. The

tubes were reloaded as soon as possible because we knew we would
have to start all over again. For breakfast we took out the best
provisions we could find – *Knackwurst* (highly seasoned sausages)
and strawberries and cream. (Oberleutnant Herbert Zeissler)

While U.338's crew ate their celebration breakfast, forty
British and Dutch seamen were drowning or dying of exposure in the bleak wastes above, but this was how the Battle
of the Atlantic was fought – success for one side, destruction
and lonely death for the other.

There were two interesting U-boat sequels to U.338's
attack on SC.122. U.641, another Stürmer boat that had been
making for HX.229, later reported that it had been subjected
to an attack by an aircraft which was showing a red light and
which attacked with three aerial depth charges but caused no
damage. The position and time of this 'aerial attack' coincides
with the torpedoing by U.338 of the SC.122 merchant ships.
The attacking 'aircraft's' red light was the masthead distress
light which was immediately switched on by one of the
torpedoed merchant ships; the 'depth charges' were the
explosion of the first three of U.338's torpedoes. This somewhat surprised U-boat crew, also on its first patrol, resumed
its journey to HX.229 completely unaware that it had been
so close to another convoy.

A third Stürmer boat, U.598, had been a few miles to the
south and had seen the many Snowflakes fired off by the
SC.122 merchant ships when U.338 was seen on the surface.
Kapitänleutnant Gottfried Holtorf consulted his navigation
charts but decided that there could not be a convoy at this
position and that the displays of lights must be a decoy to
draw U-boats away from the true position of the convoy. This
boat, too, continued her journey to the south-west in search
of HX.229.

These three U-boats were the only ones to come near
SC.122 on this night. *Zamalek* and *Saxifrage* picked up the
last survivors and set off to catch up the convoy. *Fort Cedar
Lake* was left, still afloat but well down by the head and giving
off dense clouds of smoke.

So ended the first night of the battle. Twelve Allied merchant
ships had been lost in exchange for just two U-boats slightly
damaged. This round in the battle for merchant shipping had
clearly gone to the U-boat men.

10

Review at Dawn
of 17 March

Dawn broke on a somewhat decimated convoy and a scattered
and rather embittered escort who felt that they had been beaten
by facts outside their control and by pure weight of numbers.

This was the entry Lieutenant-Commander Luther later made
in his Report of Proceedings, describing the coming of daylight
on the morning of 17 March. The escorts had indeed been
overcome by events outside their control and were mightily
relieved that the night had finally passed. Luther's sentiment
would hold true for most of the men on the escort vessels.

But the escorts had at least known more or less what was
happening; the majority of the merchant sailors in both
convoys had not. They had heard dull explosions but could
not tell whether these were torpedoes or depth charges. Some
of the explosions had been followed by a flash which was
obviously a ship being hit, and then by distress rockets. Sometimes a burning ship could be seen dropping astern of the
convoy with its crew taking to their boats, but no one knew
for certain whether the survivors had been picked up. Those
men not on duty had been able to do little more than lie on
their bunks fully dressed with lifejacket and 'panic bags' at
hand and hope for the best. Many men had refused to go
below and had spent most of the night on deck.

We had been in the second column from starboard (in HX.229).
The first ship to be torpedoed had been the ship ahead of us; the
next to go had been the ship astern. Then, for the rest of the night,
it had just been hell; everything seemed to have been happening
around my ship. My crew were all in the galley for safety and I
cannot speak too highly of their discipline. I remember a young
crew man, who was on the monkey island keeping watch, coming
and asking permission to remain on the bridge as he was so
frightened. I told him to stay and he would be all right. My second
officer also asked me if I thought we would ever get home; my

reply is not printable. During this time I had often heard hissing noises in the water similar to placing a white-hot poker into the water. I was later informed that these were torpedoes. I think, if my memory is correct, seven ships were sunk around us that night. (Captain W. Luckey, M.V. *Luculus*)

Captain Luckey's memory is quite accurate. His tanker was one of only four ships left in the three starboard columns of HX.229.

When dawn arrived the first action on most merchant ships had been a big count-up of the ships still in the convoy, and this was the first time the merchant seamen knew of the night's losses.

After things had quietened down for a while, we looked at the cruising sheet, marked off the unfortunate missing numbers and that, with a short prayer, was that. There was always an intense feeling of helpless anger that good men and ships would be seen no more and, eventually, a guilty sense of relief that 'we were still afloat'. (Chief Officer M. MacLellan, S.S. *Baron Stranraer*)

The rumours were rife. 'The convoy was followed by a pack of submarines.' 'There were raiders ahead.' 'The convoy was to be sacrificed for bigger things.' 'The enemy is using a new weapon.' 'Three subs have already been sunk.' There was no panic on board. We had been through it all before. One must bear in mind that our country was occupied and closed to us. We could do nothing but sail and had done so for the last three years. (Chief Cook Torbjørn Saga, M.V. *Abraham Lincoln*)

An escort vessel, probably H.M.S. *Volunteer*, came along-side the *Abraham Lincoln* and reported the full details of the night's losses to the Convoy Commodore. There was intense disappointment among the crew of the *Abraham Lincoln* when they heard that one of the lost ships was the *Elin K*, the only other Norwegian ship in the convoy, and it was understood that there were no survivors. The mood changed to joy when it turned out that the message had been misunderstood and that there had been no losses among the *Elin K*'s crew.

Dawn of 17 March ended the first phase of this battle, and before describing the fortunes of the convoys during the day-light hours of that day, another look will be taken at the various headquarters ashore, at how they received the news of the events of the past night, and at what decisions they were able to take that might influence the future course of the battle.

Both Commander Boyle and Lieutenant-Commander Luther had sent out signals giving details of the attacks upon their convoys; there was now no point in keeping radio silence and it was important that the shore headquarters were given as much up-to-date information as possible on the convoys' progress and condition. Duty officers at several places on both sides of the Atlantic had been plotting the melancholy events of the past few hours. Washington, Halifax and St John's were powerless to help; the embattled convoys were well beyond the range of any of their available aircraft and, now that the CHOP Line had been passed, control of the convoys' fortunes was completely in British hands. By coincidence, however, many of the senior Allied officers involved in the war against the U-boat were at that time in Washington to discuss the future policy for the Battle of the Atlantic. This 'Atlantic Convoy Conference', under the chairmanship of the American Admiral Ernest J. King, was being pressed by the British to implement more fully the decision taken at the Casablanca Conference two months earlier that the defeat of the U-boats must be the first Allied priority, and that V.L.R. aircraft cover should be provided for the American and Canadian side of the Atlantic. It can be assumed that details of the shadowing of HX.229 by U-boats the previous day, well within the radius of a V.L.R. aircraft if such an aircraft had been based at Newfoundland, and the resultant torpedoing of twelve merchant ships that night, were given to the conference sittings that morning by officers from Convoy and Routing, and this information undoubtedly concentrated the minds of those present on the necessity of closing the Air Gap. The Atlantic Convoy Conference will be referred to later.

At the Admiralty's Operational Intelligence Centre under The Citadel in London, Commander Hall's main Trade Convoy Plot had been brought up to date by the civilian ladies of the night watch. One of the ladies working here was Mrs Guendoline Boyle, wife of Commander Boyle. She knew her husband's ship was with the escort of SC.122. As far as the officers of the Trade Division were concerned, there was little that could be done to help HX.229 and SC.122; the future routes of both convoys already showed the direct run for England and there was no question of any diversion being ordered. It was up to Western Approaches and Coastal

Command to fight the convoys through with their sea and air escorts. The signals received during the night were, however, valuable intelligence raw material. Now that the position of the main U-boat strength had definitely been revealed, Commander Winn could mark up his Submarine Tracking Room plot with some certainty, and this information was used to ensure that other convoys and the many independent ships in the North Atlantic were diverted well clear of the U-boats.

There was considerable interest in the past night's events at the Headquarters of Western Approaches Command at Derby House, Liverpool. Unfortunately for historians, communications by telephone and teleprinter between Western Approaches and the Admiralty were so efficient that no formal War Diary was kept at Western Approaches, nor were the usual Monthly Command Reports sent to the Admiralty. For this reason no direct documentary evidence exists to show what action Admiral Horton and his staff took on receipt of the convoys' attack signals. However, there is enough information from other sources to work out what happened at Derby House.

The huge wall plot in the Operations Room had been maintained by the duty watch of Wrens. Leading Wren Mary Carlisle, one of the watch-keepers, describes her work.

Each convoy's route was marked by a different coloured elastic tape pinned to the plot. Small white submarine symbols represented the estimated positions of U-boats – they were like measles on the green-painted cork of the plot – but they were replaced by black submarines when the position of the U-boat became definitely known. Small red merchant ships at an angle represented torpedoed ships. It was terribly depressing and so frustrating because we could do nothing about it. We knew some of the escorts well and we had seen the merchant ships in the Mersey. We never lost the sense of shock when ships were torpedoed but you couldn't brood on it; you just had to get on with your work.

The Wrens, in skirts not trousers, had to mount high ladders to keep the plot and were very conscious of the male watchers in their glass-fronted offices opposite the plot. Some months earlier a member of the W.A.A.F.s looking after the Coastal Command section of the plot had fallen from her ladder and been killed and all the girls now had to wear safety harnesses.

Once again the help that could be given to the escort commanders was limited. No less than five escort vessels were already on their way to reinforce the escorts. *High-*

lander, Abelia and the Canadian corvette *Sherbrooke* were making their best speeds from St John's to catch up HX.229, and the two American ships, *Ingham* and *Babbitt*, were making for SC.122 from Iceland. *Vimy* was still under repair in Iceland and would not sail until early on the 18th to join HX.229. None of these reinforcements could be expected to join for at least twenty-four hours. The taking of one escort from SC.122 may have been considered; this convoy had six escorts in company and two more rejoining fast and was only 120 miles ahead of the desperately pressed HX.229, but SC.122, with its slower merchant ships, was potentially as vulnerable to further attack and this solution was not adopted. The earlier diversions to the east of these two convoys had left them in a vacuum with no other convoys nearby that might have been able to lend escorts.

The only further action that Western Approaches could take was to order that the merchant ships due to be detached from SC.122 and sent under escort to Iceland should remain with the main convoy and complete the voyage to England. This move was an unusual one but quite practical, and it meant that the two American escorts then coming down from Iceland could stay as permanent reinforcements. One of these ships, the destroyer *Babbitt*, was ordered not to join SC.122 but to take a new course and reinforce HX.229's escort, but both *Babbitt* and *Ingham* were still two days' steaming away, having been delayed by storms. Despite this limited help Luther and Boyle would have to fight on with what ships they had for the time being. At least no one suggested that the convoys turned back; that would have been the supreme German success.

Admiral Horton left Liverpool that evening on the night train to London in order to attend the monthly meeting of the Prime Minister's Anti-Submarine Committee the next morning. Horton took with him to London a report on a 'war game' that his staff was currently playing out on the North Atlantic plot at Derby House. The details of the latest sinkings would be sombre news for Mr Churchill's committee but the war game promised some hope for the future. This will be covered in a later chapter.

Turning from the theoretical to the actual, there was one man at Derby House who could provide vital help to the convoys: this was Air Vice-Marshal Sir Leonard Slatter, Air

Officer Commanding 15 Group of Coastal Command. Slatter's headquarters were also in Derby House and there was close cooperation between 15 Group and Western Approaches. Captain Ravenhill, who was in charge of day-to-day operations at Western Approaches, had the previous day asked for the earliest possible help that could be given for HX.229 and SC.122. It would have been much easier for 15 Group to help if the two convoys had not been diverted so far to the south and put into a wide part of the Air Gap. There were V.L.R. Liberators available at two airfields – Aldergrove in Northern Ireland and Reykjavik in Iceland; Reykjavik was slightly nearer to the two convoys. Orders had been sent to both airfields that Liberators were to fly to each convoy as soon as it was within range.

Reykjavik had laid on two aircraft of 120 Squadron to take off early in the morning of the 17th and attempt to reach HX.229, but a Force 7 wind blowing across the runways had delayed and then finally forced the cancellation of all operations from Reykjavik for that day. Bad weather was not unusual at Reykjavik, but this was a tragedy for HX.229 because there had just been a chance that these aircraft might have been able to reach the convoy that morning. Aldergrove, although further from the two convoys, had prepared two Liberators for each convoy. The first was already in the air, having taken off soon after midnight, and it hoped to meet SC.122 at dawn (of the 17th), but the two aircraft allocated to HX.229 would have to wait until late morning before taking off and even then the convoy might prove to be beyond their range and might get no air cover on this day.

There had naturally been intense activity at the Hotel Am Steinplatz in Berlin. A stream of signals had been arriving during the night and, when Admiral Dönitz held his daily situation conference at 09.30 (Central European Time), the total claims for the night had risen to fourteen merchant ships totalling 90,000 tons sunk and a further six ships damaged, although only twelve ships had actually been hit. It had been a long time since one night had produced so much success and there was every prospect of more to come. Dönitz ordered a signal to be sent to the U-boats: ' *Bravo. Dranbleiben. Weiter so.*' – 'Well done. Stick to it. Let's have some more.'

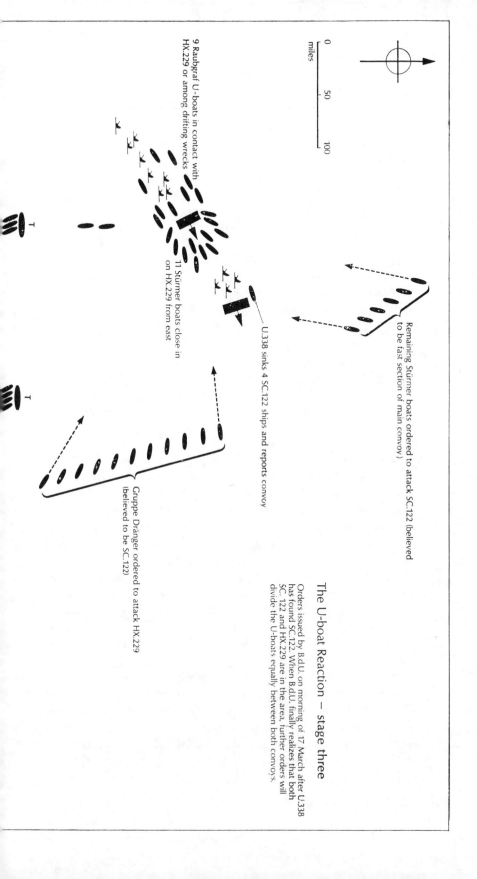

The U-boat Reaction — stage three

Orders issued by B.d.U. on morning of 17 March after U.338 has found SC.122. When B.d.U. finally realizes that both SC. 122 and HX.229 are in the area, further orders will divide the U-boats equally between both convoys.

Remaining Stürmer boats ordered to attack SC.122 (believed to be fast section of main convoy)

9 Raubgraf U-boats in contact with HX.229 or among drifting wrecks

11 Stürmer boats close in on HX.229 from east

U.338 sinks 4 SC.122 ships and reports convoy

Gruppe Dränger ordered to attack HX.229 (believed to be SC.122)

0   50   100
miles

But Dönitz and his staff officers had serious work to do. One of the success signals, the one from Kapitänleutnant Manfred Kinzel in U.338, gave the position of his attack as being in the AK 85 square of the U-boat chart, while all other U-boats had been reporting their attacks in BD 15 or BD 13. There was a difference of 120 sea miles between U.338's position and those of the other successful U-boats. Had Kinzel stumbled upon another convoy or was his navigation in error? It was soon calculated that U.338, a Stürmer boat, could not possibly have reached the convoy being attacked by the Raubgraf boats – still believed to be SC.122 – and that Kinzel had found another convoy. It was assumed that this was a faster section of ships that had been detached and sent ahead of the convoy under attack and the plan of the previous day was now changed.

Two hours later the new orders were sent to the U-boats. The six most northerly boats of the Stürmer group were ordered to make their top speed for this supposed advanced section of the main convoy, but if they could not find these ships by mid-afternoon they were to carry on to the main convoy for the second night's attacks; all other Stürmer and Dränger boats were to continue towards what was still considered the main convoy. But further signals from U.338 showed that the ships Kinzel was now shadowing were steaming at 7 knots while the 'main' convoy had been making a steady 8½ or 9 knots. At last the German officers realized that the convoy sighted by U.653 the previous morning and attacked so successfully the previous night was HX.229 and not SC.122 as they had thought, and that SC.122 was actually 120 miles ahead and on a roughly similar course.

By early afternoon two more Stürmer boats had contacted SC.122 and confirmed its position, and also sent the news that it contained more merchant ships than HX.229. Yet again the orders to the U-boats were changed. The northern Stürmer boats were now to stay searching for SC.122 and the five southerly Stürmer boats and the northern half of the Dränger group were also to operate against this convoy. It should be stated that these were not the only instructions sent to the U-boats on that day, but the details given above and shown on the maps cover the main instructions given to the U-boats during the day of 17 March. The effects of the many signals received from U-boats during the day meant that Dönitz and his staff had been confused for many hours as to which convoys were actually in the area. This had resulted in some U-boats receiving several different orders, and this caused them to lose valuable time in closing the convoys for the coming night's operations. But by late afternoon the situation had clarified and the twenty-eight U-boats of the Stürmer and Dränger groups were now disposed in such a way that during the second phase of the convoy battle both convoys would be threatened by roughly equal numbers of U-boats. As far as U-boat Headquarters was concerned it would remain one operation: its War Diary continued to refer merely to *Geleitzug Nummer 19* to the end.

The U-boat Headquarters staff at the Am Steinplatz would continue to be very busy for several days. Kapitänleutnant Adalbert Schnee, one of the operations officers, remembers that he did not sleep for four nights and Admiral Dönitz probably did not visit the main Kriegsmarine Headquarters at the Tirpitz-Ufer at all during this period, preferring to concentrate on this big U-boat battle. One officer required by Dönitz to attend in person at the daily situation conference at Am Steinplatz was Kapitän Heinz Bonatz, head of the B-Dienst which was now picking up the merchant-ship distress signals, and Bonatz was able to identify many of the torpedoed ships by name. Bonatz looked into the U-boat Operations Room on each visit for a quick look to see what a certain U-boat was doing; his son, Leutnant Heinz Bonatz, was First Watchofficer in U.*336*, which was one of the Dränger boats hurrying into action against SC.122 on that morning.*

It is appropriate to include details here of the more long-term decisions that were taken at U-boat Headquarters on this day. Many fresh boats were in the process of coming out into the Atlantic and it may have been a temptation to throw these also into this promising convoy battle. But it was decided that there were enough boats already allocated to HX.229 and SC.122 and that these fresh boats could be better employed elsewhere. The B-Dienst had intercepted a signal revealing that the next westbound convoy, ONS.1 (the

---

* After returning safely from this patrol, Leutnant Bonatz was offered a place in a U-boat commander's training course but his captain, Kapitänleutnant Hans Hunger, persuaded him to stay on in U.*336*. They were all lost on 4 October 1943 when the boat was sunk by an American aircraft based in Iceland.

Admiralty started numbering the ONS. convoys again after
ONS.171 sailed), would be at a position off Iceland on the
morning of 21 March to link up with merchant ships joining
from Iceland. Five new U-boats from German ports and three
more fresh out from Biscay were ordered to form a new
group, Seeteufel (Sea Devil), and trap ONS.1. The three
Seeteufel boats from Biscay passed very close to the SC.122/
HX.229 battle but were ordered not to join in unless any
chance targets presented themselves. The main implication of
the Seeteufel group for this story is that the position it would
take up to search for ONS.1 was also very close to the route
still to be taken by HX.229A.

It was a peculiar battle. The sailors of both sides could not
have felt more isolated and lonely out in the Atlantic and
could hardly have fought under conditions of more danger and
discomfort. Yet those who directed the battle worked from
comfortable offices in great cities – Washington, Liverpool,
London, Berlin; their greatest danger was that they might
become chance victims of a traffic accident or an air raid, and
not even the last of these threatened Washington. This
comment implies no criticism of these 'directors of operations';
it was just the nature of that battle.

11

Six Hearses
Bearing 180 Degrees

It is time to return to the continuing story of the two convoys which were under U-boat attack. On this occasion it will be more convenient to deal first with the slower SC.122 which, at dawn on 17 March, was still 110 miles ahead of HX.229. There were now forty-four merchant ships in SC.122 with six escorts in attendance – the destroyers *Havelock* and *Upshur*, the frigate *Swale* and corvettes *Buttercup*, *Lavender* and *Pimpernel*. *Godetia* was soon to rejoin after her long absence picking up the crew of the trawler *Campobello*, but *Saxifrage* and the rescue ship *Zamalek* would take another twenty-four hours to catch up after their rescue work of the previous night. The weather was much improved with calmer seas and only a light wind, but with brief snow showers. U.358, which had attacked the convoy the previous night, was following at some distance on the surface and soon after dawn was joined by a second boat, U.666. The two U-boats settled down to the long slog around the convoy on the surface which would bring them to an attack position ahead of the convoy.

Twenty miles behind SC.122, U.439 was also cruising on the surface in the half light of dawn when her lookouts sighted a burning ship. Although U.439 was on her seventh patrol, it was the first trip in command for her captain, Oberleutnant Helmut von Tippelskirch, and this was his first sight of any surface ship since sailing from Brest six weeks earlier. He turned towards the ship which soon turned out to be a large cargo vessel, well afire and settling by the bows. U.439 had come upon the derelict *Fort Cedar Lake* just after *Zamalek* had picked up her crew. Suddenly, an alert lookout spotted something that absolutely amazed the Germans – a large aircraft approaching from the east. They had thought themselves quite safe from aircraft so far out in the Atlantic. Tippelskirch, the officer of the watch, and the four lookouts shot down the conning tower ladder and the crew of U.439 took the boat down in a *Schnelltauchen* – a crash dive.

The aircraft whose appearance had so surprised the Germans was an 86 Squadron Liberator captained by Flying Officer Cyril Burcher, a New South Wales man. 86 Squadron was the recently reformed squadron at Aldergrove equipped exclusively with the Very Long Range version of the Liberator. This aircraft had taken off from Aldergrove more than eight hours earlier to join SC.122 soon after dawn, but Burcher had flown past the convoy in the dark and had just started the homing procedure by which *Havelock* would guide the aircraft to the convoy by wireless when Burcher himself had spotted the surfaced U-boat in his path. It was the first U-boat he or his crew had seen.

The Liberator came in fast to attack but the U-boat disappeared just before the aircraft was in a position to drop its depth charges. Sergeant Jack Lloyd, the New Zealand second pilot, called off the seconds following the time the U-boat had submerged so that Burcher could calculate how far ahead of the 'swirl' he had to aim. Four depth charges were released and a few seconds later there appeared the usual eruptions and cascades of white water as they exploded. The Liberator circled with its crew watching for the tell-tale wreckage or oil that would show that their attack had been successful, but U.439's well-drilled and experienced crew had got her down in time to frustrate Burcher's aim; the Australian later estimated that he had overshot. The U-boat was not damaged but she was shaken up and spent much of the remainder of that day submerged. It is an interesting point that Tippelskirch recorded the attacking aircraft as being a Sunderland flying boat.

Burcher signalled his attack both to the convoy and to Coastal Command and within minutes the news of it would have been on its way to the Admiralty's Submarine Tracking Room. The signal was also picked up by the B-Dienst and Dönitz would be annoyed when he was told that aircraft had intervened so early in this battle. Twenty minutes later the Liberator sighted the convoy, a good piece of work at a distance of 910 sea miles from its base; not all aircraft found their convoys at this range. Commander Boyle asked the aircraft to carry out a Cobra patrol at ten miles. Cobra was a standard search ahead of and on each flank of the convoy with the intention of catching any U-boat coming in to a position from which it could carry out an underwater attack.

On his first sweep on the port flank of the convoy Burcher's crew sighted a second U-boat and attacked it in the same manner as before. The Liberator's crew were delighted to see a patch of oil spreading across the surface soon afterwards and had high hopes that they had at least seriously damaged the U-boat. This boat was Manfred Kinzel's U.338, but once again the U-boat had dived too quickly and no damage had been caused; the oil was probably a small quantity deliberately released as a bluff – a trick often played by U-boats when attacked from the air. The Liberator had also been seen by U.666 which dived for half an hour. Although Burcher had no depth charges remaining, he offered to stay with the convoy as long as his fuel allowed and continued to patrol ahead until 09.15. Seven hours later, with insufficient fuel to reach its own base, the Liberator landed at the Fleet Air Arm airfield at Eglington near Londonderry. It had been in the air for eighteen hours and twenty minutes.

The cramped and tired crew of Flying Officer Burcher were to play no further part in this convoy battle but their appearance over SC.122 had now changed the entire complexion of that battle. The U-boat men were to curse the Coastal Command aircraft many times in the next few days. One useful result of Flying Officer Burcher's activities was the great uplift of morale in the convoy on the Liberator's appearance; this may be exemplified in the comments of Lieutenant Herbert Gravely, one of the officers on the American destroyer *Upshur*.

I recall the joy when we first saw an aircraft with us. At about the time they came out to cover us we had decided that this convoy experience would last forever and that the Atlantic really had no 'other side'. We gave a real cheer when that first aircraft was spotted.*

This Liberator had undoubtedly frustrated the intentions of three U-boats, but the respite gained was not permanent.

---

* In this flight in support of SC.122 and in their next five trips, Flying Officer Burcher and his crew sighted no less than ten U-boats and attacked nine of them. This intense activity was most unusual for Coastal Command aircraft and was believed to be a record at that time. The crew flew thirty-seven operational flights with 86 Squadron and attacked four more U-boats and one surface vessel before being rested a year later. It is probable that they sank U.632 on 6 April 1943 and U.419 on 8 October 1943.

Although Burcher had stayed as long as he could, there was a gap of more than two hours before the next aircraft arrived. During this interval U.338 and U.666 both came back to the surface and another boat, U.305, spotted the convoy's smoke. The three U-boats started again to work their way round to the front of the convoy.

The two-hour gap in the air cover caused by the high winds at Reykjavik airfield was unfortunate. Just on midday, a time remembered well by many of the merchant ships' officers who were taking their noon sights, there was sudden confusion on the port side of the convoy. The Belgian corvette *Godetia* swung hard to starboard and several men on the merchant ships in the convoy's port column saw the track of a torpedo. The torpedo crossed just ahead of the American ship *Cartago*; this was a narrow escape for the ten Red Cross ladies on board that ship, but a few seconds later there was a huge explosion as the torpedo struck the leading ship in the second column, throwing up a great cloud of smoke and debris.

Captain White, the Convoy Commodore, saw the torpedo explosion and immediately ordered *Glenapp*'s siren to be sounded as a signal to the convoy to make an emergency turn to starboard, away from the probable position of the U-boat. Every merchant ship turned in answer to the signal but, even as they were turning, a second torpedo was seen approaching the ships in the port column; it was running so shallow that its nose kept appearing as it passed through each wave. This torpedo passed between the Icelandic ships *Godafoss* and *Fjalfoss*, then on through the second column on a line that would take it to the Dutch ship *Parkhaven* in the third column, but her master had seen the torpedo and ordered his helmsman to put the ship even harder to starboard. This just saved *Parkhaven*, and the torpedo then ran out of the back of the convoy without hitting anything.

The ship that had been hit was the old Panamanian steamer *Granville*, loaded with military stores for Iceland and also carrying 500 bags of mail for the American garrison there; several bags of this mail were later picked up by U.603. The torpedo had torn a great hole in one of *Granville*'s holds and in the deck above. The crew had been having a roast-chicken lunch when their ship was hit. Several men were killed in the explosion and the remainder could be seen rushing around the deck trying to get the only two lifeboats away before the ship sank. One boat had been damaged in the explosion and the second capsized on launching. The survivors got away as best they could on these boats or on liferafts. A cargo of petrol in one of the holds caught fire and the ship started to break her back where the torpedo had hit. Captain Fridrich Matzen, the Danish master, was the last away and had to cling to the upturned boat with another man and the ship's dog. The *Granville* then broke in two and, after bow and stern had hung a few seconds in the air in a huge V, the two parts disappeared quietly beneath the sea.

The corvette *Lavender* had been detailed to pick up the men in the water, a task that did not take long, for the conditions were good and the crew of *Granville* was not a large one. Thirty-three men were soon safely on board the corvette, including the dog and the only passenger, a U.S. Army colonel taking passage to Iceland. *Lavender* was just about to leave when another man was seen in the sea, alive but making no effort to swim. It would have been a slow and difficult task to manoeuvre the corvette to this man so her first lieutenant, Lieutenant William Weller, R.N.V.R., dived into the water, swam out to the man and towed him to the ship's side. The rescued man turned out to be *Granville*'s second mate and, only feebly alive, he was taken to *Lavender*'s sick bay. Twelve men were missing, including the fireman who had looked after the dog.

The U-boat that had made this attack was U.338, the same boat that had sunk four ships in her surprise attack the previous night. On this occasion, it had been a deliberate, set-piece underwater attack. The temporary absence of an aircraft had allowed Kapitänleutnant Kinzel to get ahead of the convoy on the surface, and then he had submerged and allowed the convoy to approach. Four torpedoes had been fired at four overlapping ships. The first had hit *Granville*, the second passed through the convoy, but Captain White's prompt action in turning the convoy had probably caused the other two torpedoes to miss completely. Kinzel claimed hits on two ships. It had been a brave attack in broad daylight on a well-defended convoy, an attack typical of the bolder breed of U-boat commander.

The destroyer *Upshur* had been in the port screen when the *Granville* had been hit. Lieutenant John White, although an engineering officer, was officer of the watch.

The *Granville* had been torpedoed just as the other officers were
sitting down for lunch. I well remember the huge cloud of dust –
rather than smoke – rising amidships in broad daylight. I rang the
General Alarm; I well remember Lieutenant-Commander Ellsworth rushing to the bridge and, with some exasperation, asking,
'What's the matter now, Johnny?' I felt like answering, 'It's not my
fault. It's those goddam Germans.' But I just pointed to the
merchantman breaking up about half a mile away.

Godetia and *Upshur* turned outwards and searched for the
U-boat. Both escorts gained an Asdic contact almost at once,
but *Godetia* was nearest and attacked first. The Belgians and
Americans made three attacks in all and dropped twenty-
seven depth charges, but no oil or debris came to the surface
and the contact was then lost. U.338 had escaped the first
salvo from *Godetia*, probably because *Godetia*'s Asdic dome
was later found to have been dented by ice on the previous
voyage and was leaking. Kinzel had then gone deep, levelling
out at 200 metres (654 feet) – 100 feet lower than the
deepest settings of the escorts' depth charges. When contact
was lost *Upshur* was called back to the convoy, but *Godetia* was
allowed to stay in the area for the next three hours. By the
time U.338 dared to surface, the convoy was well away, but
U.666, another U-boat that had been in contact, also reported
being depth-charged and being subjected to a long Asdic
search (the 'ping' of an Asdic beam could be heard striking a
submarine's hull), so she must have been very close to U.338
and she too lost contact with the convoy because of *Godetia*'s
presence.

While these events had been taking place, the third U-boat
that had been in touch with SC.122, U.305, had been ten
miles out on the convoy's port side, but this boat was
surprised by the next aircraft coming in to cover the convoy.
Fähnrich Wolfgang Jacobsen was one of the lookouts.

We were attacked by a Sunderland and it came in from my search
sector. I was very new on board and I hadn't seen this big aeroplane
although it was only one mile away and about 200 metres up. I was
looking too high. The captain was the first to see it. We dived and
were bombed but not damaged. He gave me a real good telling off
and it was explained to me that every single man on board was
responsible for the rest of the crew. I was always the first to see an
attacking plane after this.

This aircraft was not a Sunderland, which could not fly out
this far, but the Liberator of Flight-Sergeant 'Smokey'
Stoves of 120 Squadron. After attacking U.305 he reported to

the convoy and for the next two hours patrolled the convoy at a range of ten miles. This circling Liberator forced U.305 to dive twice more but Kapitänleutnant Rudolf Bahr, another forceful commander, managed to keep the convoy in sight all afternoon and continued to report its progress to other boats.

The remainder of the daylight hours were quiet. *Lavender's* sick berth attendant worked for three hours trying to revive the unconscious second mate of the *Granville*, but without success, and the poor man died. The body was sewn up in canvas with spare firebars from the engine-room at the feet. As many of *Lavender's* crew as could be spared and all the *Granville* survivors gathered on the corvette's deck for the brief funeral service. The 'Still' was piped, *Lavender* lowered her ensign and all nearby ships that could see what was happening did the same. *Lavender's* captain, Lieutenant L. G. Pilcher, read a short prayer; the engines were stopped for a few moments; one end of the burial board was raised and the body of Second Officer Carmel Micallef, from Sliema in Malta, slid quietly from under the Union Jack into the sea. At least one victim of the battle for merchant shipping had received a decent burial.

On board most of the other ships in the convoy the afternoon was uneventful. No doubt many of the crews of both the escorts and the merchant ships took the opportunity to sleep. U.305 was still hanging on and her signals were picked up in *Havelock*. Commander Boyle several times ordered his faster escorts out during the afternoon to make offensive sweeps but they never caught U.305. When Flight-Sergeant Stoves had to leave the convoy in mid-afternoon, he should have been relieved by another Liberator which was to have stayed with the convoy until dusk, but this aircraft failed to appear. Convoy SC.122 was left without air escort during these last, vital hours of daylight.

While SC.122 had managed to survive the daylight hours of 17 March with the loss of just one merchant ship, HX.229 had continued to plug along about 100 miles behind. At least four and possibly five of the HX.229 merchant ships torpedoed during the night were still afloat at dawn and most of the morning's action was around these ships. To summarize these in the order in which they had been hit: *James Ogle-thorpe* with part of her crew on board was attempting to return to St John's, *William Eustis* may still have been afloat despite *Volunteer*'s four depth charges, and *Nariva*, *Irénée du Pont* and *Southern Princess*, which had been hit together, were definitely afloat although *Southern Princess* had turned turtle and was no more than a blazing hulk.

The corvette *Anemone* had remained behind with these last three ships. She had on board the entire crew of the *Nariva* which, except for a small fire in the foc's'le and being down by the bows, appeared to be in good condition. Captain Dodds was hopeful of getting his crew back on board at daybreak and of saving *Nariva* and her valuable cargo of frozen meat. *Anemone*'s captain, Lieutenant-Commander Pat King, R.N.R., who had served many years as a merchant officer in the Orient Line, was quite willing to wait and make the attempt. At dawn Captain Dodds, Second Officer Williams and Second Engineer Brophy were rowed across to their derelict ship in one of *Anemone*'s boats.

She looked a bit pathetic with her bows deep in the water and the foredeck awash as she rose and fell in the heavy swell. I noted the battery-powered red distress light was still burning and there was still a wisp of steam and smoke pouring out of the funnel. The surface of the sea all around the ship was littered with still-frozen carcasses of lamb and mutton which had obviously been washed out of the lower hold spaces through the enormous hole blown open by the torpedo. (Second Officer G. D. Williams)

The three officers split up, Williams and Brophy going down to examine the engine-room – a weird and frightening visit. The boilers were found to contain no water, only steam, and the furnace crowns were red hot; any attempt to refill the boilers would cause them to collapse. There was no hope of raising steam for a considerable time. Brophy went to his cabin to retrieve a new raincoat which he still possesses today, and Williams had time to pick up a large Pinocchio doll that the radio officer had bought in New York for his niece. They had only just got back into *Anemone*'s boat when, to their dismay, they saw the corvette suddenly move off at top speed.

Oberleutnant Siegfried Koitschka, captain of U.616, had been nearby for some time.

I sighted a steamer which was drifting but I couldn't seem to get into position for an attack at periscope depth, partly because of the heavy swell but partly because the steamer was drifting so aimlessly that she was difficult to plot. I tried twice to get into position but couldn't, although we were so close to it at one time that the steamer nearly collided with us and we had to go deeper to avoid it.

There was a corvette in attendance all the time, stopped only 400 metres away. I felt that such a simple target as this corvette only needed one torpedo set at three or four metres and to explode with a magnetic pistol. We used the torpedo in the stern tube; it was a G7e which ran comparatively slowly on an electric motor and I hoped the corvette would not see its track. I was surprised when it missed, and swore.

Anemone's lookouts had seen the track of the torpedo and the sudden starting of her engines had caused the torpedo to miss astern by half a ship's length. *Anemone* then swept along the torpedo's track searching for the U-boat but found nothing and soon returned to pick up the three men in the boat. Lieutenant-Commander King was still willing to transfer *Nariva*'s crew, despite the U-boat's presence, but Captain Dodds, after consulting his officers, decided that salvage was only possible if a tug was sent for and *Anemone* remained standing by. Such conditions were clearly impossible. It was felt by many of *Anemone*'s crew that *Nariva* could have been reboarded and could have sailed on but that *Nariva*'s officers knew that they would never have been able to persuade their men to return. In view of the numbers of U-boats operating in that area it is unlikely that *Nariva* would have stood much chance alone and it is as well that she was left.

While all this had been happening the *Southern Princess* had finally sunk. *Anemone* tried to sink both *Nariva* and *Irénée du Pont* by putting a few rounds of 4-inch into their waterlines, but the shells only made small holes. A depth charge was then dropped under each ship but they still refused to sink. *Anemone* could remain no longer and set off after the convoy. Oberleutnant Koitschka watched the corvette go and then came in and fired off another torpedo at *Nariva*. This too missed and Koitschka, on his first Atlantic trip after a long spell in command of a training U-boat, gave up, thoroughly disgusted with three unsuccessful attacks during the night and morning. He too set off to catch up the convoy again.

Of all the ships involved in this convoy battle, the corvette *Anemone* seemed destined to be involved in more incidents and to see more action than any other. Within only fifteen minutes of leaving the two derelict merchant ships, one of *Anemone*'s lookouts sighted a surfaced U-boat about six miles to the north and *Anemone* immediately gave chase. The U-boat dived at once and it took the corvette thirty minutes to reach the spot where the U-boat had been seen, but a firm Asdic contact was immediately gained and ten depth charges were dropped, set to explode between 150 and 385 feet. Unfortunately the depth-charge explosions once more caused *Anemone*'s Asdic set to give trouble and contact was lost. The corvette searched for another half hour but, when the U-boat failed to surface again, King resumed his journey back to the convoy. It is not known which U-boat had been attacked and these depth charges probably caused no damage. *Anemone* was to see plenty more action yet.

It is very difficult to be precise over what happened to the various torpedoed merchant ships left afloat behind Convoy HX.229. It was an established German procedure for any U-boat that was low on fuel and not able to continue with the high-speed work involved in direct attacks on a convoy to use their discretion and 'sweep up' behind the convoy, finishing off any stragglers or derelict ships they found there. Several U-boats operated in this way on 17 March, although they were not all short of fuel; some were still searching for the main convoy and happened to be in that area. U.665 (Ober-leutnant Hans-Jürgen Haupt) fired two torpedoes at a ship and claimed one hit but no sinking and U.228 claimed two hits on another ship seen on its own. Kapitänleutnant Heinz Walkerling in U.91 was also in this area and made several attacks on merchant ships; post-war reference books credit his boat with the *Fangschuss* (finishing-off shot) of four ships – *James Oglethorpe*, *William Eustis*, *Irénée du Pont* and *Nariva*. But a study of the positions in which these ships were torpedoed and of U.91's movements leads to the conclusion that Walkerling probably sank only *Nariva* and *Irénée du Pont*.*

There was no one aboard three of these ships when they sank, so exact details of their end are now academic. But there was another ship whose fate is of more importance. The American *James Oglethorpe* had been left with part of her crew on board and with steam up, attempting to make St

---

* This is the first of several disagreements with reference books, but the reader is asked to remember that the compilers of these tackled the mammoth task of covering the U-boat sinkings of the entire war while I have been able to concentrate on a five-day period.

John's. She never arrived and no signal was ever received from her. This ship was too far west to be one of U.91's 'sweeping-up' victims, and no other U-boat claimed to have sunk a ship between *James Oglethorpe*'s last-known position and Newfoundland during the next few days. Perhaps this Liberty ship had been damaged by the original torpedo explosion more than had been suspected and just fell apart, or perhaps she foundered in a storm that blew through on the following day. It can only be concluded that the *James Oglethorpe* was lost at some unknown position east of Newfoundland and that the gallant Captain Long and those of his crew who had stayed with their ship were drowned. The discipline of American merchant seamen is not always shown to advantage, but the brave men who were lost attempting to save the *James Oglethorpe* should also be remembered.

One of the U-boats operating behind HX.229 on the morning of the 17th was Oberleutnant Adolf Graef's U.664, and Graef became involved in one small development in the technical side of the U-boat war.

We had a Radione, a special short-wave wireless receiver for U-boats. We could not pick up our own German stations so far out in the Atlantic but we could tune into Gustav Siegfried Eins and Deutscher Sender Atlantik; these were English propaganda programmes that pretended to be German stations but we knew they were broadcast from England by pre-war refugees. We were forbidden to listen to them but most of us did so and Gustav Siegfried Eins became a well-known character amongst U-boat men.

Quite by chance I picked up on the Radione the conversation of the escorts. I reported this when we returned to base and on our next trip we were given a new man who could understand English and he made a note of everything he heard. Although the escorts spoke in a rough code, he could work out how many escorts were with the convoy. We handed the record of these conversations in at the end of that next patrol and I asked if my U-boat could be fitted with a direction finder so that we could find the convoys. This was done and for the next patrol we carried a second English-speaking man. I then asked if more U-boats could be fitted with the direction finder so that we could get not only a bearing on the convoy but a definite fix by triangulation.

On that first occasion Graef had been listening to Lieutenant-Commander Luther's radio-telephone link with the other escort vessels and with the Convoy Commodore. Graef's experiments came to an abrupt end on 9 August 1943, when U.664 was stranded out in the Atlantic almost without

fuel following the sinking of one of the German tanker
U-boats and was sunk by aircraft of the escort carrier U.S.S.
Card. Most of the crew, including the two English-speaking
specialists, became prisoners-of-war.*

It is time to rejoin the main part of Convoy HX.229. At dawn
of 17 March, only twenty-eight out of the original forty
merchant ships that had left New York were still in the
convoy, although the *Tekoa* was catching up from her rescue
work. The escorts *Mansfield, Pennywort* and *Anemone* were
also busy astern, leaving only *Volunteer* and *Beverley* protecting the convoy. The merchant ships were still pressing on
at their best speed, which was between 9 and 10 knots on the
same course of 053 degrees. Some of the merchant ship
officers would later complain that, although several small
tactical turns were ordered, the convoy always returned to its
course of 053 degrees and maintained this general direction
throughout the worst of the U-boat attacks; this certainly
helped the U-boats to keep contact. There is no doubt that
Lieutenant-Commander Luther considered this, but he
decided that the straight run to get within air cover was more
important.

Two events occurred affecting the escorts – one was good
news for the convoy, one bad news. At dawn on the 17th, the
convoy reached 35 degrees West, which was normally the
limit of the destroyer *Mansfield*'s operations from St John's
and she usually turned back at this point. On this occasion
Mansfield had received a signal, probably from Halifax, telling her to attempt to refuel from one of the escort oilers and
stay with the convoy as long as possible. The weather and the
U-boat action had not allowed *Mansfield* to refuel since
receipt of the signal but Lieutenant-Commander Hill decided

* The story of these 'black propaganda' programmes is well covered in
Sefton Delmer's book *Black Boomerang*, Secker & Warburg, 1962. There were
three programmes aimed at German servicemen – Gustav Siegfried Eins,
Deutscher Kurzwellensender Atlantik, which was mainly for U-boat men, and
Soldatensender Calais which was mainly for soldiers. They put out lively and
modern dance music and a first-class and accurate news service but mixed with
subtle propaganda. The Kurzwellensender Atlantik's signature tune, '*Es war
in Schöneberg im Monat Mai,*' and Vicky, the sailors' sweetheart, were well
known to most U-boat men. The station often broadcast details of U-boat
sinkings and the names of captured U-boat men long before the Germans could
release such information, although the British sometimes made mistakes –
Kapitänleutnant Kurt Baberg in U.618 heard the news of his own boat's destruction and of his own death. 'On this day we celebrated our second birth.'

to stay on with the convoy, hoping that an opportunity would occur later. When *Mansfield* rejoined the convoy that morning from her rescue work, Lieutenant-Commander Luther was told of Hill's decision and so HX.229's escort was not reduced at this critical time.

This welcome advantage was partly cancelled out by the failure to rejoin of *Pennywort*. Through a misunderstanding, *Pennywort* had not received details of the course change the previous evening just before the first U-boat attack. *Pennywort* had been in the process of changing to a new position in the screen at the time. Ever since then *Pennywort* had been engaged in rescue work at various positions behind the convoy and when this was over she had steamed to catch up, but on the old convoy course of 028 degrees instead of the new course of 053 degrees. The error was eventually discovered but *Pennywort* would not rejoin until just before dusk, twelve hours later than expected.

During the first few hours of daylight, while the convoy only had two destroyers as escort, the German U-boats were building up their strength again. Although many of the Raubgraf boats were dropping out, having shot off all their torpedoes or being too low on fuel, the first of the Stürmer and Dränger boats were now arriving in the area. U.603 was the official contact keeper; she was having no trouble in keeping the convoy in sight and was sending out a steady stream of signals. Six U-boats were in contact with HX.229 soon after dawn, eight by 09.00, ten by 10.30. Although SC.122 was receiving some air cover that morning, HX.229 was still in the Air Gap and there were no Liberators to drive down the U-boats around the convoy.

In mid-morning Commodore Mayall decided to fill in the many gaps in the battered starboard side of his convoy by reducing the number of columns from eleven to nine. Third Officer R. McRae was on watch in one of the merchant ships affected, the Donaldson Line ship *Coracero*.

The *Coracero* was instructed to take the leadership of a new column on the starboard side of the convoy which we proceeded to do and duly arrived on station, then two or three other ships were instructed to fall in column astern of us. I remember looking out to starboard with nothing between us and the horizon and feeling decidedly naked and uncomfortable but, within half an hour, felt much better when another ship proceeded to head a new column to starboard of us. She was of Dutch nationality and from the build of

her superstructure looked as if she was built for trading in the East
or East Indies.

It was standard practice of the apprentice to disappear about
10.30 and procure coffee and buns. The apprentice on my watch was
named Young; he was a nice youngster and, being well trained, he
proceeded at 10.30 on this priority-plus task. Before I had a chance
to drink my coffee a tremendous explosion burst forth from starboard of us and, on wheeling round, all I could see of our Dutch
colleague was a cloud of smoke, spray and pieces, nothing solid. I
shot to the bridge and swept my eyes to starboard and, Hell's bells,
there it was, a torpedo track heading for us about 200 yards away.

I immediately shouted helm orders 'hard a-starboard' (as the
book instructs always turn into a torpedo track) and then hung onto
the parapet of the bridge and watched the torpedo track converging
with the bow swinging towards it. The period of watching for the
results must have been only thirty or forty seconds but they were
elongated ones and it is true that under stress one thinks and does
strange things. While waiting, my mind did a turn back to a book
about World War One I had read as a boy, written by I think
'Taffrail', and it opened with the captain of a destroyer in exactly
the same position as I was now in, his ship was turning into the
path of a torpedo and facing certain destruction and, in the remaining seconds, his whole life flashed before his eyes. Well, no such
quick show came my way and, whether it was disappointment or a
vent of anger or what, I remember thinking it was a load of old
rubbish. It was a tense moment as our bow swung and covered the
torpedo track and, for a fraction of a second, everything stopped –
thoughts, blood, heart-beat – and then the track emerged on the
port side. We had made it.

I immediately gave the helm order 'amidships' and then grasped
the lanyard of the ship's siren and commenced to sound six short
blasts to warn the convoy that a torpedo was approaching from
starboard. But, between the second and third blast, all hell broke
loose as a further torpedo hit us on the starboard side at the after
end of the engine-room.

The Dutch ship torpedoed just before *Coracero* was the
Terkoelei. She was an East Indiaman, but had been the
German-owned *Essen* until May 1940 when she had been
taken as a prize by the Dutch at Sourabaya; she now had a
mixed crew of Dutchmen, Lascars and British gunners. Most
of her crew got away in four lifeboats but the Lascars were
completely shocked and nothing could persuade them to man
the oars and row clear of their rapidly sinking ship. Two of
the lifeboats were caught when *Terkoelei* capsized – one by the
mainmast and one by the funnel and many of the Lascars soon
died in the cold water. The second torpedoed ship, *Coracero*,
did not go down at once and her crew got away safely except

for the five men on duty in the engine-room, who were probably killed instantly when the torpedo struck.

Unknown to each other, two U-boats had made submerged torpedo attacks on the starboard side of the convoy at almost the same time. U.384 fired three torpedoes from the starboard bow at 11.05 and U.631 fired one torpedo from the starboard quarter at 11.06; both claimed hits. U.384 was later credited with the sinking of the *Coracero* and U.631 with that of the *Terkoelei* but these were conclusions reached on the assumption that the merchant ships were in their original, widely separated, positions before the Commodore put them together in the front corner of the convoy. It is possible that U.384's torpedoes hit both ships and that U.631's single torpedo failed to score, but the truth of this will never be known. If U.631 had sunk the *Terkoelei*, however, it is ironical that Blohm and Voss had built this merchant ship in their Hamburg shipyards in 1923 and that U.631 was built in the same yards in 1942.

The escorts carried out their usual sweep after this attack but found no trace of the attacking U-boats. Lieutenant-Commander Luther was faced once more with the problem of survivors; again no merchant ship had stopped to carry out the rescue work, perhaps understandably so in broad daylight with U-boats obviously about. Luther decided to detail *Mansfield* for rescue work but remained himself in *Volunteer* to screen the operation.

Lieutenant-Commander Hill brought his destroyer in, but first he dropped two single depth charges set at 100 feet and 300 feet 'more to make a noise for the submarine's benefit than with any hope of being in contact'. Then, for nearly two hours, the destroyer worked at this hazardous, and harrowing, task of picking up shocked and exhausted men from the sea.

Some particular survivors were on a capsized lifeboat. All managed to negotiate the scrambling nets except one who was in the water and clinging to the bilge grab rail. He didn't have the strength to pull himself from the water and he stared with unblinking eyes at the people safe above him. One of our leading stokers scrambled down and, with Lieutenant-Commander Hill doing his nut to get underway, the survivor was hauled to safety only to die within five minutes of being rescued. (Engine Room Artificer G. T. Smith)

The dead man was a young English gunner from *Terkoelei*.

Lieutenant-Commander Hill himself describes the rescue of the last man picked up, one of *Coracero*'s crew.

Two men were on a raft; one was picked up easily but the other was too weak to jump to the net or catch a lifeline. I manœuvred the ship gently to avoid injuring him but we were on his leeward side and the strong wind blew *Mansfield* away from the raft. I made another attempt to get him but the same thing happened again. The crew were getting very nervous at being stopped so long and my first lieutenant pointed out that the man had already had two chances and that our crew plus the 130 or so survivors on board should be considered. I decided to have one more try and proceeded to take *Mansfield* out and bring her in on the raft's weather side. The man on the raft waved desperately as we steamed away – he thought we were abandoning him – but we came in all right this time and picked him up. I considered that I had learnt a useful lesson in seamanship.

Thirty-eight men were lost from *Terkoelei*, mostly Lascars who had been in the capsized lifeboats. One of this ship's Dutch casualties was Chief Cook Samuel Schaap, who had been torpedoed twice before and had always said that the third torpedoing would be fatal. All of the *Coracero* men who had escaped the torpedo explosion had been saved. *Coracero* was still afloat and *Mansfield* was ordered to finish it off by torpedo.

As torpedo officer, I am somewhat apprehensive about scoring a hit and decided to fire over open sights from the torpedo tubes. We steamed up slowly and were about to fire when the poor vessel sank, thus relieving me from any embarrassment. (Sub-Lieutenant A. R. Guy)

Mansfield steamed away to catch up the convoy.

Once again the reader might spare a thought for the predicament of Gordon John Luther. As *Mansfield* and *Volunteer* steamed back to the convoy Luther heard that *Beverley* was in action with another U-boat and had left the merchant ships without any protection at all until *Volunteer* and *Mansfield* could rejoin. Luther's Report of Proceedings reflects the anguish he was now in.

It appeared to me that the time had now come when help must be asked for in the form of reinforcements as, with attacks by day and night and the escorts performing the dual role of escorting and rescue work, there was little hope of saving more than a fraction of the Convoy. Moreover, fuelling had been out of the question due to the weather and the persistent attacks on the Convoy and H.M.S. *Mansfield* was getting very low in both fuel and water. This help was asked for in my 13.10 and 14.37 to A.I.G. 303.

The relevant signals read:

HX.229 ATTACKED, TWO SHIPS TORPEDOED.
REQUEST EARLY REINFORCEMENT OF ESCORT.
51,45 NORTH, 32.36 WEST.
HAVE BEVERLEY AND MANSFIELD IN COM-
PANY, PENNYWORT AND ANEMONE OVER-
TAKING ASTERN. PERSISTENT ATTACKS WILL
NOT PERMIT FUELLING AND SITUATION IS
BECOMING CRITICAL. D/F AND SIGHTING
INDICATES MANY U-BOATS IN CONTACT.

The second signal was probably sent in answer to an
Admiralty signal requesting amplification of the first. Unfortunately, as has been described, there was little immediate
help that either the Admiralty or Western Approaches could
give without robbing SC.122 of part of her escort, and this
course had been decided against. Luther's hard-pressed escort
would have to fight on without help.

The merchant ships in the convoy were also well aware of
the danger they were in. The Ellerman Line ship *City of Agra*
had been next to *Coracero* and her master, Captain Nancollis,
now signalled to the Commodore and asked permission to
press on independently. His ship was capable of making 14
knots and he preferred to take his lone chance rather than stay
in the 9-knot convoy. Commodore Mayall refused the request.

Although it was not obvious to anyone at the time, the
fortunes of Convoy HX.229 were actually taking a turn for
the better at this time. Three quarters of an hour after
Coracero and *Terkoelei* were torpedoed, the destroyer *Beverley*
was taking up position ahead of the convoy when she spotted
a U-boat eight miles away and directly in the convoy's path.
Beverley turned towards the U-boat and a few moments later
the masthead lookout reported a second U-boat in the same
direction. *Beverley* gave chase at her top speed and both
U-boats dived, but when speed was reduced to start sweeping
with Asdic *Beverley* was immediately rewarded with a firm
contact. *Beverley* was an experienced North Atlantic escort
which, with H.M.S. *Vimy*, had sunk a Type IXC U-boat,
U.187, the previous month.* The previous night she had had

* On 18 March, the day after the present action, English newspapers
published a description of the sinking of U.187. When they reached England,
the *Beverley* men were disappointed to find that *Vimy* had received most of the
publicity, probably because *Vimy*'s captain, Lieutenant-Commander R. B.
Stannard, R.N.R., was a Victoria Cross holder.

a brush with another U-boat and had damaged this with the only salvo of depth charges dropped before having to rush back to the convoy. Now Lieutenant-Commander Price decided to carry out a proper series of careful attacks, although it meant leaving the convoy to sail on without escort. Over the next two and a half hours *Beverley* made six attacks.

Beverley's Asdic contact was U.530, another Type IXC boat which was on its first patrol although she had already scored, sinking a straggler from SC.121 a week earlier. Her captain, Kapitänleutnant Kurt Lange, was a pre-war merchant officer who was now thirty-nine years old and was believed by his crew to be the second oldest captain in the U-boat fleet. He had never served in an operational U-boat before but during training had often told his crew that he intended to bring them back safely from every patrol, even though this might mean not always pressing home his attacks in dangerous conditions. Lange had just found HX.229 and sent out his sighting report when *Beverley* forced him to dive.

U.530 was hunted by *Beverley* for the next two and a half hours and the early depth-charge attacks were very accurate. All the lights in the U-boat went out, and water started to force its way in through the torpedo tube hatches. The torpedo-men tightened the clamps but were unable to stop a steady inflow. Containers between the outer deck and the pressure hull were damaged and these too filled with water. With this growing weight of water in the boat itself and in the outer deck containers, the U-boat sank steadily. Lange decided to pump out the midships diesel tanks to get more buoyancy, but the pumps were also damaged and this could not be done. One man noticed that 'the younger men were very steady but the married ones looked scared', but another man says that 'we were all in terrible fear'. The U-boat went down to 240 metres and the crew, believing that 200 metres was their boat's limit, expected to be crushed at any moment. It was at this stage that the German sailors, with only the dim emergency lighting, with water swilling around their feet, with the hull of their boat creaking under pressure, heard quite clearly the passage of the destroyer directly overhead and believed they were about to meet their end.

Many destroyers carried one Mark X depth charge. This was a huge tubular charge, ten feet long and loaded with 2,000 pounds of explosive. The Mark X, sometimes called

Little Hector, was carried in a torpedo tube and was fired over the side by a cordite charge. When *Beverley's* Asdic officer felt that he had a really good contact on the U-boat below, the captain ordered the Mark X to be fired; it would be the first occasion *Beverley* had done so and the event caused much interest. The destroyer made a slow, careful approach, the Mark X flopped over the side and *Beverley* put on speed to get well clear of the anticipated explosion. Nothing happened – the firing mechanism had failed. It is probable that this run by *Beverley* was the one heard directly overhead by U.530's crew. Almost one ton of explosive should have detonated above the U-boat, which would surely have been further damaged by the shock waves or forced deeper. In either case, the pressure hull would finally have given way.

Twenty minutes later *Beverley* made her last attack, but it may be that contact with U.530 had been lost, and this attack was on a second U-boat, U.600, which was not damaged. Lieutenant-Commander Price now decided that he must give up the hunt and get back to the convoy before dark, and *Beverley* left the scene.

Kapitänleutnant Lange was able to stop his boat sinking further by violent use of his electric engines, and that evening U.530 surfaced with the entire crew fearful that the destroyer would be quietly lying in wait to finish them off. The U-boat's upper deck was found to be crushed and the steel cladding of the conning tower was 'rolled up like a piece of paper'. The crew had much cause to be thankful for the workmanship of the Deutsche Werft at Hamburg which had built U.530. This crippled U-boat took no further part in the convoy battle, and *Beverley's* activities had also caused at least two and possibly three other U-boats to take fright and lose contact with the convoy.

Kapitänleutnant Lange kept his promise to bring U.530's crew back safely from every patrol until January 1945, when he went to a shore position. In nearly two years he had sunk two stragglers and damaged a third ship – a fairly average score. U.530's luck still held; in May 1945, under the command of Oberleutnant Otto Wermuth, it was off the East Coast of the United States but ignored the order to surrender and two months later put into the River Plate and asked to be interned by the Argentine Navy. Although the crew were later handed over to the U.S. Navy, several of them later returned and married Argentine girls.

Volunteer had caught up the convoy again while *Beverley* had been doing such good work, and Lieutenant-Commander Luther was relieved to find that nothing untoward had happened in the period that the merchant ships had been left without escort. It has already been told how HX.229 had been allocated three V.L.R. Liberator aircraft as cover during the day and how the first two of these could not take off from Iceland due to high winds. There was bitter disappointment on *Volunteer* when the third aircraft, from Northern Ireland, failed to arrive on time and it looked as though HX.229 would again have no air cover in the important hours leading up to nightfall. But this mood changed to relief an hour later when not one but two Liberators were spotted approaching the convoy and their signal lamps were soon blinking messages to *Volunteer*.

The extra Liberator was flown by Flying Officer Chas Hammond of 86 Squadron; he had been detailed to provide SC.122's pre-dusk patrol but had failed to find that convoy and flown well past it. *Volunteer* signalled what little information it could about SC.122's whereabouts and the Liberator flew away but, despite a four-hour search, it never found its own convoy. It was for this reason that SC.122 failed to receive its air cover in the last hours of daylight. (Hammond's crew had further trouble on their homeward flight. They were diverted to Benbecula because of bad weather at Aldergrove but had great difficulty in making a landfall in the dark. After two engines cut out through lack of petrol, a landing ground was finally found at Eglinton near Londonderry and the Liberator landed after having been in the air for twenty hours and thirty minutes. Then the arrival at Eglinton was not reported to their home airfield and, while the exhausted crew slept all through 18 March, six Coastal Command aircraft were searching over the sea for them.)

The other aircraft reporting to *Volunteer* was the Liberator allocated to HX.229. It was flown by a veteran 120 Squadron crew captained by Flying Officer S. E. Esler, a red-haired Northern Irishman known on the squadron as 'Red' Esler. His crew were on their thirty-second operational flight and had already made seven attacks on U-boats, scoring one 'kill'

(U.661) on 15 October 1942. Esler had arrived earlier at the convoy's estimated position but no ships were in sight, and he had then tried to contact the convoy by radio but without success. Esler's navigator was Flight Sergeant T. J. Kempton.

The Skipper had asked me what we should do. The normal procedure would have been to fly a C.L.A. (Creeping Line Ahead) search, that is a series of wide sweeps down the expected track of the convoy, but I was very confident about my navigation – navigation conditions had been good – so I suggested that the convoy had been in difficulties and had been delayed, that we should assume that it was still on its original course and that we should fly straight down the track without wasting time with the C.L.A. We did this and picked up the convoy on radar twenty minutes later. We met it in position 51.29 North, 32.51 West, nearly 1,000 miles from base. The convoy consisted of far fewer ships than our briefing had indicated.

A combination of good navigation by Kempton and Luther's insistence that the convoy stick to its course resulted in this happy meeting. This aircraft's activities around HX.229 were at a greater range than had been achieved on any previous Liberator sortie.

Esler was asked to patrol at visibility distance and for the next two hours the Liberator flew around the convoy without seeing anything, although it is probable that several U-boats saw the aircraft and submerged. It is obvious that the U-boats became less cautious as the hours passed and were determined to work their way round to the front of the convoy before dark. At 17.05 Esler's crew spotted two surfaced U-boats ten miles away. It was the start of a furious hour of action. Esler was able to approach from out of the sun and came to within 3,000 yards of one of the U-boats before it dived. Five depth charges were dropped in a line ahead of the swirl and the R.A.F. men were delighted to see the bows and conning tower of the U-boat surge up and then sink again beneath the surface. This was Oberleutnant Hans Trojer's U.221; the U-boat was badly shaken and one man was injured.

No less than three U-boats were next sighted in a group and two of these dived hurriedly as the Liberator came in. The third U-boat unwisely stayed on the surface and opened an inaccurate machine-gun fire which was returned by Esler's gunners. Esler dropped his last depth charge alongside the German boat; it was U.608 (Kapitänleutnant Rolf Struckmeier), which later recorded that she had been attacked by a

'Sunderland', and it too was shaken sufficiently to keep out of the way for many hours. Then, almost immediately, another U-boat was sighted, but the Liberator could do no more than attack this with machine-gun fire and drop marine-marker-flares for scare effect; one of these scored a 'direct hit' on the U-boat.

Esler's patrol time was almost up, so he returned to the convoy and flashed the following signal to *Volunteer*:

SIX HEARSES IN SIGHT BEARING 180 DEGREES TWENTY-FIVE MILES. I GO.

A 'hearse' was the Coastal Command code for a U-boat; 'I go' was the usual way of saying that the aircraft had to leave. This signal, however, was read in *Volunteer* as 'five miles' and not 'twenty-five miles' and it caused absolute horror – the Liberator leaving, darkness closing in and already six U-boats almost within torpedo range of the convoy. A signal was flashed back to the aircraft saying that no escorts could be spared to deal with these U-boats and could the aircraft help. Esler sent no reply to this but disappeared behind the convoy. In fact, he stayed a little longer and made one more sweep before setting off for home. Flying Officer Esler's signal, as received by *Volunteer*, 'Six hearses in sight, bearing 180 degrees, five miles', became quite a celebrated signal among Western Approaches escorts, representing just about the worst news an escort commander could receive just before dark. We shall meet Red Esler and his crew again.

The nine U-boats sighted and the six attacks made on this day by three crews were easily a new record for Coastal Command, and were justification of the recently introduced policy of concentrating the few available V.L.R. airacrft over threatened convoys at the expense of 'offensive' sweeps and the escorting of other convoys not under the threat of U-boat attack.

During the day that was just closing, *Volunteer*'s HF/DF operators had picked up no less than thirty-four signals from U-boats around HX.229 but, thanks to *Beverley*'s vigorous depth-charge attacks and even more so to the work of Flying Officer Esler's Liberator, it is probable that only the contact keeper U.603, which had no torpedoes, and no more than two other U-boats were in firm contact with HX.229 at this time, and only one U-boat, U.305, was in contact with SC.122. Twelve U-boats – U.89, 91, 435, 468, 530, 600, 616, 638, 653, 664, 665 and 758 – had reported to their headquarters that they had been forced to drop out of the operation for various reasons – no torpedoes, shortage of fuel, mechanical trouble, depth-charge damage – but this still left thirty-one U-boats, mostly Stürmer and Dränger boats, to continue the fight; many of these were quite near the two target convoys. Everything would now depend upon how many of these could find the convoys during the coming night.

At dusk on 17 March, SC.122 had just passed the half-way mark of the ocean crossing from Newfoundland to Northern Ireland while HX.229 was now only ninety miles behind the slow convoy. The valuable ships of HX.229A were 600 miles away to the north-west on their tedious northern diversion route; the only U-boat within 400 miles of this convoy was U.229, which had been banished to her remote weather reporting position off Greenland, but the first boats of the new Seeteufel group were beginning to assemble off Iceland, more than 800 miles ahead of HX.229A but very close to its future route.

The nature of the German operations against the convoys was now changing. The weather was worsening again with strengthening winds and frequent snow squalls – not the best of weather for U-boats. The surprise air patrols of the past day had frustrated many of the German boats and they knew that there would be even more intense air activity on the following day. The previous firm orders allocating certain U-boats to specific convoys would now increasingly be ignored by the Germans out in the Atlantic, with captains listening carefully to the reports of other boats and making their own decisions. It would become a 'catch-as-catch-can' type of operation from now to the end of the convoy battle, but this was not contrary to Dönitz's philosophy – the sinking of any Allied merchant ship by any means and in any place was what the battle for merchant shipping was all about.

The seamen in the ships of SC.122 and HX.229 and the crews of thirty-one U-boats prepared for the second night of the battle.

12

The Second Round

The corvette *Pimpernel* was steaming in Position C of SC. 122's screen, which meant that she was guarding the convoy from the approach of a U-boat from the starboard bow. At 20.06 *Pimpernel's* radar operator picked up a clear contact and almost immediately afterwards the officer of the watch and one of the lookouts saw a U-boat 6,000 yards away on the surface, not out in *Pimpernel's* sector, but between *Pimpernel* and the frigate *Swale*, which was the next ship back in the screen and only 1,000 yards from the U-boat. *Pimpernel* immediately reported the sighting to Commander Boyle in *Havelock* and turned towards the U-boat but the enemy soon disappeared, diving towards the convoy. *Pimpernel's* warning was too late; four torpedoes were on their way. *Swale* never did see the U-boat.

The U-boat was U.305, which had earlier in the day been depth-charged by Flight-Sergeant Stoves' Liberator but without being damaged. The failure of SC.122's pre-dusk Liberator to appear had enabled U.305 to regain touch with the convoy. Fähnrich Wolfgang Jacobsen, who had earlier been so seasick and had then failed to spot the attacking Liberator, remembers this attack on the convoy.

We had steamed hard at 16 knots to get round to the front of the convoy then we stopped about seven miles ahead and waited. We could see the convoy coming down on to us quite clearly in the bright moonlight and watched carefully to pick up which targets we should go for. When the four ships that we were aiming for were overlapping, we fired. I think we hit two 8,000-ton ships of the *Port Adelaide* class. We dived then and got underneath the convoy. One of the ships we had hit sank almost at once and we could hear the boilers blowing up under water. We had the feeling that it was coming down all around us.

We considered that this was the standard U-boat attack and we were doing no more than following the B.d.U. Tactics Book.

U.305's torpedoes had, indeed, struck two ships. They were the 8,789-ton refrigerated ship *Port Auckland* – the

Germans' identification was good – and the smaller cargo ship *Zouave*; both ships were British. They were in the fifth and sixth of the convoy's ten columns, which shows that no part of a convoy could be completely safe. The *Port Auckland* was hit first, in the starboard side, and several minutes later the *Zouave* was hit on the port side; this may well be a case of a FAT torpedo turning round and making its second run through the convoy before finding a target.

Zouave was not a very elegant or well-liked ship, being one of the so-called 'Burntisland Economy Jobs' of the early 1930s, and was sailing with a mixed crew of Welshmen, Maltese, Africans and a few others. She was loaded with 7,000 tons of iron filings and sank very quickly; several men were trapped in the engine-room but most of the remainder of the crew got away in a lifeboat. It was probably the sinking *Zouave* that could be heard by the Germans in U.305 'coming down all around us'. In his Survivors Report, Captain W. H. Cambridge* wrote that 'the vessel then gracefully slid down by the stern, almost as if she was being launched for the first time', but another man did not put it quite so graciously:

> The ship sank in a very few minutes, not only because we had an extra heavy cargo but also because she was a rattling old tub and I can hear her to this day heaving a sigh of relief as she sank. She literally fell to bits – there were rivets flying about like machine-gun bullets. There were no real regrets at her going down; none of us were aiming to rejoin after this voyage. (Cook S. Banda)

The *Port Auckland* stayed afloat for some time but she too had been hit in the engine-room; one of the engineers, W. P. Shevlin, describes the terror of the engine-room men:

> The fourth engineer and the greaser both made a dash for the ladder with myself a slow starter in the rear. The other greaser was somewhere at the back of the engine and never stood a chance. I never reached the foot of the ladder which was only a few feet away. The greaser and the fourth did and the greaser made it to the top but I don't think the fourth got very far at all; I heard him shout almost immediately as though he had fallen. The engines had stopped themselves as the water came pouring in probably putting out the fires, or perhaps a main steam pipe had carried away in the stokehold which is where she must have been hit. The lights had gone out and everything was in utter darkness.

* Captain Cambridge of the *Zouave* and Kapitänleutnant Rudolf Bahr of U.603 had a common interest in the destruction of the battle cruiser H.M.S. *Hood* by the *Bismarck* and the *Prinz Eugen* in May 1941. Bahr had been serving in the *Prinz Eugen* and Captain Cambridge's son, Sub-Lieutenant John Cambridge, was an engineering officer aboard the *Hood* and was lost.

I must have been swept away from the manœuvring platform around the forward end of the port engine and towards the port-side engine-room bilges. I checked myself en route by grabbing at anything I could feel and fought back but, by now, I was lost in the absolute darkness. I thought it was 'finish'. I kept struggling but really did not know where I was. I was quite calm and did not panic but struggled frantically enough. I thought sadly of my wife and my family and thought how sad would be the news to them. I had heard the trapped foreman in the stokehold screaming but now it seemed all quiet. I did not know how high the water was; I remember one fleeting thought – that I was already drowned and this was death.

It was a torch of the search party, shone into the engine-room, that finally enabled the struggling Shevlin to gain the ladder and escape. One trimmer, badly burnt on face and hands, managed to reach the deck and two more trimmers were rescued from their stokeholds. The fourth engineer had fallen back into the engine-room and been drowned and five firemen and a greaser were also lost, as was the officers' steward who went for a stroll along the boat-deck while waiting for his lifeboat to be launched and probably fell through an open hatch. He was not seen again.

One lifeboat had been smashed by the torpedo explosion and here are two views of the launching of the three remaining boats, the first by one of the ship's officers:

The boats were lowered and scramble-nets dropped overside in way of them and it was found that, owing to the heavy rolling, it was only necessary for several people at a time to stand in the top of the scramble-nets holding on to the lifelines and step or jump into the lifeboat at the appropriate moment – that is when the lifeboat was level with the upper-deck on the downward roll of the ship. This needed very nice judgement as it must be remembered that it was a pretty dark night and no light, other than an odd torch or two, was available. In the circumstances it was quite remarkable that exactly 100 people embarked in the three lifeboats with no one being injured or even falling into the ditch! (Second Officer J. G. A. Dunn)

But Sergeant Barry Ware, one of the ten Australian pilots aboard as passengers, did not see it through a seaman's eyes:

My turn came and my cobber already in the boat was yelling to me to grab the rope. On the first two attempts it was out of reach, on the third I had it. I kicked myself out from the side of the ship and then felt my own weight; my hands were numb from the cold and then red hot as the rope was streaming through my hands. I let go and in those split seconds not knowing whether it was to be the sea or the boat then suddenly I was amongst a sea of arms and legs. I had made it. The next quarter of an hour was chaotic. We were way overloaded and they were still coming down; unless we cleared the ship shortly, the lifeboat would surely founder as she was being unmercifully smashed against the ship's side. Then suddenly the ropes were limp; there were no more to come. The davit lines were freed and we were clear.

The seas were mountainous, sweeping down the side of the waves into the trough then screaming up the other side to reach the crest breaking over the top of us. We became violently ill, vomiting over one another with no apologies – we all understood.

Not all were as sick as Sergeant Ware; shouts were heard from the seamen handling different boats, 'Race you to Liverpool', and men in one boat were singing – 'Roll out the Barrel'.*

Commander Boyle had ordered a Half Raspberry search after the two ships had been torpedoed, but this had produced nothing. The rescue ship *Zamalek* had still not caught up from her previous night's work, but Boyle was not prepared to abandon the survivors and he ordered the corvette *Godetia* to carry on with rescue work. The Belgians were often given this type of duty and already had the crew of H.M.S. *Campobello* on board.

The first survivors that were found were two men shouting from the sea among some drifting debris which was all that was left of the *Zouave*. One man was picked up easily but the second could not hold onto the line he was thrown; he shouted that his legs were broken but before the Belgians could help him the line went slack. *Zouave*'s lifeboat was then found but there was much dismay in the boat when they were hailed from out of the dark in English with a heavy foreign accent; the *Zouave* men thought the U-boat had come back to machine-gun them until they realized that the voices belonged to friends. One of *Port Auckland*'s boats was found next but, while its occupants were being picked up, something quite frightening occurred, described here by Lieutenant G. B. Rogers, one of *Campobello*'s officers who was helping out by acting as anti-submarine officer in *Godetia*'s Asdic hut.

Suddenly the Asdic operator called out 'Torpedo fired, Sir'. I repeated the message and listened with my earphones. I could hear

* None of the ten Australians was lost, but three were killed later while flying, Flight Sergeant John Underhill in a training accident in North Wales on 25 July 1943, Pilot Officer Ernest Skillen in a Lancaster bomber of 50 Squadron in a raid on Aachen on 12 April 1944, and Flying Officer Russell Tickner also in a Lancaster, of 196 Squadron, over Norway on 25 February 1945.

the distinctive noise of a 'tin fish' running and bent over the receiver to study the bearing. The bearing remained steady. I passed on this information and kept checking. With the large lifeboat alongside we could not move and if we did start up probably would not get away in time. I thought our only hope was in our comparatively shallow draught but did not put much faith in that. Suddenly there was a terrific noise in my earphones followed by another as I snatched them off. The sinking ship had been right on the bearing and had caught the torpedo which I had been anticipating. Being below the rim of the bridge I had not been able to relate my knowledge to the situation outside. That 'fish' had only been running for a couple of minutes but every second had seemed like hours as I realized they were coming straight for us.

After the convoy had passed over him, Kapitänleutnant Bahr had deemed it safe to surface again and his lookouts had immediately spotted the still-floating *Port Auckland*. Bahr decided to finish the ship off and fired this one torpedo. He was completely unaware that a corvette was rescuing survivors on the other side of *Port Auckland*. The Belgians pulled the last few men out of the lifeboat and *Godetia* got under way at once. A firm radar contact was picked up 3,000 yards away and *Godetia*'s captain, Lieutenant Larose, ordered the U-boat to be chased. *Godetia* worked up to $15\frac{1}{2}$ knots but the radar showed that the Germans were running slightly faster on the surface. Larose ordered the 4-inch gun to open fire but, before it could do so, the radar 'blip' disappeared. U.305 had dived.

Godetia could not do much more because her Asdic was out of operation. Larose ordered one depth charge only to be dropped in the area where the U-boat had dived in order to discourage the Germans from coming up again and then returned to his rescue work. The remaining *Port Auckland* survivors were soon found and picked up. The *Port Auckland* itself was seen to be sinking fast and after making one more search for stray men in the sea *Godetia*, now packed with 165 survivors, set off to catch up the convoy.

This quite typical incident of the Battle of the Atlantic had lasted just three hours from the time U.305 fired the first torpedoes. Two ships had been lost with 15,000 tons of cargo and 129 bags of mail. Twenty merchant seamen and one Royal Maritime Artillery gunner had died but 140 men had been saved to sail or fly again.

Meanwhile Convoy SC.122 had continued on course with its six escorts with every reason to believe that the convoy was

to be in for a bad night with further attacks to come. Two
hours after the first torpedoing, *Havelock* did pick up a strong
HF/DF contact on the convoy's port side and swept out on
the bearing at high speed, but nothing was found. This
U-boat signal was a sighting report transmitted by Oberleutnant Helmut von Tippelskirch in U.439, which had just
made its first contact with the convoy. *Havelock* had been seen
by the U-boat's lookouts and U.439 had dived.

As the hours passed away without incident, men began to
hope that the U-boats had been shaken off. The optimists
were, on this occasion, correct. U.305 remained under water
for three hours after *Godetia*'s single depth charge, and
Havelock's prompt move when she picked up U.439's signal
persuaded that boat to stay down long enough to lose the
convoy. These were the only U-boats to have been in contact
with SC.122 on the second night of the battle. The convoy
steamed safely on and dawn of 18 March was greeted with
much relief.

Only eighty miles behind SC.122 and still steadily catching
up was Convoy HX.229. The 'six hearses' signal from the
Liberator the previous evening had 'nearly given a heart
attack' to the officers on *Volunteer*'s bridge. Lieutenant-
Commander Luther's Report of Proceedings again reveals
his fears for the coming night.

> This shattering piece of information caused me very seriously to
> think. H.M.S. *Mansfield*'s fuel condition precluded her being used
> for any high-speed work and to send one escort back to deal with
> this pack could only defer their attack. I therefore decided to do
> nothing but await events and deal with them as best as I could as
> they arose. I myself foresaw the worst and thought that we were in
> for a night very much worse than last – and only three escorts (one
> of them a lame duck) to fight the battle *and* do what we could for
> survivors.

The reader knows something that Lieutenant-Commander
Luther and every man in *Volunteer* did not know – that the
six 'hearses' or U-boats had been twenty-five miles from the
convoy and not five.

Volunteer's first lieutenant, Lieutenant Leslie, had earlier
in the day 'experimented with the recently-issued benzedrine'
to stay awake while his captain had a sleep, so Luther was
well rested to face the coming night. Luther had then broadcast over the ship's loudspeaker system telling his crew that

they were faced with a large number of U-boats manned by brave and determined men. He warned them that there was every indication that the coming night would be a repeat of the heavy attacks of the first night of the battle. It is probable that in many ships of the convoy captains and masters were saying much the same to their crews.

As darkness closed in, the convoy had only the three destroyers – *Volunteer*, *Beverley* and *Mansfield* – to form a screen. This time Luther placed his own ship ahead of the convoy, making wide sweeps out to both bows, never dropping below 18 knots, the maximum speed at which Asdic could still operate. Lieutenant Leslie says this 'covered the maximum ground and we were also less likely to be hit. Everyone expected the worst.'

The first event of the night was the rejoining of the corvettes *Anemone* and *Pennywort* which had both been absent for twenty-four hours. Luther was delighted to see them and allocated them positions in the screen. For the first time since midday of the 16th, thirty-six hours earlier, and for the first time since the U-boat attacks had started, the escort commander now had his entire group of five escort vessels at hand. Nothing happened after this until 02.30* on the 18th when *Volunteer*'s HF/DF picked up a U-boat signal close ahead of the convoy. *Volunteer* ran out on the bearing but could not find the U-boat, although she dropped one depth charge for scare effect.

Happily for the men sailing in Convoy HX.229, *Volunteer*'s sally out to put down the signalling U-boat at 02.30 was the only contact this convoy had with their enemies during that second night of the battle. This U-boat had not spotted the convoy. The U-boat Headquarters War Diary shows that its last signal from a U-boat in touch with HX.229 had been at 14.30 the previous afternoon, but firm contact had then been lost and only three uncertain hydrophone contacts were reported during the night.

It may seem difficult to understand how thirty aggressive U-boats could have completely lost two large convoys that had made no major course changes. But there are at least three explanations – the air patrols of the previous afternoon,

---

* Convoy Time changed at midnight with clocks being put forward one hour, so times quoted here will now be one hour slow on Greenwich Mean Time and two hours slow on Central European Time.

although not covering the two convoys all the time, had been in an area that was not often visited by Allied aircraft and had thoroughly disorganized the Germans; the weather, although not yet really rough, had been bad enough to further discomfort the Germans. And, then, luck, that unpredictable gift of the Gods, had this time favoured the seamen in the convoy ships; many U-boats must have passed very close to the convoy during that night but just not close enough to gain a sighting and bring other U-boats in to continue the massacre of the first night. Finally, it should not be forgotten how vast are the ocean wastes of the Atlantic, so that even a large convoy became a difficult quarry for a U-boat lookout only a few feet above sea level, especially at night. Radar would have helped, but the U-boats had no radar.

The most serious event, aboard *Volunteer* at least, during those hours of darkness, was a medical case. When Flying Officer Esler's Liberator had flown over and flashed his dramatic signal the previous evening, *Volunteer*'s captain had actually been talking to his ship's doctor, Surgeon-Lieutenant Humphrey Osmond, about an American merchant seaman on board who was complaining of 'a pain in the belly'.

Then my personal dilemma – not as appalling as the captain's – began. I found the 'bellyache' was an acute appendicitis. There was much keenness to operate 'to give him a chance'. Chiefy Ward [Warrant Engineer G. H. Ward] was very keen to help. Now it was rough, we couldn't heave to. I am no surgeon. I totted up the score, looked at my Hamilton Bailey [a surgical textbook] and found the Ochsner–Sherrin conservative technique for adults; it only had a 1 per cent mortality rate in Swedish lumber camps. The patient is placed in a semi-upright position in bed supported by pillows, is given very small quantities of liquid by mouth and, if necessary, fed by vein. The whole point of the Ochsner–Sherrin technique is that it prevents any immediate need for operation. What Chiefy Ward wanted was a repetition of an operation done on a previous commission when my predecessor, a surgical registrar I think from King's or University College, had indeed removed an appendix with the help of the Chief who gave the anaesthetic. This had whetted his appetite for these activities; I think engineers do have a strong inclination to help doctors in these matters. Chiefy was a great friend and I think this was about the only disagreement we ever had, but he grumbled about it for the next three months. The patient was later landed at Londonderry and operated on successfully.

Dawn of 18 March brought in the third day of the battle. The snow showers had become heavier and more frequent and a full north-north-west gale was now blowing around both convoys. But the weather was a blessing for the convoy men; these were not conditions favourable for a U-boat lookout searching for the lost convoys. The remainder of this chapter will describe the fortunes of SC.122 and HX.229 during the stormy daylight hours of 18 March. SC.122, still ahead of HX.229, will be covered first.

The weather may have been bad around the convoys but it was good at two other important places – the airfields at Reykjavik and Aldergrove – and one Liberator had taken off from both of these airfields in the early hours to provide the first patrols for SC.122. Both aircraft found the convoy, reported to *Havelock*, and patrolled for the next four hours, dodging in and out of the snow showers around the convoy; neither aircraft found U-boats. But SC.122 was being followed. After U.305 had sunk *Port Auckland* and *Zouave* early the previous night and been put down by *Godetia*, Kapitän-leutnant Rudolf Bahr had surfaced and steamed hard through the remainder of the night along SC.122's track. Soon after dawn he had spotted the convoy ahead and got off a good sighting report to U-boat Headquarters. It was not long before other boats were making use of Bahr's signals and calculating courses to intercept this convoy.

The weather over SC.122 improved slightly as the day wore on, and the Liberators kept turning up regularly to provide cover. Soon after midday, the crew of Flying Officer J. K. Moffat spotted a U-boat and put in a good attack with four depth charges. The U-boat was reported to Commander Boyle who sent out a corvette, but it found nothing. The next two Liberators to come on patrol also met U-boats. Pilot Officer A. W. Fraser's crew made two attacks, using up all their depth charges, and then flew right over a fully surfaced Type VIIC U-boat but could do no more than make sure it dived. Flying Officer R. T. F. Turner's crew then came on the scene and they soon sighted a U-boat.

We were flying at 600 feet, just under the cloud base, hoping to surprise the U-boats. It was getting a little dark but we could see the open areas of ocean quite easily. We sighted it three miles away but he dived too quickly to be worth depth charges. The U-boat had been in a good position to attack the convoy so I adopted what we called 'baiting tactics'. We drew off to a reasonable distance into the dark part of the horizon, made a wide turn and then came back. We sighted him again on the surface and again he dived quickly. I wouldn't normally have used depth charges in such unfavourable conditions but we were nearly at the end of our patrol so we attacked.

Turner's U-boat was U.642 (Kapitänleutnant Herbert Brünning); the attack failed to cause any damage but Brünning stayed below for the next hour.

This concluded a highly efficient and successful day for the Liberator crews, at least as far as SC.122 was concerned. Despite the poor weather, five crews – all from 120 Squadron – had found the convoy and air cover had been continuous for more than ten hours. There had been six U-boat sightings and four attacks made on them; no U-boat had been able to attack the convoy all day. It was a seemingly perfect demonstration of air power. But this fine work had not been as effective as might have appeared. It is probable that three of the four attacks – Moffat's and Fraser's two – had all been on the same U-boat; Kapitänleutnant Bahr's U.305 had been the target on all three occasions. But, after surviving each attack without damage, Bahr had resurfaced and continued shadowing the convoy and getting off his sighting reports. One must admire the courage of Bahr and his crew, on their first patrol, in sticking to their task so tenaciously. They were now exhausted and the batteries of the boat were almost run flat with the frequent dives. Bahr withdrew after the last attack but, partly as a result of this U-boat's efforts, five other U-boats had come either within sight of the convoy or so close that they would make contact that night.*

The convoy's surface escorts also came into action during the evening. One of the Liberators had left a flare floating over the position in which a U-boat had been seen diving, and *Upshur* was sent out to investigate. After half an hour of searching, the Americans picked up a contact on Asdic (Sonar to the Americans) and gave it seven depth charges, but no further contacts resulted and *Upshur* was called back to the convoy an hour later. Her contact was probably U.642 which was not damaged. Then *Lavender* picked up a radar contact on the convoy's port side and gave chase. The U-boat was soon sighted and *Lavender* opened fire with her 4-inch gun

* U.305 failed to sink any more merchant ships after *Port Auckland* and *Zouave*, but on 20 September 1943 she sank the Canadian destroyer *St Croix* and, on 7 January 1944, the frigate H.M.S. *Tweed*. U.305, with Kapitänleutnant Bahr still in command, was sunk ten days later by the destroyer H.M.S. *Wanderer* and the frigate H.M.S. *Glenarm*. There were no survivors.

before the U-boat dived. *Lavender* soon made Asdic contact and carried out two good depth-charge attacks. The second of these was followed by underwater rumblings and an oil slick appeared on the surface. These attacks by *Lavender* had been very effective; the U-boat was Kapitänleutnant Manfred Kinzel's U.338 and there is no doubt that Kinzel was preparing to press home an attack as soon as it became dark. *Lavender*'s depth charges had caused several serious leaks and put this boat out of the battle for the coming night. Finally, just as it was getting dark, *Saxifrage* gained a 'doubtful' contact and dropped seven depth charges for luck but without apparent result.

And so SC.122's day ended with U-boats being attacked by aircraft and escort vessels all around and this vigorous defence preventing any attacks being made on the convoy. The *Empire Morn*, a 7,000-ton cargo ship, had developed boiler trouble during the afternoon and gradually dropped back from the convoy. By dark she was several miles behind and not being escorted. Stragglers had to take their chance; the main convoy had to be protected. It is not certain whether SC.122 was being shadowed and reported at dusk but three undamaged U-boats were very close if not actually in touch.

Convoy HX.229's day had not been as trouble-free as had that of SC.122. At dawn, Lieutenant-Commander Hill in *Mansfield* had decided that his short-endurance destroyer could no longer stay with the convoy. There had been no opportunity to refuel and the gale then blowing would obviously not allow this now. Hill had been ordered to stay with the convoy as long as he could and the decision on when to leave was one that only he could make. The previous night had been quiet, air patrols were expected shortly, and *Highlander* and possibly other reinforcements were also expected to join that day. Hill reported his decision to *Volunteer*, and *Mansfield* left the convoy. She had done much sound and sensible work in recent days, attacking two U-boats and damaging one of them, filling a gap in the escort screen, and rescuing 130 men from three torpedoed ships.

Lieutenant-Commander Hill had a second decision to make – where was *Mansfield* to go? His regular base was St John's but this was now 960 miles away and there were two Allied bases that were nearer – Hvalfjordur in Iceland, 700 miles, and Londonderry, 800 miles. But to make Hvalfjordur would mean steaming mainly into the gale, while a course for Londonderry would mean a faster passage with the gale pushing the destroyer along. Hill chose Londonderry – 'it was a decision about which there was no disagreement on board'; after six months on the other side of the Atlantic, *Mansfield*'s crew were certain to get home leave. The destroyer steamed off to the east at an economical 12 knots. This happy return was marred by only one incident. Only a few hours after leaving the convoy, a heavy sea swept *Mansfield*'s afterdeck and washed overboard one of the crew, Able Seaman Alfred Thomas. Although a thorough search was made, the poor man was not found.

Mansfield's departure left the convoy with only four escorts – *Volunteer*, *Beverley*, *Anemone* and *Pennywort*. *Beverley* was the same class as *Mansfield* but had been fitted with extra fuel tanks so her endurance was not a problem. There were soon other troubles for Lieutenant-Commander Luther in addition to this reduced strength of escorts. Five Liberator aircraft had been allocated to his convoy and were due to provide continuous air cover throughout the daylight hours. The first of these was flown by Flight Sergeant J. H. Bookless, an Australian pilot with 86 Squadron. When Bookless approached the area he found poor weather conditions and asked for bearings from *Volunteer*. For two hours nothing was heard and, even when radio contact was made and bearings received, the crew could see no ships in the poor visibility. After five frustrating hours of searching, Bookless gave up and flew back to Aldergrove where he landed with nothing whatever to show for more than sixteen hours in the air. This crew were only on their second operational flight and their squadron had only recently commenced Atlantic operations. It might be thought that this contributed to the failure to find the convoy but the next four aircraft dispatched to HX.229 on this day were all flown by experienced crews of 120 Squadron and not one of these found the convoy either.

There was later an investigation into these 'convoy not met' flights and *Volunteer* was asked to explain herself. *Volunteer*'s subsequent report pointed out that, not being a regular senior officer's ship, there was only a small communications department and it had been necessary to call in HF/DF operators to handle unfamiliar equipment. Also, HX.229 had been so close to SC.122, which was homing aircraft on the same frequency, that this caused confusion in the poor visibility of that day.

And so HX.229 had to go without any direct air escort all that day. Two of the aircraft searching for the convoy did find U-boats and one of these, captained by Squadron Leader Desmond Isted, put in a very good attack on U.610, whose captain was the aristocratic Kapitänleutnant Walter von Freyberg-Eisenberg-Allmendingen (known more simply as von Freyberg). The attacking aircraft, again recorded by the Germans as being a Sunderland, had been seen some distance away and von Freyberg had thought about fighting it out on the surface but, with the sea so rough, his gunners would not have stood much chance and von Freyberg wisely dived. At least one of Isted's depth charges exploded right over the U-boat, damaging both periscopes, both compasses, a compressor and many instruments and fuses. It took five hours to carry out essential repairs and, although this boat continued searching for the convoys, she never came into contact again.*
The second U-boat contact was by Flying Officer R. Goodfellow's crew, and it was the only occasion in the SC.122/ HX.229 operation that an aircraft detected a U-boat by radar; all other sightings were visual. Goodfellow spotted his U-boat in time to see it dive but too late to make an attack.

One of the reasons why *Volunteer*'s wireless office was so busy was that *Highlander* and possibly also *Abelia* called up and signalled that they expected to join early that afternoon. The arrival of the regular escort commander in a destroyer and of an extra corvette would have been of immense value to the convoy but, alas, it was not to be. When asked what the convoy's position was, *Volunteer* signalled the Convoy Commodore's estimated noon position which later turned out to be forty miles east of the convoy's true position. This resulted in *Highlander* and *Abelia* steering for a position ahead of the convoy and both sailed past HX.229 without seeing the convoy. Commander Day did not ask for a homing signal, which would have corrected the error at once, because he could tell from the radio traffic that *Volunteer* was having difficulty homing in the Liberators. The signal position was

* Kapitänleutnant von Freyberg's U.610 ended her career dramatically in operations against Convoy SC.143 on 8 October 1943. She sank the Polish destroyer *Orkan* but was then depth-charged and sunk by a Sunderland of 423 (Canadian) Squadron. There were no survivors.

further confused when one of *Highlander*'s coders was washed off his feet while walking to the cipher officer's cabin; the man was saved but two cipher books he was carrying were washed overboard. This was reported to the Admiralty and all involved soon changed to another cipher but *Highlander* could not follow what was happening during a crucial period.

This written description of events may seem a little dry and the reader may like to try again to picture a harassed and tired Lieutenant-Commander Luther, working mainly from the open bridge of *Volunteer*, trying to protect his convoy with only four escorts, trying to sort out the signals muddle and bring Liberators in, trying to achieve meetings with *Highlander* and *Abelia*, all this in the shrieking, howling gale that brought flurries of snow and caused *Volunteer* to pitch and roll heavily over every wave.

U.221 was a boat with a hectic career behind her and Ober-leutnant Hans Trojer was one of the forceful type of captains; he had been in U-boats since the very first day of the war, although part of the intervening time had been spent in command of a training boat. Among U.221's past exploits had been a collision with another U-boat the previous December, which had resulted in the sinking of the other boat, U.254, with the loss of all but three men. Trojer and his crew had also been in action only a week ago against Convoy HX.228 and had torpedoed two merchant ships including one of the Luckenbach ships, *Andrea F. Luckenbach*, which was carrying ammunition and had disappeared in a spectacular explosion. U.221 had then been severely depth-charged by the French corvette *Aconit*. Now, here she was on the afternoon of 18 March, the only U-boat in contact with HX.229.

The failure of the Liberators to find and cover HX.229 had allowed Trojer to work his way steadily round to the front of the convoy. At 13.50 he dived to periscope depth and after allowing the convoy to come right up to him he had easily found a gap in the widely-spaced escort's screen, the rough weather concealing his periscope. Third Officer R. H. Keyworth was on duty in *Canadian Star*, which was in the third column from the port side of the convoy.

I was watching a Liberty ship at the head of the next column to port when it appeared to be struck by a heavy sea throwing spray over the entire ship as high as the funnel but, when it lost way, I realized that it had been torpedoed. I rang the alarm bells and the men came up to action stations. The gunner on the port wing of the bridge was tearing the cover off his Oerlikon gun when I suddenly spotted, coming up between our ship and the next column to port, a periscope about a yard clear of the water; the surface of the sea was a little calmer in the lee of *Canadian Star*. The captain and I called out at the same time and the gunner started to swing his Oerlikon onto it, but, almost at once, we were hit. It felt as though the whole ship had blown up underneath me. There was a tremendous amount of cordite, I can still almost taste it. The captain ordered Abandon Ship. I dashed into the chart house to get my little 'get-away bag' with my sextant and some of my navigation books. I had ideas of a long voyage in an open boat.

Canadian Star was the fast Blue Star Line refrigerated ship that normally sailed independently but had been ordered to join HX.229 because her gun had been damaged in an accident. She was carrying twenty-four passengers, mostly service officers and their families from India, the Far East and Australia. Some of these had been forced to flee from Singapore when the Japanese invaded and after a long wait in Australia had been at last looking forward to reaching England. Two torpedoes had hit, both at the after end of the ship, and she was settling fast by the stern. The explosions had wrecked two of the four lifeboats but there was room in the other two boats for all on board. The Liberty ship torpedoed at the same time was the *Walter Q. Gresham* – another new ship on its maiden voyage – loaded with 9,000 tons of sugar and powdered milk. The torpedo had blown off the propeller and torn a huge hole in the side of the ship. The powdered milk in the cargo was washed out of the aft hold and was being whipped up by the gale into an enormous white froth.

When the two ships had been torpedoed, Lieutenant-Commander Luther had ordered the escorts to perform an Artichoke, the usual move following a daylight torpedo attack, in which the escorts searched for the U-boat astern of the convoy. *Volunteer*, *Anemone* and *Pennywort* steamed right past the sinking ships. Surgeon-Lieutenant Osmond was again on *Volunteer*'s bridge.

I stood well back to keep out of the way. The captain was following the Asdic and we swept round to the sinking ships at the rear of the convoy. There was a Liberty going down in the gale-slashed sea with the Stars and Stripes at the stern; her guncrew were manning their after gun – futile but gallant. There was a lifeboat – a man clinging to a spar. It was that appalling dilemma which a brave, decent and humane young captain had to face and decide on in about five minutes – the survivors or the submarine? We wallowed past them, the captain shouting from the loud-hailer, 'We'll be back.' We had to get the submarine before it came round again. Our survivors (the *William Eustis* crew picked up earlier) seemed angry and resentful. I believe that decision hung around the captain's neck for the rest of his life. He could never be sure that he had done the right thing because we didn't get the submarine.

The Artichoke sweep only produced an uncertain contact by *Pennywort* on which she dropped six depth charges for luck and, after half an hour, Luther decided to give up searching for the U-boat. He detailed *Anemone* and *Pennywort* to rescue the survivors.

Back at the scene of the torpedoing that half hour had been a desperate time, especially for the passengers on the *Canadian Star*. They had started to take their places in the two remaining lifeboats but, when only a few women and two children were in one of these, a young seaman was caught unawares by a sudden lurch and a rope slipped through his hands. The women and children were tipped into the sea and swept away. On *Canadian Star*'s deck a horrified Royal Artillery colonel watched this tragedy; his wife and young son had been in the boat. Another woman had nearly reached this boat but had been delayed by her husband, who had been asleep when the torpedo struck, not being able to find his trousers and the wife, quite cross with him at the delay, was later to admit that this had saved her life. The passengers were very calm and one noted that 'there was a stiff-upper-lip attitude; we were all very polite to each other'. Most of the passengers managed to get into the remaining lifeboat, the women and children being packed 'heads and tails' into the bottom for protection from the weather. This boat was then got away safely with Chief Officer Percy Hunt in charge.

The movement of the lifeboat was at that time completely at the mercy of the sea; she was alongside one second and twenty feet off the next. I remember pulling the Fourth Engineer out of the water and seeing the Doctor trying to get down the pilot ladder. He would not let go of the ladder and we could not get near him. I don't know what eventually happened to him. [Colonel Alexander Craighead, Indian Medical Service, was lost.] In the meantime I issued orders to pay out the sea anchor and four oars were manned to assist in keeping the boat head-on to the wave tops. It was the first time I had seen a sea anchor in a sea that size but it worked marvellously.

The *Canadian Star* was now left with most of her crew on board but with only one capsized boat, still attached to its fall ropes, and several liferafts. Second Officer Clarke-Hunt was one of those who made use of the rafts.

It slipped into the sea and swung at the end of a rope about ten feet from the ship's side. I could see that the men were stunned and shocked and they just waited for the raft to come back. I urged them to jump in and get on it but no one would do; the non-swimmers would have had to rely on their lifejackets. I set off first and the others soon followed. We soon had twenty-two men on a ten-man raft, most of them had to hang onto the side somehow. We lost six of these fairly quickly; you would see them getting cold, a certain look came into their eyes and then they just gave up. The army officer who had seen his wife and child drown was the first to go . . . Only nine of us were eventually picked up.

One man decided that he had no chance of surviving. The fifty-eight-year-old ship's carpenter, a Highland Scot who had settled in New Zealand, described as 'a rough but fine man', was seen on the *Canadian Star*'s stern, over which huge waves were now breaking. He called out to one of the ship's officers, 'Goodbye, Sir. It was a good life while it lasted,' waved and then calmly 'walked right into the path of a wave pounding across the after deck. It was like a minnow being swallowed by a whale.'

There still remained a few crew members aboard. Third Officer Keyworth managed to right the capsized lifeboat at the end of its fall and was hoping to get the water out of it, but the *Canadian Star* was now going so quickly that Keyworth and his helpers had to get into the boat, waterlogged as it was. The only man now left on board was Captain Miller, who had made sure that everyone possible had got away. He refused to go in any of the boats or rafts and many think that he deliberately went down with his ship, although this young officer had everything to live for, having married just before this voyage. It is believed that if all his passengers and crew could have been saved Captain Miller would have attempted to save himself but would not do so after seeing some of his passengers drown. One of his officers later regarded this decision as 'the ultimate devotion to duty' another regarded it as 'damned stupid, what use was a dead man to his country?'

The lifeboats and rafts of *Canadian Star* and of the *Walter*

Q. Gresham had become well scattered by the storm before the escorts came back. For the next two hours the rescue work went on in atrocious weather conditions. Both rescuing corvettes had survivors from previously torpedoed ships on board and these helped with the rescue work. The following quotations from both Royal Navy men and the previous survivors on the two escorts reveal much of the bravery, suffering and horror of the Battle of the Atlantic – *Pennywort's* work first.

One survivor, each time the wave brought him high enough for us to reach him, would grab the rescuer's arm and, when the wave moved out from under him, he would fall back into the sea. Then the same thing would happen again; he would be a little further toward the stern each time he was close enough to grab but at last he was hauled aboard. There were two survivors next. One was holding onto a raft with his left hand and to another man with his right. As they drifted alongside we could see the one he was holding was dead; his head was leaning back and the lifejacket pushed up under his chin. His mouth and eyes were open and, as each wave broke over him, his eyes and mouth would not move. (Able Seaman T. Napier, ex *James Oglethorpe*)

We had, of course, scrambling nets over, also ropes and lines, and the anchors had been lowered a little. Two Americans managed to get on an anchor and one managed to climb up the chain and inboard but the other one fell back and was swept away. The one that was saved was terribly upset at this; he said that they had been buddies for years. (Wireman R. A. White)

The American captain was in a lifeboat with a coloured steward who was in a very bad way, mainly from exposure and swallowing oil. The boat was on the weather side of the ship and one moment was level with the rail, the next thirty feet below us. We managed to grab the captain eventually when the boat was on an upsurge but couldn't get the steward. We steamed in a circle, not easy in the weather we were having, and got the boat on our lee side. I was about to jump in with a line, as the man was virtually unconscious, but was beaten to it by one of our Norwegian survivors. We got the coloured man in but he died shortly afterwards. (Sub-Lieutenant L. M. Maude-Roxby)

And on *Anemone*, which was working mainly among the *Canadian Star*'s survivors:

The skipper of the *Anemone* won our admiration by his superb handling of the little escort vessel as, in spite of very rough seas, he manoeuvred his ship close to a raft which had been sighted with about eighteen soaked, cold and shivering survivors clinging desperately as the raft teetered perilously on the crest of each wave, threatening to hurl its occupants into the wild, spume-streaked sea.

And as the raft got closer we thought we could hear these poor, soaked U-boat victims shouting or crying for help until we could hear, borne down to us on the howling gale, the voices were lustily singing, 'She's a lassie from Lancashire, just a lassie from Lancashire'. We didn't know whether to laugh or cry. (Second Officer G. D. Williams, ex *Nariva*)

The survivors were getting onto the scrambling nets and our sailors were getting hold of both sides of the net and flinging them inboard in great heaps. One of them was an army colonel who stood up and announced himself by name. A Geordie seaman said, 'What the . . . . . . do you expect me to do about it. Get forward!' The crew called him Colonel Blimp after that.

There was a little boy; an R.A.F. officer threw him like a rugger ball up onto our deck and one of our stokers caught him. He was still breathing but completely numb. The little boy's mother came up all right but the R.A.F. officer's wife got trapped between the lifeboat and *Anemone*'s side and was badly crushed. Then there was a very pretty girl with long hair; she grabbed the net but slipped back. A sailor, with split-second timing, leant over, grabbed her hair and swung her right up and onto *Anemone*'s deck. (Lieutenant D. C. Christopherson)

Third Officer Keyworth had got away from the *Canadian Star* in his waterlogged lifeboat with several seamen. Their boat had overturned once and some men had been lost, but Keyworth and the remainder had, by a desperate effort, turned the boat over again although it remained full of water. Keyworth tells of his subsequent fight with the elements.

It was useless to bale; the sea just swept through the boat from end to end. I could see the men, one by one, their eyes glazing and eventually losing their grip and being washed up and down the boat and eventually out of it altogether. Then I started to get a feeling of cosiness, ready to relax just as though I had come in and sat by a warm fire and just couldn't keep awake.

I had a perfectly clear vision of my mother outside my home in Wellington, New Zealand, and at the gate, at the bottom of the long garden path, was the postman with an envelope in his hand. I knew very well that this told her that I had been lost at sea. As I lapsed further into this subconscious state, the postman started to move. I couldn't bear the thought of my mother being distressed and I managed to rouse myself from this drifting-away feeling. The postman stopped. I started going again and the postman started walking up the path again so, once more, I pulled myself together and he stopped. I should think that this process was repeated seven or eight times. By the time he was three quarters up the path, I saw a corvette. I think now that if I had gone right into the coma I would have seen the postman reach my mother and that the vision I had been seeing enabled me to reach to the bottom of my endurance and it saved my life.

The corvette eventually picked us up; there were four of us left. The *Anemone* men reached down and pulled me over the rail; a sea threw the lifeboat against my legs and I heard one of the navy men shout, 'Stretcher! His legs have gone.' I thought they had been cut off. The last I can remember is lying on the stretcher and seeing my sea boots still there so I knew my legs were O.K. Then I passed out.

Lieutenant Christopherson describes how *Anemone* picked up her last survivor.

We took a last look round and saw a man on a hatch cover, spreadeagled as though riding a surfboard. He was pretty far gone so we got our best heaving-line man and he managed to throw the line right across the survivor's neck and the man calmly reached up, caught the line, and took one turn around his wrist. He could do no more than this but it was enough to allow him to be dragged to the net and taken aboard. He was completely numb from the waist down. It took a lot of work to bring him round but we were delighted that he was able to walk off the ship when we reached port.

Despite the hard work of *Anemone*'s young sick-berth attendant and of many volunteer helpers, two survivors died after being picked up – the R.A.F. officer's wife who had crushed her back and one of the men in Keyworth's lifeboat. Two-year-old Noel Wright, the youngest survivor, made a good recovery. The *Canadian Star*'s total death roll was thirty out of the eighty-seven who had been on board. Seven of the dead were passengers, including one complete family, also two children, and an Australian businessman who had survived Gallipoli and the worst of the First World War only to die here in the Atlantic. The *Walter Q. Gresham*'s losses were twenty-seven out of sixty-nine and included a very popular second mate, Walter Farragut from New Orleans, described by his union's newspaper as 'one of the men who help to clear out the Gulf phonies and build up the Union (the N.M.U.) there'. Two fortunate survivors were technicians employed by Lockheeds who had been travelling as passengers.

Gordon John Luther did not live for long after the war but it is known that he bitterly regretted his initial decision to leave the survivors and take all the escorts on the abortive U-boat hunt and, to his death, he felt responsible for the fifty-seven lives lost from the *Canadian Star* and the *Walter Q. Gresham*.

Three of *Canadian Star*'s officers were later decorated.

Captain David Miller was posthumously awarded Lloyd's
War Medal for ensuring that every possible person aboard
was saved; Chief Officer Hunt and Third Officer Keyworth
were both awarded M.B.E.s for fine work in launching and
handling lifeboats. Oberleutnant Trojer was informed by
radio four days later that he had been awarded the Knight's
Cross for his recent successes, but he and all his crew were
lost to a Coastal Command Halifax in Biscay in September of
that year.

An important development in Convoy HX.229's fortunes
occurred while the rescue work was taking place. After
steaming at high speed for just 1,000 sea miles from St John's,
H.M.S. *Highlander* finally joined and Commander E. C. L.
Day took over command of his escort group from the weary
Lieutenant-Commander Luther. Day was not well pleased to
find that the position he had been signalled for the convoy had
been forty miles in error. The position given had been the
Convoy Commodore's estimated noon position, and Day's
Report of Proceedings later contained a sharp criticism of the
Abraham Lincoln's navigation and urged that this ship should
not be used to carry a Convoy Commodore again. In theory
the Commodore's noon position had been the average of the
estimated positions of all the merchant ships in the convoy
but the vile weather had probably made navigation difficult.
Whatever had gone wrong, the mistake had deprived the
convoy of the services of this destroyer during the recent
attack by U.221. When he did join, Commander Day was
also slightly annoyed to find that Luther had detailed two
ships from the escort to rescue work.

The second of HX.229's hoped-for reinforcements, the
corvette *Abelia*, failed to appear that day. She too had sailed
all the way from St John's and had also suffered from the
confusion over the convoy's position. She actually passed
HX.229 that night and steamed on well ahead of the convoy.
HX.229's other reinforcements – U.S.S. *Babbitt*, H.M.C.S.
Sherbrooke and H.M.S. *Vimy* – were all still some distance
from the convoy.

The experienced Commander Day planned his moves for
the third night of the battle.

I hoped the convoy would be seen by U-boats just before nightfall and reported on its existing course of 053 degrees and that they

would attempt to get ahead of the convoy. I gave orders that the convoy was to turn on to 080 degrees just after dark, hoping to leave the U-boats ahead of the old route. After four hours on 080 degrees the convoy was to turn on to 040 degrees so that we should be back on our original route at dawn.

Not too far away, Commander Boyle was making similar plans for SC.122.

Commander Day had another problem to face just before dark. The American freighter *Mathew Luckenbach* had been between *Canadian Star* and *Walter Q. Gresham* when those ships had been torpedoed, and her crew had watched their occupants attempting to launch lifeboats in the gale conditions. Able Seaman Pat Civitillo was aboard the *Mathew Luckenbach*.

After it was all over we were in the mess hall discussing the attack. We were surprised when the Captain and the Gunnery Officer appeared. The Captain requested all crew members off watch to be present. Then he gave his opinion of our situation. He explained that the *Mathew Luckenbach* was a 15-knot ship placed in a 9-knot convoy. Eighteen ships had been sunk so far. The escorts were ineffective against the subs. It was more than likely the submarine pack would sink most of the remaining ships. The *Mathew Luckenbach* had run alone on other trips without mishap and he was confident he could do it again, providing we lost the submarine pack. He said he could do this under the cover of darkness by slipping out of the convoy and steaming ahead, leaving the submarine pack to go after the convoy.

He asked for a show of hands by those in favour of it. We all raised our hands. He also had a statement written out to this effect, which he asked us all to sign. We all signed.

Commander Day spotted *Mathew Luckenbach* steaming away from the convoy.

I saw him sneaking off, went alongside and shouted through the loudhailer, 'Obey the last order of the Convoy Commodore.' He replied, 'I can steam 15 knots. I'm making my best way home.' I told him he was steaming into danger – I didn't know that he was but I thought it might turn him back.

The American ship steamed on into the dark to take her chance alone. She had become a 'romper' or 'runner' - a ship that deliberately sailed on from a convoy.

Dusk of 18 March marks the end of the second phase of this convoy battle. It had been an inconclusive phase, with bad weather and air patrols forcing many U-boats to lose contact and allowing the convoys a little respite, but four more merchant ships had been sunk in the past twenty-four hours, bringing the total lost so far to nineteen – twelve from HX.229 and seven from SC.122. In return, five U-boats had been damaged – four by surface escorts and one by aircraft – but only one of these had been forced out of the battle.

HX.229's superior speed had enabled that convoy almost to catch up SC.122 during the past day. The naval Official History states that 'the two convoys were gradually closing each other and so ultimately formed a large mass of shipping in a relatively confined space of ocean'.* This has often been taken literally but Captain Roskill was probably writing figuratively. The two convoys were still seventy miles apart on this evening and would not come much closer than this. But even this proximity was now to affect the course of the encounter between convoy and U-boat. SC.122 was at that time being shadowed by U.642, although the escorts had detected the signals and chased the U-boat away. The weather was moderating rapidly and well over twenty U-boats were still in the chase, many of them making for SC.122 in response to U.642's signals. Neither of the convoy escort commanders knew that the other convoy was now so close. It is probable that Convoy HX.229 was not being shadowed at this time but Commander Day, in ordering HX.229 to turn on to 080 degrees in an hour's time, would actually steam his convoy for four hours straight at the U-boats coming up astern of SC.122.

As darkness came down in the late evening of 18 March, the final phase of the convoy battle was about to open.

* Roskill, op. cit., Vol. II, p. 365.

13

The Final Stage

Although SC.122 and HX.229 were still to make several small alterations to their courses, the two convoys would maintain roughly parallel courses from the night of 18 March until reaching England. HX.229 would always be to the north and, although it would soon pass SC.122, it would never be very far from the slower convoy. It had always been one operation for the Germans but the two convoys were so close in this last stage that the Germans could rarely be certain on which convoy any U-boat was operating and U-boat captains were content if they could gain contact with either. The Coastal Command air operations too would increasingly benefit both convoys irrespective of which convoy an aircraft had been detailed to cover. For these reasons, events will now more often be described as they occur and the exact circumstances of each convoy are not so important.

Twenty-four U-boats were still in the fight. Most of these were well spread out astern of the convoys, having been harassed during the day by aircraft and bad weather, but were now trying to catch up and take advantage of the calmer conditions and bright moonlight to score in this third night of the battle. SC.122 had two U-boats – U.338 and U.642 – in contact by dusk and U.666 joined about midnight. But Commander Boyle had a strong escort around this convoy and these boats found it difficult to penetrate his screen. Oberleutnant Herbert Engel brought U.666 in at 23.17 and fired a full salvo at three overlapping cargo ships, then turned away and fired his stern tube torpedo as well. These were probably long-range shots from outside the screen and no ships were hit. The convoy was not aware of the attack and Engel was able to reload his tubes undisturbed and start planning another attempt.

Convoy HX.229 was steering into the area astern of SC.122 which contained many U-boats. U.406 was the first boat to stumble upon HX.229 at about midnight. This boat, commanded by Kapitänleutnant Horst Dieterichs, was suffering from mechanical trouble. Her engineering officer, Oberleutnant Rudi Toepfer, says this was caused by 'carbon deposits and thick sludge in both crank cases; this choked up the filters whenever we ran at high speeds, thereby causing a dangerously low oil pressure. Sabotage would have been an easy explanation but poor quality fuel seemed to be more likely.' U.406 could only limp along behind HX.229, but Dieterichs plotted the convoy's course and speed and his signals soon brought other boats into contact. Once again, the battered HX.229 convoy was threatened with a pack attack as at least ten more U-boats answered the signals and sighted the convoy during the next few hours.* The convoy's two absent escorts – *Anemone* and *Pennywort* – turned up just before midnight from their recent rescue work, bringing Commander Day's strength to five ships. The familiar U-boat signals were soon being picked up from nearby and the weary escorts prepared themselves once more for action.

But the first incident did not occur until 03.10 when *Anemone* sighted a U-boat astern of the convoy. There was surprise in *Anemone* at this because the U-boat was in the unfavourable position of being up-moon and down-wind to the convoy. *Anemone*, always in the action, set off to attack and Commander Day sent *Volunteer*, which had hardly used any depth charges while Luther had been in command of the escort, to join in also.

Anemone's U-boat was U.615, which was the boat that had muffed its chance on the first night of attacks on HX.229 because its captain had insisted on giving up a good position on the convoy's flank to come round behind the convoy. Leutnant von Egan-Krieger tells how his captain repeated the error now.

The weather was clear with a bright moon and the convoy looked a fine picture with the illumination of the moon behind it. Again the captain made a decisive mistake in deciding to get between the moon and the convoy against all theoretical teaching and practical experience. Again we wasted time in changing over our position

---

* U.406 was sunk by the frigate, H.M.S. *Spey*, on 18 February 1944. Kapitänleutnant Dieterichs was lost, but most of the crew were picked up by an American ship together with three radar technicians who were carrying out trials in the U-boat and whose capture, according to the Official History, yielded valuable technical information. Oberleutnant Toepfer settled in Los Angeles after the war.

and finished up in a far less advantageous one. When we were at last in the attack position, an escort immediately came upon us at high speed. Our only salvation was in the cellar.

Anemone came up just after U.615 dived, gained an Asdic contact and dropped ten depth charges 'set to 150 feet and 385 feet as it was thought that the U-boat would probably try to go deep'. Some lady passengers recently rescued from *Canadian Star* were in *Anemone*'s petty officers' sleeping quarters at the extreme rear of the corvette and the depth-charge explosions terrified them, although the little boy who had been picked up thought it was all great fun. No doubt to improve their morale, the passengers were later told that the U-boat had been sunk but this was not so.

Anemone's first depth charges signified the opening of a furious ninety minutes of action around the convoy. *Volunteer* settled down with *Anemone* to hunt U.615 and these two escorts made seven attacks in the next ninety minutes. So thick were the U-boats that three were probably being hunted during this period and two of them, U.134 (Oberleutnant Hans-Günther Brosin) and U.440 (Kapitänleutnant Hans Geissler), were damaged; the original target, U.615, was unharmed. Soon after *Anemone*'s first contact, *Pennywort* also gained a contact on the convoy's port beam and she dashed off to attack this. But, while three of the five escorts were engaged in these actions, a fresh U-boat, U.441, came in and fired five torpedoes at three merchant ships on the convoy's port side. Kapitänleutnant Klaus Hartmann was delighted to hear explosions and put in an elaborate claim for a 'passenger-freighter' sunk, a second ship set on fire and left sinking, and two more ships hit. Such successes would have been disastrous for this already hard-hit convoy, but all his torpedoes had missed. The only explosions Hartmann could have heard were depth charges and his claims to have seen burning and sinking ships can only have been a figment of his imagination.

On top of all this action, *Highlander* spotted a further U-boat; Lieutenant D. G. M. Gardner describes *Highlander*'s attack on it.

It was very close and, from our bridge, the U-boat captain could be seen on the conning tower watching the convoy through his binoculars; his stern lookout must have been asleep. We increased speed, but didn't reduce again quickly enough when he did see us and dived, with the result that there was too much noise for the Asdic to work as it should have done and we lost him. I think the captain was rather annoyed with himself afterwards when he realized his mistake.

Highlander had a full complement of depth charges and ten of these were dropped despite the loss of Asdic contact.

No one aboard *Highlander* knew that she was in great danger while she tried to regain contact with the U-boat. Kapitänleutnant Rolf Struckmeier in U.608 was only 4,000 metres away and had *Highlander* in his sights. He quickly fired three torpedoes and then dived. After the torpedoes had run for five and a half minutes, two initial explosions were heard and then larger underwater explosions. Struckmeier believed the first two explosions were his torpedoes hitting the destroyer and that the underwater ones were the sinking victim's stock of depth charges going off. He consulted his warship recognition books, concluded that he had torpedoed a Defender-class destroyer and claimed it as sunk. This was yet again an optimistic claim; *Highlander* never even saw the torpedoes. She had regained contact with her quarry and all the explosions Struckmeier heard were the results of *Highlander*'s depth-charge attacks.

The U-boat that *Highlander* was hunting was U.439, commanded by Oberleutnant Helmut von Tippelskirch. Commander Day put in three more attacks, one of which was a failure due to the Hedgehog not firing, and then decided to return to the convoy. U.439 was slightly damaged and one of her crew, Maschinenobergefreiter Gerhard Schmeling, pays this tribute to *Highlander*. 'This was our seventh patrol and we had been depth-charged many times before but never as seriously as on this occasion.'

Gradually the furious action died down. Four of HX.229's five escorts had made attacks on U-boats. Seventy-one depth charges had been dropped and three Hedgehog attacks made. Three U-boats had been damaged, although not seriously, but many other U-boats had been near enough to take prudent flight and had submerged and stayed deep. By dawn, not a single U-boat was in sure touch with the convoy but these encounters demonstrate yet again how difficult it was to destroy a U-boat when escorts had to keep returning to their convoy.*

* The three U-boats damaged in these attacks all returned safely to their bases but were all sunk in the following months. Soon after setting out for her

Anemone was not finished with the U-boats yet. Lieutenant-Commander King tells of his return to the convoy.

With only two depth charges left and once again far astern of the convoy, course was set to rejoin. Just as daylight was breaking, two U-boats were sighted ahead on the surface, one on each bow. We headed towards one of these with the intention of opening fire with our 4-inch gun but both U-boats dived simultaneously as we approached. This was perhaps just as well because had both U-boats remained on the surface and fought it out with their guns it would have been two against one and *Anemone*, with her single gun, could only have engaged one U-boat at a time leaving the other to shoot unmolested. This could have been disastrous with all our survivors on board as we were entirely alone at the time.

When *Anemone* did rejoin the convoy, she was placed in the screen near *Volunteer*. *Anemone*, with only two depth charges remaining, flashed a signal to the destroyer offering to exchange some of her female survivors in return for some of *Volunteer*'s depth charges.

While Commander Day's escorts had been beating off the U-boats near HX.229, other Germans had been nosing around SC.122. Three U-boats – U.338, U.642 and U.666 – were in firm contact with the convoy and possibly an unknown fourth boat also. U.642 was soon forced to dive when *Havelock* picked up an HF/DF bearing on one of its signals, and U.338, which had been heavily depth-charged the previous evening by *Lavender*, was in a poor mechanical condition and also had to give up. This left U.666 and a possible fourth U-boat both on the starboard side of the convoy and in the favourable down-moon position. At 04.48 the frigate *Swale* was in the starboard-beam position of the screen and Sub-Lieutenant L. A. Spicer was officer of the watch.

The R.D.F. (radar) reported a contact at about 4,000 yards. There had been several such contacts during the night but they hadn't come to anything and I didn't want to bring the crew to Action Stations again unnecessarily. I was looking in the direction of the contact through my binoculars when, at that very moment, a merchant ship went up behind me. I called Action Stations.

next patrol, U.439 collided with another U-boat in the Bay of Biscay during a storm in the night of 3 May and both boats sank with much loss of life. U.440 was caught in the Bay of Biscay by a Sunderland of 201 Squadron on 31 May 1943; her new captain obeyed recent orders to stay up and fight and U.440 was sunk. U.134 shot down U.S. Navy Blimp K34 off Florida on 18 July 1943 but was then sunk by a 179 Squadron Wellington off Spain on 24 August. There were no survivors from these last two boats.

Swale's Report of Proceedings shows that the torpedoing occurred only one minute after her radar contact and that the U-boat immediately disappeared but was replaced at once by an Asdic contact. *Swale* ran down on the submerged U-boat which was then only 200 yards from the starboard column of the convoy and just had time to drop the big fourteen-charge pattern that was this modern escort's big punch before the Asdic contact disappeared in the confused water left by the wake of the merchant ships.

The merchant ship that had been hit was the old Greek cargo ship *Carras* and she was the leading ship in SC.122's starboard column. She had been hit by one torpedo and Captain Mazavinos ordered his crew to abandon ship at once. But the U-boat that had torpedoed *Carras* was not the one sighted and attacked by *Swale*. U.666, which had earlier fired five torpedoes at this convoy with success, had reloaded her tubes, caught up again and managed to enter the screen by passing between two escorts in the dark sector away from the moon. Oberleutnant Engel had then taken his time about aiming three torpedoes at ships in the starboard columns, had fired and then turned and ran to the rear of the convoy on the surface without being detected. Engel had claimed hits on three ships but *Carras* was his only success – the first of this new U-boat. The identity of the U-boat attacked by *Swale* is not known.

The rescue ship *Zamalek* had no difficulty in picking up the thirty-four-man crew of *Carras* and no one was lost or injured. The Greeks brought plenty of drink with them and were quite willing to share this with the other survivors on *Zamalek*; a Welsh fireman already aboard says 'there was great rejoicing, but quietly'. These events took SC.122 up to dawn. The third night of the U-boat battle had brought four separate torpedo attacks on the two convoys but these had resulted in only one old merchant ship being hit – although it was the twentieth ship lost in the battle – and no lives had been lost. It had been a night of good sound defence and sturdy counter-attack by the escort vessels of both convoys and the merchant ships had been able to steam many valuable miles nearer to England and safety.

Before leaving SC.122 in the dawn light of 19 March, the story of *Carras* should be completed. The Greek ship showed no signs of sinking and Lieutenant-Commander Crosby

Dawson, R.N.R., the English officer on his last voyage in command of the Belgians in H.M.S. *Buttercup*, radioed to Commander Boyle that the ship might be salvaged. Boyle ordered that a crew be put aboard *Carras* if possible and signalled the situation to the Admiralty who sent a naval tug, H.M.S. *Saucy*, covered by the corvette *Jonquil*, from London-derry 570 miles away, a good indication of how valuable even one old Greek freighter was to the Allied cause. Captain Morris of *Zamalek* went down to the survivors' quarters on his ship and appealed for volunteers to man *Carras*; there was little interest in his appeal and the idea was abandoned. *Buttercup* then sent a party of Belgians, under Lieutenant P. H. V. M. van Waesburghe, across to the drifting merchant ship in order to dump the codes which the Greek master had left on board in his haste. Van Waesburghe tells his own story.

We got aboard quite easily; she was low in the water but not getting worse. I looked for the heavy box so that I could dump the Confidential Books but couldn't find it so I had to take the books back with me. Our men had found a live sheep and wanted to take it back also so that we could have a change from the eternal corned-beef stew but I wouldn't allow it and forbade them to take anything else.

When I got back to our boat I found that they had taken a few things like cigarettes and chocolate; it was the old pirate instinct they had. One man, Raymond Beuren our coxswain and formerly an Ostend fisherman, was wearing the Greek captain's jacket and he often wore it later on in *Buttercup*.

The Belgians returned to their own ship. Lieutenant-Commander Dawson decided to sink *Carras* and a few shots of 4-inch were fired and one depth charge dropped. The Belgians are convinced that *Carras* was sinking when they left, but U.333 found a drifting cargo ship in this position twelve hours later and was credited with the sinking of it; however, this may have been another derelict ship. *Buttercup* neglected to inform the escort commander that *Carras* could not be saved, with the result that the tug and corvette from Londonderry spent several wasted hours at sea before the situation was cleared up.

It was now three days and nights since U.653 had sighted HX.229 and started this encounter between convoy and U-boats. The morning of 19 March brought far better weather than had been experienced for several days, with calm seas, light winds and good visibility although there were still occasional showers of rain or sleet. The vigorous counter-attacks by the escorts of both convoys had forced every German U-boat to break off and, for the second time in the battle, both convoys were free from shadowers. Within a short time, too, both convoys received their first air patrols, Fortresses (B-17s) this time, for the convoys had now come within range of these aircraft. An increasing sense of security now pervaded the Allied ships with the friendly shores of Britain only four days' steaming away and the prospect of air cover all the way. But the Germans were far from giving up. Dönitz knew that Coastal Command's patrols would be more intense than ever but he sent orders that the operation was to continue for another twenty-four hours and urged his tired boats to one more last effort.

SC.122 received a boost soon after dawn when her first escort reinforcement arrived. This was the U.S. Coast Guard Cutter *Ingham* which had taken four storm-tossed days to come down from Ireland. This powerful, modern ship was half as big again as any of the regular escorts with the convoy, and her commander, Captain A. M. Martinson, U.S.C.G., was actually senior in rank to Commander Boyle, but Martinson placed himself unreservedly under Boyle's orders.

Another American reinforcement had come down from Iceland with *Ingham*. This was the destroyer U.S.S. *Babbitt*, believed by her crew to be the fourth oldest ship in service with the U.S. Navy and reputed to have been refused by the British earlier in the war. But *Babbitt* was very experienced in North Atlantic convoy work; indeed her crew had just earned themselves a Battle Star for their involvement in the fight for SC.121. *Babbitt* was steaming to reinforce HX.229 and her captain, Lieutenant-Commander Samuel F. Quarles, describes her approach to HX.229 during the early hours of that day.

We plotted the position of the convoy on our chart along with the last reported positions of many German submarines. There were so many between us and the convoy that we felt as if we were about to take on the whole German submarine Fleet. We were quite thrilled because we were hungry for more action. The likelihood of finding a submarine on the surface by radar while making 22 knots during the night was so great that many persons stayed up so as to get to their battle stations more quickly.

At 04.25, when about ten miles from the convoy, *Babbitt's* radar picked up a good contact dead ahead and 2,200 yards away. The sub must have seen us immediately because she dived at once. We picked her up on Sonar soon and proceeded to attack with full patterns of depth charges. We could detect the evasive tactics of the sub while we were proceeding towards the 'Drop' position. We would head towards the sub initially and then put the rudder over full as soon as it became evident that the sub had turned towards either our port or starboard. In one of these attacks, the Executive Officer and I made a small wager as to which way the sub would turn. I do not remember who won.

While turning to make the third attack we picked the sub up on our radar, indicating that she had surfaced. We carried out a standard plan to illuminate the sector of horizon in the direction of the sub. This was done but due to her low profile we could not see her in spite of the fact that our Star Shells almost turned night into day. The fact that there were many 'stars' coming down on their parachutes with their metal-clad burdens turned out to be a most unfortunate development; our radar picked all of these up and the screen was so flooded that we could not find the sub by this means again.

We gained further good Sonar contacts during which we dropped many depth charges. By this time, we had used all we had topside and replacements had to be hoisted up from the magazines by block and tackle. This was so slow that at times we would simply point the ship's bow towards the sub contact, slow down and wait for one or two more depth charges to be gotten ready. Then we would go to standard attack speed and fire those.

Babbitt's attack continued for a little over five hours and she dropped fifty-three depth charges in eleven separate attacks. Some diesel oil and grease were found floating on the surface and the Americans collected samples which were later handed over to the Royal Navy for analysis. After the eleventh attack no more Sonar contact could be gained and after a further hour of searching Lieutenant-Commander Quarles set off to join the convoy.

There is no doubt that the U-boat attacked by *Babbitt* was the new Type IXC boat, U.190, and it had been a harsh baptism of fire for her crew. Kapitänleutnant Max Wintermeyer had been trailing HX.229 intermittently for two days without making any attacks. The first depth-charge patterns had been very accurate and had forced U.190 to plunge down out of control to a depth of 310 metres, a depth which was claimed as a record one for U-boats at that time. When the boat had been brought under control by its apprehensive crew, Wintermeyer ordered her up very carefully ten metres at a time. When 180 metres was reached, *Babbitt* attacked again and this caused serious damage – all lights went out and water started to enter the U-boat. The engine-room managed to produce some power from the electric motors and Wintermeyer went deep again. Several more depth-charge patterns were dropped, although no further damage was caused and U.190 was able to surface later in the day, but she was out of this battle and was ordered to base for repairs. It is probable that the U-boat that was detected on the surface by *Babbitt's* radar in the middle of the action was not U.190 but another boat, U.406, which dived when *Babbitt's* starshells were fired and was not directly attacked by the depth charges.

Both Wintermeyer and his First Watchofficer, Oberleutnant Hans-Edwin Reith, are still alive, and it is interesting that it is the opinion of both of them that they had been attacked not by one escort vessel but by a Support Group, this belief being based on the fact that *Babbitt* had approached from the open sea away from the convoy and on the intensity and persistency of the depth-charge attacks, so full marks here for *Babbitt* in attacking so furiously that the Germans believed she was more than one ship.*

The *Mathew Luckenbach*, which had 'romped' from HX.229, had steamed on during the night and drawn forty or so miles ahead of its convoy. The crew had slept well and were pleased with their decision to leave. On the fine morning of the 19th they saw a friendly aircraft and then, on the southern horizon, two small warships. All seemed well.

After eating a good breakfast, the bosun turned my watch to on the lifeboats. They had flooded during the storm and he set us to baling them out. We were busy with this when someone on lookout yelled 'TORPEDOooooo'. Then silence. I stood there frozen for a few seconds. Then I realized that I had not taken my lifebelt with me. I started to make a mad dash to my quarters for it but I took no more than ten steps when the lookouts' shouts became more urgent and excited, now yelling, 'Torpedo. Torpedo. Torpedo.' I realized I would never make it to my quarters. I grabbed on to a guy wire and held on with all my might. There was a deafening roar and I was wrenched from the guy wire by tons of water and tossed head over heels across the deck until I was caught by the chain railing on the starboard side. Just as I felt that the pressure of the water was about to crush me, it subsided.

* U.190 survived the war safely and, under the command of Oberleutnant Reith, surrendered at Halifax, Nova Scotia. She was used as an experimental boat by the Royal Canadian Navy before being ingloriously scuttled in October 1947.

Able Seaman Pat Civitillo, the eager Maritime Service seaman, was about to be shipwrecked on his first voyage.

When he left HX.229, *Mathew Luckenbach*'s master, Captain Atwood H. Borden, had unwittingly steamed into the area of ocean near Convoy SC.122; the two small warships to the south were that convoy's port escort screen. Several U-boats were also in that area searching for the convoy and one of these had sighted this merchant ship steaming innocently along on its own. The lucky Germans were Kapitänleutnant Herbert Uhlig and the crew of U.527, another Type IXC boat on its first patrol; *Mathew Luckenbach* was their first victim. As he had seen patrolling aircraft in the vicinity, Uhlig had made a submerged attack and had nearly been outpaced by the fast merchant ship; his torpedoes had caught up only slowly from astern. These had been spotted by the American lookouts but, although the *Mathew Luckenbach*'s rudder had been put hard-a-starboard, this action had been too late and the ship was hit by two of U.527's well-aimed torpedoes.

Captain Borden ordered distress rockets to be fired and his crew to take to the boats. The Chinese chief steward caused some surprise when he appeared 'with a large laundry bag over his shoulder and dressed in his rubber lifesaving suit. He looked like someone from another planet. We all looked at him in awe.' No one had been hurt by the torpedo explosions and the crew all got away safely in boats or rafts. *Mathew Luckenbach* remained floating, low in the water but on an even keel, and showed no signs of sinking. Her crew now felt very lonely indeed – no convoy, no escorts, no rescue ships.

But the Americans were not destined to be in the water long and owed their safety to a fellow American. When the *Ingham* joined SC.122 she was placed in the port quarter of the screen and was the nearest ship to *Mathew Luckenbach*. Lieutenant John Waters was Officer of the Deck and just happened to catch a glimpse of a column of water on the horizon and then the distress rockets. Commander Boyle received the report and, as it was a quiet period, Boyle sent *Ingham* out to pick up survivors and *Upshur* to cover her. The rescue of the *Mathew Luckenbach*'s crew thus became an all-American affair. The weather was good and *Ingham* had no difficulty with the rescue. As one sailor of *Mathew Luckenbach*'s Armed Guard was hauled over *Ingham*'s side, he caught sight of 'U.S. Coast Guard' stencilled on his rescuer's lifejacket and said 'What in hell is the Coast Guard doing picking me up out here?' *Ingham's* tough bosun's mate grabbed him, turned him round and nearly threw him back into the ocean 'to wait for the Navy to get you'. All the survivors were soon on board *Ingham,* including Chief Mate Willy Heyme who had served in the Imperial German Navy in the First World War and who now commented that the U-boat men were 'a very efficient crowd'.

Mathew Luckenbach was still afloat and her engines were known to be capable of raising steam. Captain Martinson of *Ingham* suggested that the merchant ship's crew be returned to save their ship but there were no volunteers. Lieutenant J. A. Martin, a regular Coast Guard officer, then suggested that he and two other of *Ingham's* officers and a party of men should board *Mathew Luckenbach* and attempt to save the ship. Lieutenant John Waters was one of the other officers.

I will frankly admit that I did not relish any such assignment. Having spent the entire winter in almost continual combat I had no illusions about how safe the North Atlantic was. To have attempted to run some 600 miles with the ocean swarming with U-boats in a ship with two torpedoes already in it would have been little less than suicidal. Even so, I think the captain was giving it some serious thoughts.

Fortunately, Captain Martinson decided against the salvage attempt and *Ingham* returned to her convoy.*

After she had torpedoed the American ship, U.527 spotted the two warships and also a patrolling aircraft. Kapitän-leutnant Uhlig thought it better to remain submerged and wait before finishing off the *Mathew Luckenbach.*

I decided to withdraw underwater to a distance where I could surface. There were, among others, three considerations in my mind – this ship could not escape because she was heavily damaged; the aircraft could protect her only for a certain time then must

---

* Before he could go off watch, Lieutenant Waters had to follow Coast Guard regulations and record the names and addresses of all the sixty-seven survivors in *Ingham's* log. This had two interesting sequels. As a result of having to spend so much time after his watch should have ended, Waters later suggested a new system of log-keeping which was accepted by the Coast Guard and for which Waters received an 'incentive award' of 1,500 dollars. The second result is that thirty-one years later I found the names and addresses in *Ingham's* log and this enabled me to trace four members of *Mathew Luckenbach's* crew.

Lieutenant (later Captain) Waters later wrote an excellent book, *Bloody Winter* (D. Van Nostrand, 1967) on the convoy battles in the North Atlantic during the winter of 1942–3.

return to base; there was some hope that the ship would sink without another attack within an hour or two and I could save my torpedoes.

After spending several hours at a distance of ten nautical miles, I surfaced and saw the ship still on the surface without any escort. So I submerged again in order to approach her and sink her by a *Fangschuss*. When I had her at the right position and distance, only a few moments before my command to fire, she was hit by the torpedo of another U-boat and sank rapidly. I surfaced and so did the other submarine. It was U.523 and Kapitänleutnant Werner Pietzsch was the commander.

In her last moments *Mathew Luckenbach* had performed a useful duty. Not only were U.523 and U.527 queuing up to finish off this ship in the late afternoon of that day, but no less than three other boats – U.336, U.598 and U.618 – were preparing to torpedo this easy target and were disappointed to be cheated by U.523's torpedoes.*

The two convoys had come so close to Britain by the morning of that day, 19 March, that Coastal Command were now able to offer the services of seven squadrons containing three different aircraft types – Liberators, Fortresses and Sunderlands – and a careful plan had been worked out to make the best use of each type of aircraft. The Benbecula-based Fortresses of 206 and 220 Squadrons were to provide the close cover of the convoys from dawn onwards, 206 Squadron being allocated to HX.229 and 220 Squadron to SC.122. At the same time Sunderlands from the squadrons based at Castle Archdale on Lough Erne in Northern Ireland were to take off and fly a 'parallel track sweep' to cover a large area of ocean around the general position of the two convoys. Then, in the later afternoon, Liberators of 86 and 120 Squadrons were to take over the close escort of each convoy and provide cover as far into the night as possible. With the good flying weather and with Dönitz's orders to the U-boats to keep up with the convoys, there was every promise of a rich day for Coastal Command.

The first aircraft of the day to take off from Benbecula was Fortress 'N' of 220 Squadron. The regular captain of this

---

* The two new Type IX C boats that shared the sinking of *Mathew Luckenbach* never gained any further success. U.523 was sunk on 25 August 1943, by H.M.S. *Wanderer* and H.M.S. *Wallflower* near a Gibraltar convoy; Kapitänleutnant Pietzsch and thirty-six of his crew were rescued. U.527's end will be described later. This sinking of one old American merchant ship represents a poor return for the loss to the Germans of two Type IX C U-boats.

crew was Warrant Officer M. Bortoft, but his flight commander, Squadron-Leader W. E. Edser, had decided to fly
this operation. Flight-Sergeant T. E. Kynnersley, the regular
second pilot, had thus been forced out of the cockpit and sat in
the nose of the Fortress.

Benbecula at three in the morning gave an impression of great
cleanliness and of climatological coolth. Winkles on the sea shore,
some rocks, a lot of sand and cold, chilling sea.

After daylight I was twiddling my thumbs, looking at the sea and
noting that the pilot was remaining below descending cloud, i.e. we
flew lower and lower until we passed below the lowest part of a
fairly inactive front which had a classic example of roll-cloud, like a
bolster, detached from the main base, rolling round and round,
stretching away, left and right, as far as we could see. As we went
under, like a limbo dancer, the substantial Yagi (radar) array above
my head, protruding through the nose into the air ahead of the
aircraft, sprouted a great silent, persistent magenta-coloured spark,
fifteen to thirty feet in length. The pilot, startled, gyrated the aircraft in some curious way but then, quite suddenly, we were in clear
skies and sunshine and the spark had gone. We laughed and I said
it was nothing to do with me.

SC.122 lay ahead. From the air, ships in convoy always reminded
me of a herd of grazing cattle – not really going anywhere.

This aircraft found no U-boats around SC.122 but HX.
229's first Fortress was a 206 Squadron aircraft flown by
Pilot Officer L. G. Clark and his crew. The convoy was soon
found and the Fortress was asked by Commander Day to
patrol at a distance of thirty miles. Within only twenty
minutes, the Fortress came upon a squall astern of the convoy
and Clark decided to fly into it, hoping to catch a U-boat
unawares. This move was completely successful and a fully
surfaced U-boat was found there. Clark was able to carry out
an attack before the U-boat had time to dive; four depth
charges were dropped and two explosions were seen on either
side of the U-boat. On its next pass the Fortress crew found
that the U-boat had disappeared, but heavy black gulps of oil
could be seen coming to the surface and soon a large circle of
oil lay above the U-boat's last position.

It had been a perfect attack on this U-boat which had
thought itself safe, at least for a few minutes, in the squall.
There seems little doubt that this U-boat was U.384 and that
it was sunk by Pilot Officer Clark's attack. U.384 failed to
return to base after this patrol and sent no signals after this
day. The U-boat Headquarters War Diary assumes the boat

to have been '*am* 19.3.43 *in AK* 69 *verloren*' – 'lost on 19 March 1943 in AK 69'. The scene of Clark's attack was actually in AL 47, which is the neighbouring square of the U-boat chart to AK 69, and all other attacks on U-boats at this time can be accounted for in the records of other U-boats.

In the absence of survivors from U.384 the reader can only try to imagine the scene in the U-boat, possibly crushed early by the depth charges and the crew swiftly engulfed or possibly forced down into the depths to be crushed more slowly by the pressure of 1,000 feet of sea water. Less than two days earlier Oberleutnant von Rosenberg-Gruszczynski and his men had probably feasted in celebration of their sinking of the *Coracero*; now their boat had become the traditional 'iron coffin' on the ocean bed.*

The next attack by a Fortress can also be linked to a U-boat. The second aircraft to come on patrol for SC.122 was captained by Flying Officer Bill Knowles, and a U-boat was soon spotted on the surface. The Fortress came in to attack; the U-boat was slow to dive and four depth charges ahead of the swirl brought firstly oil, then the U-boat itself to the surface where it lay motionless with its decks awash. The Fortress was too close to attack again and the U-boat managed to dive again while the Fortress was circling to make a fresh approach. The second depth-charge attack was made just as the U-boat disappeared and more oil came to the surface after this. Knowles and his crew had sunk U.633 twelve days earlier and were now hopeful that they had sunk a second U-boat.

The time of this attack coincides exactly with the experiences of Oberleutnant Herbert Engel's U.666, which had torpedoed the *Carras* four hours earlier. The depth charges had caused serious damage and, after the second attack, the boat plunged down to 190 metres before the Germans could bring it under control. U.666 survived, but this was another U-boat that would take no further part in this convoy battle.

Five Sunderlands had taken off from Lough Erne at dawn. The Irish Republic allowed the Sunderlands to fly through a narrow corridor up the valley of the River Erne in the county of Donegal and out over the sandy beach of Bundoran to the Atlantic. This concession, not available to the Liberators at

---

* The names and home towns of U.384's crew are recorded in Appendix 5 alongside the crews of the torpedoed merchant ships.

Aldergrove, saved a fuel-consuming flight around the north coast of Donegal and greatly extended the Sunderlands' effective range. The Sunderland crews also had the locations of various suppliers of fuel in bays on the west coast of Ireland where they could land and refuel with no questions asked, but only if they were in serious trouble.

On this morning the five flying boats set course for their allocated search areas. The intention of these parallel track sweeps was certainly to attack and destroy U-boats if possible but also to force U-boats over a wide area to submerge and prevent them working round to positions ahead of the convoy before nightfall. It was boring work and a Sunderland crew might fly a complete sortie without seeing a U-boat, but the aircraft may have been seen several times by U-boats which had then been forced to dive.

On this day the Sunderlands did more than this and two of the five Sunderlands caught U-boats on the surface. The first attack was by Flight Lieutenant G. A. Church's crew of 228 Squadron who depth-charged U.608 at a position well astern of SC.122 although the U-boat was not damaged by this attack. It was left to Flight Lieutenant Clare Bradley's crew flying one of 423 (Canadian) Squadron's aircraft to have a really exciting flight. A periscope was sighted only five miles behind SC.122 but the Sunderland was too late to attack. Nearby was the merchant ship *Empire Morn* which had straggled from this convoy with boiler trouble and was still trying to catch up. Bradley flew over *Empire Morn* to warn her of the nearby U-boat and then flew back to find that the U-boat had resurfaced. This time Bradley was able to make an attack but only two out of the intended six depth charges released. Bradley stayed in the area for some time, hoping to keep the U-boat away from the *Empire Morn* and finally dropped his remaining depth charges one by one for 'scare effect' before turning for home.

Bradley and his crew had undoubtedly saved the *Empire Morn* from torpedo attack. The U-boat was U.338 which, under the determined Manfred Kinzel, had already sunk five merchant ships in SC.122 and survived a damaging depth-charge attack by *Upshur* and *Lavender*. Kinzel had been getting into an attack position against the merchant ship when first sighted by the Sunderland and forced to submerge. If Bradley's six depth charges had all released on the second encounter, these might well have finished off this U-boat; the two that did drop caused a serious leak in the pressure hull and much minor damage. When the Sunderland finally left, Kinzel tried to catch *Empire Morn* again but, despite thrashing his engines, he could not catch up.* This was one more U-boat out of the battle, but we shall meet Kinzel and his boat, the 'Wild Donkey', once more.

It had been a good morning for Coastal Command with one U-boat probably sunk, two damaged and many more put down. The close cover lasted for the remainder of the day with Liberators taking over from the Fortresses, but there were no more encounters before evening. No torpedo attacks were made on the convoys during that day. It was an effective demonstration of well-planned air cover completely dominating the U-boats.

The convoys had steamed hard all day and made further valuable progress towards complete safety. The only change in their composition was the arrival of the corvette *Abelia* to join the escort of HX.229. *Abelia*'s long haul from St John's of almost 1,200 miles effectively demonstrates the huge area covered by this operation.

That evening, U-boat Headquarters broadcast orders that the coming night was to be the last of the battle. At least fifteen U-boats were still in the general area of the two convoys and available for operations but many of these were now suffering from mechanical defects or had dropped well back because of the intensive air patrols of the day. The weary captains of some boats probably felt in their bones that this convoy battle was now as good as over and may have found good reasons for not pressing too hard. U-boat men tell of the great mental strain of a prolonged operation such as this one. The merchant ship, escort and aircraft crews may have felt that their enemies were invisible, safe and always full of confidence, but a U-boat man, who may not have seen his home base or another friendly U-boat for several weeks, felt immensely lonely and isolated. After the four days of this convoy battle he was

---

* *Empire Morn* was one of the C.A.M. ships (Catapult Armed Merchant Ship) but was not carrying a fighter on this voyage. After this narrow escape on SC.122, she met serious trouble on her next voyage. She had just landed her cargo of thirty-two Typhoon fighters at Casablanca when there was a severe explosion against her stern causing twenty-six deaths. It was never established whether the explosion was caused by a mine or a torpedo. *Empire Morn* was repaired and sailed on until scrapped in 1972.

physically and mentally exhausted, never free from the fear
that he was about to be pounced upon by ship or aircraft and
depth-charged into eternity.

Despite this, Dönitz's orders on this last night met with
some response and both convoys were probably being
reported by evening. Kapitänleutnant Herbert Brünning, in
U.642, had been in touch with SC.122 that morning and,
despite being forced down several times by aircraft, had kept
resurfacing and had managed to catch occasional glimpses of
smoke and follow the convoy's course. When darkness came,
he picked up the convoy on his hydrophones and sent off
further reporting signals. Another boat, U.631 under Oberleutnant Jürgen Krüger, is recorded as being able to do the
same with HX.229 so that this convoy was also being reported before dark but again by hydrophone and not by visual
contact. The weather on this night is recorded as being
exceptionally calm with a full moon and the surface of the
ocean as being a glassy calm. HX.229's escort was further
reinforced during the early part of the night when the British
destroyer *Vimy* at last arrived from Iceland and the American
Babbitt turned up after its long depth-charge encounter with
U.190. HX.229 now had eight escorts in its screen – a far
cry from the position when Lieutenant-Commander Luther
had faced the first night of the battle in the Air Gap with half
that number.

SC.122 also had eight escorts present and also an unexpected bonus in the form of night-time air cover. Flying
Officer Red Esler, who had provided HX.229's first air cover
two days earlier, was now detailed to cover SC.122 for the
last two hours of daylight and to remain with the convoy after
dusk if he thought it possible to do so. Esler had soon used
two of his depth charges on a U-boat caught while diving;
this was U.590 (Kapitänleutnant Heinrich Müller-Edzards)
but no damage had been caused. Visibility was so good that
Esler decided to stay on after dark. *Havelock* was picking up
HF/DF contacts and the Liberator was asked to investigate
these. While Esler was following up one of these leads from
Havelock the following incident occurred; it is described by
Flight Sergeant T. J. Kempton, whose navigator's position
was in the nose of the Liberator.

We were approaching the convoy from behind when I spotted the
wake of a vessel ahead. Then a dark shape became visible. It was

moving fast and seemed to be rather big for a U-boat; in that position it could have been one of the escort ships. As we overtook the vessel, the skipper had to decide whether to attack. I was the only one with an uninterrupted view in the darkness and the identification had to rest with me. The vessel proceeded onwards, quite unaware of the aircraft's presence. If it was a U-boat, it was an excellent target – but supposing it was an escort – we could not risk an attack until we were sure. This was an agonizing decision and we had to err on the side of safety.

We were nearly on the vessel before I was able to positively identify a U-boat and, as the aircraft went in to attack, the bomb doors were still opening. The pilot's press-button release was only operative with bomb doors fully open and the release mechanism failed to work.

So the Liberator roared right over a very surprised U-boat and the four depth charges stayed in the bomb bay. The rear gunner fired five long bursts which scored hits around the conning tower and a flame float was dropped but the U-boat had dived by the time the Liberator came round again. It is not known which boat this was but it had certainly been very close to destruction. Esler and his crew continued this nocturnal guardianship of SC.122 for a further two and a half hours but without further incident. The Liberator signalled its farewell to *Havelock* and 'then back to Aldergrove to meet the dawn'. In their two patrols, one with HX.229 and one with SC.122, this crew had flown for thirty-six out of the past seventy hours, had sighted eight U-boats and made five attacks.*

It was an otherwise quiet night for SC.122. The only action by the surface escorts had occurred before midnight when *Ingham*, patrolling well astern of the convoy, picked up a doubtful Sonar contact and dropped a pattern of depth charges but no U-boat reported an attack at that time. The ships of HX.229 saw no action of any kind. U-boat Headquarters were no doubt disappointed by the lack of success that night and Dönitz and his staff realized that, with the prospect of intense air cover next day, the operation was as

---

* Esler was one of several 120 Squadron pilots who later became test pilots and lost their lives developing new aircraft. Of the pilots who flew in support of SC.122 and HX.229, Pilot Officer (later Flight Lieutenant) A. W. Fraser, R.A.A.F., was killed at Odiham on 4 July 1944, Red Esler flew the Avro 707 – a one-third-scale prototype of the famous Vulcan bomber built to investigate the low-speed flying characteristics of delta-winged aircraft – but was killed when the aircraft crashed near Blackbushe on 30 September 1949, and Squadron Leader Desmond Isted is also believed to have died as a post-war test pilot but I have not been able to obtain details of his death.

good as over. Orders went out that all U-boats in contact with the convoys were to break off at dawn and search back along the convoys' paths for stragglers and other targets of opportunity. It is not known whether any U-boats had made firm contact with either convoy during the night. U.631 and U.642 had continued reporting during the early part of the night but their contact was mainly through hydrophones and both boats finally lost contact. At dawn they turned back to the west. They can probably claim to have been the last U-boats in touch.*

There remained one last act of this battle to be played out. At dawn on 20 March, the aircraft of Coastal Command arrived in strength. Not surprisingly the close escorts of the convoys, provided mainly by the Fortresses from Benbecula, found nothing all day but the eight Sunderlands sent out from Castle Archdale on two separate parallel track sweeps had several encounters with the many U-boats moving away to the west.

The first sighting was by the 201 Squadron aircraft of Flight Lieutenant Dudley Hewitt and is well described by the flight-engineer, Sergeant J. H. Bunton.

The sub had been fully surfaced and on hearing our engines he made to crash dive but was still very much visible when we arrived above him. Six depth charges were dropped and these had the effect of blowing him back to the surface with the bows in the air at an angle of 45 degrees.

We circled, with the wing-tip almost touching the waves, firing the odd burst at the conning tower. After a minute or two, the sub settled on an even keel and the hatch was opened, crew members endeavoured to get out and reach their cannon on deck; these unfortunate souls were quickly dealt with by our gunners. Three were seen to fall into the sea.

After a further minute or so, no movement was visible from the conning tower so we made a wider circle and prepared to run in to kill. Our remaining charges completely straddled the sub; it was, of course, a sitting duck. When the turmoil of water had subsided, we saw bodies and oil but only for a few seconds; the wind and rain made it impossible for us to see any more.

Radio silence was broken during the engagement and, when it was over, we asked if we should return to base as we then had no

* U.642's crew turned out to be a lucky one. The boat was bombed by the U.S.A.A.F. at Toulon on 5 July 1944 and damaged beyond repair but only two men were killed. The crew, still under Kapitänleutnant Brünning, then took over a new boat, U.3518, and served until April 1945 when the boat was damaged by a mine in the Baltic but managed to make Kiel without casualties.

depth charges. The reply was that we were to continue with the patrol. The navigator said to the wireless-operator, 'Ask them if they expect us to throw the bloody anchor at the next one.' In the excitement, the wireless-operator started to send this message before he realized what he was doing.

Flight Lieutenant Hewitt's crew were later credited with a kill but this was not so. By a rare coincidence, two U-boats had been involved. The first was Kapitänleutnant Herbert Uhlig's U.527, which had torpedoed the *Mathew Luckenbach* the previous day; U.527 was damaged, but not seriously so, by Hewitt's first salvo of depth charges and had dived safely out of the way. At that very moment a second U-boat, Kapitänleutnant Gottfried Holtorf's U.598, surfaced and was attacked in the manner described by Sergeant Bunton. The Sunderland's machine-gun fire had scored many hits on the U-boat's upper works but none of the German seamen were killed and the depth charges caused no serious damage. The times of the two attacks coincide exactly in 201 Squadron's and in German records.*

One hundred miles away a 423 Squadron Sunderland had also found a target. The flying boat was manned by the part-Canadian crew of Flying Officer Brant Howell. German sailors were seen scrambling back into the conning tower as the Sunderland attacked and five depth charges straddled the U-boat before it had completely submerged. A patch of oil appeared on the surface above the spot where the U-boat had dived. Howell's crew were delighted with their work and believed they had seriously damaged, if not destroyed, the U-boat, but this was not so. This was U.631 which was badly shaken up but survived.

The first clash of the battle for the two convoys had occurred nearly four days earlier and 600 miles to the west when Oberleutnant Bertelsmann in U.603 had torpedoed the *Elin K.* Now, at a position well behind the two convoys at 15.54 Convoy Time on 20 March, the last engagement of the

---

* Both of these U-boats were later disposed of by U.S. Navy aircraft, again by a rare coincidence on the same day, 23 July 1943. U.527 was sunk by aircraft of U.S.S. *Bogue* near the Azores and U.598 off Brazil by U.S. Navy Patrol Squadron 107. There were thirteen survivors from U.527 including Kapitänleutnant Uhlig but only a junior officer and one rating from U.598.

Flight Lieutenant Dudley Hewitt received a D.F.C. for this attack, but was killed flying with 228 Squadron over the Bay of Biscay in June 1944 while attacking a Flak ship.

battle was destined to take place between Sunderland 'T' of 201 Squadron piloted by Flying Officer W. C. Robertson, a Scot, and the Type VIIC U.631 captained by Oberleutnant Jürgen Krüger. Squadron Leader D. G. T. Hayes, the squadron Gunnery Leader, was in the Sunderland's front turret when the U-boat was sighted.

It showed no sign of diving and the captain said, 'These depth charges aren't going to be much good if he stays up.' They were designed to crush a submerged U-boat, not to destroy a U-boat on the top. We weren't going to go away without attacking it but we had no offensive weapons, just these old dustbins of depth charges. They had heavier calibre guns than we had and the Sunderland, at its slow speed, was a good target for them. But we couldn't leave it and we decided to have a go with the depth charges. We dropped down to 150 feet and commenced the run in.

I had done about 1,000 hours of terribly boring flying with Coastal Command and then, when it did happen, it couldn't have been a more exciting or frightening situation. I was face to face with him and I thought that he had as good a chance of hitting us as we had of hitting him. I saw the flash of his gun. I opened fire at about 400 yards and could see my fire hitting the water and then striking around the conning tower but I don't know whether I hit anything. The Germans were looking up at me as we passed over and I was firing down at them. I could see their white faces – they looked petrified. I suddenly realized that the enemy was human and the fact that they stayed up on top and kept firing was a badge of courage.

Six depth charges were dropped and the U-boat was entirely covered by the water of the explosions. When this cleared, the Sunderland crew could see the U-boat's bows high in the air and then it disappeared stern first.

In this instance the U-boat crew had stood their ground and fired on the attacking Sunderland out of desperation. Krüger had wanted to dive but the conning tower hatch had jammed and the gunners had been forced to carry on until this was cleared. The disappearance of the U-boat stern first was the result of a somewhat frantic dive just as the blockage was cleared. Even so one and a half tons of sea water came down the conning tower. This was pumped out and some slight damage repaired after the Sunderland had left the scene.

The Sunderlands flew home after a successful day's work. The battle for Convoys SC.122 and HX.229 was over.

14

Distant Operations

While the two convoys were steaming the last part of their voyages in safely, it is an appropriate time to look briefly at what had been happening elsewhere in the North Atlantic and at relevant events in other waters.

When SC.122 and HX.229 had been diverted to the east and away from their original route on account of the Raubgraf U-boats six days earlier, the third convoy, HX.229A, with its cargo liners, tankers and fast freighters, had been diverted to the north. Almost immediately this convoy had run into blizzards and fog and then, as the convoy steamed even further north across the Davis Strait between Canada and Greenland, it encountered a frightening hazard – ice. For three days the merchant ships and escorts 'meandered through ice fields, growlers and island-like bergs'. Convoy formation was lost and the escorts were kept very busy; at night they illuminated the icebergs with their searchlights.

Among the first ships to be damaged by the ice were two of the escorts – *Waveney* and *Aberdeen*. Lieutenant-Commander A. E. Willmott, R.N.R., captain of *Waveney*, can describe both incidents.

It would appear that the ice was much further south than usual at that time of the year. During the night (of 18/19 March) the shell plating of both port and starboard bows of *Waveney* was pierced on and below the waterline, most fortunately forward of the collision bulkhead (the first watertight partition abaft the stem), and the shell plating along both sides for more than half-way aft was deeply indented between each frame but without any fractures. This was the result of ice pressure as she pushed slowly forward. I well remember when *Waveney* lifted her bows on a swell seeing the water pouring out and filling up again when she dipped her bows down.

Shortly after dawn on the 19th, *Aberdeen* was sighted to starboard 'aground' on an underwater ledge of a large, irregular shaped iceberg; she was heeling slightly to starboard, had ice on the foc's'le forward of the guns, and was down by the stern. I closed her with a view to rendering assistance and passing a towline across but was instructed to remain clear in case of other underwater ice ledges and to stand by while she made attempts to refloat. Eventually, after much effort with engines, *Aberdeen* slid stern first off the ice ledge. Although her bottom was damaged and the Asdic dome shattered, she floated and we set off in company to rejoin the convoy. During that same day, I saw three or four merchant vessels returning towards St John's because of ice damage and I seem to remember learning, much later, that they all got back safely except one.

Although the convoy course was altered more eastwards to try to get out of the ice, we continued pushing through it and, at dawn on the following morning, a huge flat-topped iceberg was sighted off *Waveney*'s starboard bow and I had to haul carefully out to port so as to give it a wide berth. After manœuvring clear, it was realized that the Starboard Bridge Lookout had never reported it. The first lieutenant ascertained he *was* awake and questioned the man but he insisted he could not see the iceberg, even when it was pointed out to him – only 500 yards away, high out of the water and long – it was incredible! On being shown the Bridge Standing Order Board, he claimed not to see that also. The doctor could find nothing wrong with the man; he was a problem all the way home and had to be carefully watched.

Five merchant ships had turned back for St John's with ice damage but one vessel was in more serious trouble. The large ex-whaling factory-ship *Svend Foyn*, at 14,795 tons the second largest ship in the convoy, was the leading ship in one of the columns. In the early morning of 19 March *Svend Foyn* steamed straight into an iceberg. She was badly holed and had to shut down her engines but remained afloat and rescue tugs were requested. The drama of the *Svend Foyn* lasted two more days and it is well told here by Lieutenant-Commander L. B. Philpott, R.N.R., whose sloop H.M.S. *Hastings* was detailed to stand by the damaged ship.

The two ships remained beset by icebergs which at times drifted perilously close to *Svend Foyn* (they were drifting faster than she was owing to their far larger draught). It was towards the end of the day that I learned that no tug was available to tow *Svend Foyn* in but that two vessels had been detailed to her for life-saving only; they hoped to arrive late the next day. Under no circumstances were they to tow.

The night brought a full gale but at daylight *Svend Foyn* was still riding it well and was no lower in the water. The wind fell away almost to a calm; the glass was high and steady and the sea was calm though, of course, a sizeable swell was running. I decided that it would be a useful exercise and good for morale to send a boat across to *Svend Foyn*. This was done with my R.N.R. lieutenant in charge. My officer conveyed the news of the possible arrival of rescue ships that evening though the captain of the *Svend Foyn* thought he could be towed in. However, the crew were apparently ready to leave, bags packed and so on. There was no light or heat aboard the *Svend Foyn*. All in all, my boat's crew were glad to be back. I was astonished and not a little perturbed to learn that *Svend Foyn* carried not only her own crew but a large number of D.B.S. (Distressed British Seamen).

Two U.S. rescue ships arrived at dusk and went alongside the *Svend Foyn*. It grew dark and they cast off but I was staggered to find that not one man had left the stricken ship. I am afraid the master was obsessed with the idea of being towed in though he had been told that these ships were for rescue only. But he would not let anyone leave. My final signal to him before night was a last effort to persuade him to leave and get aboard the U.S. ships, 'that is what they are here for'. But he was adamant and his last signal that night was 'Impossible to do anything until morning'.

A gale sprang up during the night and, quite suddenly, she sank. His last signal was to call us all back. She went so suddenly that we hardly knew she had gone for there were still many icebergs around and, in the conditions obtaining, she was not so very unlike an iceberg. The best we could do then was to save as many lives as we could. There were many lifeboats down, some with only a few men in them, some with many but there were also dozens of red specks in the water – they were the red lights on the personal life-belts. They must have just floated off the deck of the ship as she went down. But, as daylight came, the extent of the tragedy became apparent – a very large patch of oil fuel and many a corpse floating in it.

We spent many hours searching the scene and examining for more signs of life but it was all like a nightmare (and still is) and, somewhere about noon, we parted, I to go to Londonderry and they to their home port, each with our rather pathetic groups of survivors.

There had been 195 men aboard *Svend Foyn* when she sank and the rescue ships had picked up more survivors than Lieutenant-Commander Philpott suspected. The American rescue ships were the U.S. Coast Cutters *Modoc, Algonquin, Aivik* and *Frederick Lee* from the Greenland Patrol. *Modoc* did a particularly good job in rescuing 128 men from a total of 152 who were picked up in all. The dead were composed of British, Norwegians and Lascars and included the master, Captain Frank Thompson; their ages ranged from the fifteen-year-old galley boy from North Shields to a sixty-three-year-old seaman from India. It is ironical that these men, who might have been worrying about the danger of fire in the event of a torpedo hitting *Svend Foyn*'s cargo of high-flash-point fuel, met their death through an iceberg.

Convoy HX.229A made good progress after getting clear of the ice. The U.S. Coast Guard Cutter *George M. Bibb* came down from Iceland to take the place of *Hastings* in the escort. The convoy passed within about forty miles of U.229, the solitary U-boat that Dönitz had stationed off Greenland as a weather ship and also to watch out for the German blockade runner expected from Japan, but HX.229A was not seen by the U-boat. The reader may remember that a fresh group of U-boats, the Seeteufel group, was being assembled south of Iceland in a patrol line to search for the westbound convoy ONS.1 and that these U-boats would also be near the area through which HX.229A would steam. The Admiralty, by means which will be discussed later, knew of the Seeteufel U-boats and neatly diverted HX.229A round the patrol line. There were no more incidents and the ships, the crews and the many passengers in this convoy would arrive in Britain on 25 March having made a longer journey than expected due to the big diversion to the north and the convoy's experiences with the ice.

The arrival of HX.229A with just one ship lost, and that not due to U-boat action, was greeted with great relief by the Admiralty and by Washington. This was the first North Atlantic convoy to arrive in Britain without being found by U-boats after the last five such convoys had all been attacked.

There still remains one merchant ship unaccounted for from Convoy SC.122. The old Cardiff cargo ship *Clarissa Radcliffe*, loaded with iron ore, was one of several ships that had straggled from the convoy during the gale on the second night out from New York. All the other stragglers either put into port or caught up the convoy but not the *Clarissa Radcliffe*. Two days after the gale she was seen by H.M.C.S. *The Pas* eighty miles from Halifax and only an hour's steaming from SC.122. After *The Pas* gave her a course to steer to catch up the convoy, *Clarissa Radcliffe* was never seen again, at least not by Allied eyes.

On the afternoon of 9 March, U.621 (Oberleutnant Max Krushka) sighted a 6,000-ton merchant ship (*Clarissa Radcliffe* was 5,754 tons) drifting in the position AJ9346 of the U-boat map and, in several attacks on the ship, fired a total of eight torpedoes of which only two hit. U.621, which was on duty in the Raubgraf patrol line, saw this ship several times more in the next two days but refused to waste further torpedoes on it and finally watched it sink on the morning of 12 March in AK 7198. Post-war researchers decided that U.621's victim was the *Clarissa Radcliffe*, it being assumed that she had been abandoned by her crew before Oberleutnant Krushka sighted her, and this theory was accepted, although with reservations, by the Admiralty and by Lloyds. But AJ 9346 (53.15 degrees North, 41.05 degrees West) is 1,060 sea miles from the position off Halifax at which *The Pas* spoke to *Clarissa Radcliffe* on the morning of 9 March. U.621's merchant ship attack that afternoon must have been on another ship, probably a derelict.

There is another U-boat report that might be more relevant. U.663 (Kapitänleutnant Heinrich Schmid) sailed from Brest on 10 March and was ordered to Gruppe Seeteufel off Iceland. While steaming north on the afternoon of 18 March, U.663 sighted a 6,000-ton cargo ship steaming due east. Schmid fired three torpedoes from a submerged position which missed and then a single shot which struck the ship. Explosions and sinking noises were heard.

No merchant ship was reported lost or overdue from this position and the reference books do not give Schmid credit for a sinking. It can only be conjecture that *Clarissa Radcliffe* failed to find her convoy off Halifax and that her master, Captain Stuart Finnes, instead of turning back decided to take his chance by sailing directly to England. The position at which U.663 claimed its sinking, AK 9655 (52.20 North, 27.10 East) is only slightly south of the Great Circle route and coincides with the distance from Halifax that *Clarissa Radcliffe* might reasonably have made. If this ship was *Clarissa Radcliffe*, her crew had nine more days of life than has previously been estimated, but the fifty-five men aboard – merchant seamen from South Wales, Shetland and West Africa and twelve Royal Navy gunners – still perished in the North Atlantic. Sink on sight and no effort to help survivors – a far cry from the elaborate rules for submarine warfare of 1939.

The men who may have sunk *Clarissa Radcliffe* did not survive much longer. Coming out from Brest on her next patrol, U.663 was caught by a Coastal Command Halifax – sunk on sight and no survivors.

Before returning to SC.122 and HX.229 it might be of interest to see what had been happening elsewhere in and near the North Atlantic while these two convoys had been under attack. There had been no major engagements and, although the minor actions may have been of great and sometimes tragic importance to the men concerned in them, they were really no more than the small change of the naval war.

The concentration by Dönitz on SC.122 and HX.229 had helped several other Allied convoys. While the recent battle had been raging, three westbound convoys – ON.170, ONS. 171 and ON.172 – had reached safe waters off Newfoundland with 101 merchant ships and no losses and two more – ON.173 and ONS.1 – were well on their way. The next convoys of the SC. and HX. series had just sailed from New York but had yet to face the dangerous waters of the middle North Atlantic. One ship to cross safely had been the 83,675-ton Cunard liner *Queen Elizabeth* which was carrying 15,000 passengers, mainly American troops, from New York to Greenock. Her great speed of 28 knots and a diversion to the north had kept this mass of humanity well away from U-boats.

But the Germans had found and attacked one Allied convoy; this was UGS.6, a slow convoy of ships carrying military supplies from New York to Casablanca for the American armies fighting in Tunisia. The B-Dienst had been able to get this convoy's complete future route and Dönitz had sent twelve Type IX U-boats in two groups to trap and attack it. The convoy contained forty-five merchant ships with a strong escort of seven American destroyers. It was found and reported but the shadowing U-boat, U.130, was sunk by the destroyer U.S.S. *Champlin*. Three ships in the convoy and one straggler were sunk in the resulting battle. The water in this area, near the Azores, was so warm that only three crew members lost their lives from the three ships sunk in convoy.

One U-boat had been well away from all the convoy battles. Oberleutnant Robert Schetelig's U.229 had been off Greenland reporting the weather and looking out for the German ship *Regensburg*, the blockade runner that was expected in that area. The *Regensburg* was not met and U.229 had to remain on this cold station for three lonely weeks during which she found and sank one neutral Swedish ship.* *Regensburg* was one of four ships with scarce cargoes from Japan expected by the Germans at that time but they were not very

* By coincidence it was off Greenland that U.229 was rammed by H.M.S. *Keppel* in a convoy action on 22 September 1943. She sank with all hands.

lucky. The *Doggerbank* was sunk in error by a German U-boat, U.43 (Oberleutnant Hans-Joachim Schwantke), on 3 March near the Canary Islands; then the *Karin* was caught and sunk by American warships near Ascension Island and the *Regensburg* was eventually caught by H.M.S. *Glasgow* north of Iceland and sunk. The fourth blockade runner, the Italian ship *Pietro Orseolo*, was torpedoed by an American submarine but managed to reach Bordeaux.

Another cat-and-mouse game was at that time being played by the American submarines of Submarine Squadron Fifty, based in Scotland, in a secret operation to catch the German tanker U-boats which were known by the Allies to be of great importance to U-boat operations. Early on 21 March, U.S.S. *Herring* was on patrol off the north coast of Spain trying to catch a tanker U-boat suspected to be coming into its base at Bordeaux. In the moonlight *Herring* saw a U-boat running fast on the surface and, in a good attack, hit her with two torpedoes; the U-boat sank at once. This was U.163, not the tanker U-boat but a veteran Type IXC boat five days out from Lorient. The boat's captain, Kurt-Eduard Engelmann, had just been promoted to Fregattenkapitän, an unusually high rank for a U-boat captain. The supply U-boat that *Herring* was looking for might have have been the Type XIV U.461 which came safely into Bordeaux one day later.

Perhaps the greatest Atlantic drama apart from SC.122 and HX.229 had been enacted just south of the Equator where the fast liner *Empress of Canada*, 21,517 tons, was torpedoed on 14 March while sailing independently from the Middle East to England via the Cape route. Distress signals were picked up and four Royal Navy ships – *Corinthian*, *Boreas*, *Crocus* and *Petunia* – were sent to carry out one of the great sea rescues of the war. The *Empress of Canada* had been carrying 1,892 people when hit, 499 of whom were Italian prisoners-of-war. After a prolonged and difficult rescue operation 1,519 survivors were picked up, of whom some later died of shark bite or exposure. The corvettes *Crocus* and *Petunia* each saved 348 people which probably stood for all time as a record for a corvette rescue.* The submarine that sank *Empress of Canada* was ironically an Italian boat, the *Leonardo da Vinci*, which was herself sunk two months later off the Azores by H.M.S. *Ness* and H.M.S. *Active*.

* H.M.S. *Crocus* was later used for the filming of *The Cruel Sea*.

The Aftermath

When the U-boats withdrew from the battle at dawn on 20 March, the two convoys that had been subject to their attacks were left with a three-day voyage of 450 miles to the first landfall. Lookouts, radars and Asdics continued to work hard but as the hours unfolded with no further action the sailors soon came to realize from long experience that the U-boats had gone. There were thoughts of home and leave for the British, curiosity for the Americans making their first Atlantic voyage, great relief for all after days of tension, discomfort, irregular meals and lack of sleep. 'Chaps would be giving haircuts to their chums, whistling, skylarking and looking to the future and not giving a thought to the weeks now behind them. The whole ship bore a carefree air.'

But there were deeper thoughts in some minds. A radio broadcast had been picked up in which Winston Churchill had spoken of 'a struggle taking place at this very moment between two homeward-bound convoys in the North Atlantic against far superior odds'. Men had seen ships sunk and wondered if friends on those ships had survived. Some officers would already be discussing whether certain decisions taken in the heat of the battle had been correct. The captains of escort vessels retired to their cabins and composed Reports of Proceedings, probably cursing this paper war but perhaps wondering whether historians would one day study their paragraphs. The Convoy Commodores were also composing their reports for the Admiralty Trade Division; Commodore Mayall finished his report on HX.229 with the dry comment that 'apart from U-boat attacks, the voyage was fairly average'.

The last days of the voyage were anything but 'fairly average' for those ships who had taken survivors aboard. Eleven ships were carrying more than 1,100 survivors between them; over 800 of these survivors were in small naval vessels, and the remainder were in the rescue ship

Zamalek which had 165 survivors and in the merchant ship *Tekoa* which had saved 146 men. In all of these ships except *Zamalek* there was great strain on the food and accommodation of the 'hosts'. Bully-beef stew and hard biscuits were made much use of, but the survivors on the Belgian *Godetia* much appreciated their food cooked in continental style. In another ship one survivor was so hungry that he exchanged a watch that he had recently bought in New York for a bar of chocolate. Accommodation was so short on *Anemone* that some men had to sleep in sheltered corners of the upper deck.

Most survivors recovered quickly from the shock of being torpedoed, and there was ample opportunity for men of various ships and different countries to study each other's behaviour in these cramped and trying conditions. Many reports give best marks to the Dutch and Norwegian survivors who were always found to be hardworking and helpful, but the same could not always be said for some of the British and Americans. The captain of *Pennywort* had trouble with some American survivors who demanded overtime rates for working; he locked them in an ammunition store until they became more agreeable, but he found that the Norwegians and Dutchmen worked so well that '*Pennywort* became the cleanest ship in the Royal Navy'. One party of 'filthy-oil-soaked survivors' from a British merchant ship refused to clear up their quarters on *Anemone* until they were guaranteed payment – pay for a British crew stopped when their ship was sunk. *Anemone*'s first lieutenant ordered that these men were not to be fed until the quarters were cleaned. An American survivor in *Volunteer* brought 'howls of amusement when he asked where he could find the soda fountain and the laundry'.

Isolated incidents of bad feeling should not obscure the fact that most survivors were extremely grateful to the men who had saved them. The Lascars in *Mansfield* 'showed their gratitude with dog-like devotion'. The survivors of three ships aboard *Godetia* signed letters of appreciation for the Belgian crew; these signatures helped me to find several of the survivors thirty years later. When one of *Mansfield*'s sailors was lost overboard on the voyage home, Dutch survivors aboard made a collection for his family and several other ships later received substantial sums of money and

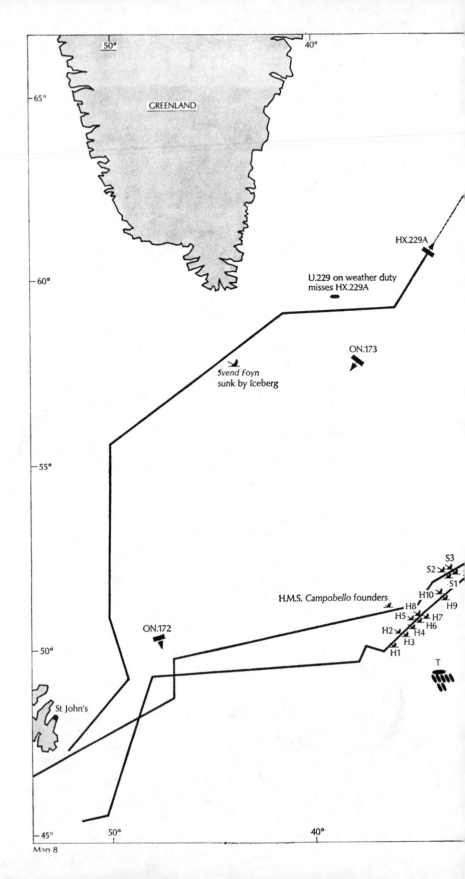

GREENLAND

50°   40°

65°

HX.229A

U.229 on weather duty
misses HX.229A

60°

ON.173

Svend Foyn
sunk by iceberg

55°

S3
S2
S1
H10
H8   H9
H.M.S. *Campobello* founders   H5   H7
H6
H2   H4
ON.172   H3
50°   H1

T

St John's

45°

50°   40°

Map 8

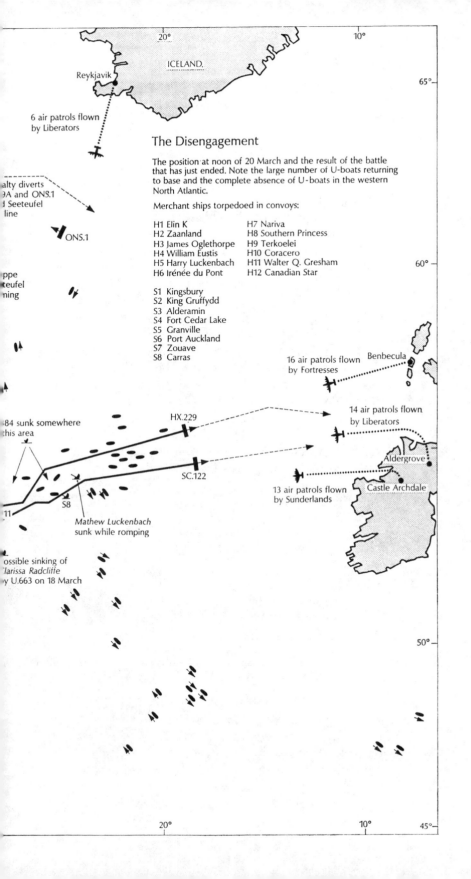

20°   10°

ICELAND.

Reykjavik   65°—

6 air patrols flown
by Liberators

The Disengagement

The position at noon of 20 March and the result of the battle
that has just ended. Note the large number of U-boats returning
to base and the complete absence of U-boats in the western
North Atlantic.

alty diverts
9A and ONS.1
d Seeteufel
line

ONS.1

Merchant ships torpedoed in convoys:

H1 Elin K	H7 Nariva
H2 Zaanland	H8 Southern Princess
H3 James Oglethorpe	H9 Terkoelei
H4 William Eustis	H10 Coracero
H5 Harry Luckenbach	H11 Walter Q. Gresham
H6 Irénée du Pont	H12 Canadian Star

60° —

ppe
teufel
ning

S1 Kingsbury
S2 King Gruffydd
S3 Alderamin
S4 Fort Cedar Lake
S5 Granville
S6 Port Auckland
S7 Zouave
S8 Carras

16 air patrols flown   Benbecula
by Fortresses

84 sunk somewhere
this area

HX.229

14 air patrols flown
by Liberators

Aldergrove

SC.122

13 air patrols flown   Castle Archdale
by Sunderlands

S8

Mathew Luckenbach
sunk while romping

11

ossible sinking of
larissa Radcliffe
y U.663 on 18 March

50° —

20°   10°   45°—

other gifts from men whose lives had been saved. Survivors of the *Canadian Star* and the *Nariva* presented every man on *Anemone* with a suitably inscribed silver ashtray; these are still much-valued souvenirs of the convoy battle.

The voyage continued in calm weather. Late on the 20th a last reinforcement turned up for HX.229; this was the Canadian corvette *Sherbrooke* which had been delayed at St John's with repairs and finally caught up when HX.229 was 1,440 miles from Newfoundland! Several escorts with large numbers of survivors on board were allowed to leave their convoys early. In this way *Anemone*, *Pennywort* and *Volunteer* all left HX.229, leaving the faithful old four-stacker *Beverley* as the only escort to remain with HX.229 from the scratch group that had originally joined from St John's. *Tekoa* and *Zamalek*, with their loads of survivors, were also allowed to steam on ahead. The three American ships, *Upshur*, *Ingham* and *Babbitt*, were released and the crews of all three were delighted to be ordered to English ports instead of to Iceland as they had expected. Although they did not know it, these Americans were finished with the North Atlantic. At the recent Atlantic Convoy Conference Admiral King insisted on withdrawing all U.S. Navy escorts from the North Atlantic trade convoys because he did not like 'mixed' groups; it is probable that he did not like his ships serving under British command. In return the U.S. Navy took over protection of convoys on the Dutch West Indies–U.K. route. The men on *Upshur*, *Ingham* and *Babbitt* therefore had Admiral King to thank for their transfer to warmer and calmer waters.

On 22 March, HX.229 and SC.122 both came into sight of land when Inishtrahull lighthouse off Northern Ireland was spotted. Commodore Mayall had a signal flashed to the merchant ships of HX.229;

CONGRATULATIONS. WE MADE IT. THANKS TO ALL OFFICERS AND CREW MEMBERS.

Local escorts came out from Londonderry and all the ocean escorts left for a fast run in to their respective bases. Sections of merchant ships started leaving both convoys to proceed to various ports. In one of the coastal convoys resulting from the splitting up of the main convoys, two of the SC.122 ships, the ex-Italian *Carso* and the Swedish ship *Atland*, were in the same column; they had been neighbours in SC.122 all the way

The brains behind the U-boat campaign. Grossadmiral Karl Dönitz (centre) in the B.d.U. Operations Room. On the right is Konteradmiral Eberhard Godt, Chief of the Operations Division, and on the left is Kapitänleutnant Adalbert Schnee, A1 Ops. Schnee was a successful U-boat commander of the early war years and returned to sea in the last months of the war.

A thoughtful Wren at work on the North Atlantic plot.

Admiral Sir Max Horton in the Operations Room at Derby House, Liverpool, with part of the North Atlantic plot as a background. The symbols of the eastbound Convoy ON.150 date the photograph as early December 1942, immediately after Horton had taken over command of Western Approaches.

German torpedo man checks up on one of 'eels'. Note the boxes of French apples ...ed in the torpedo room.

...S.S. *Babbitt* nearly gets a U-boat.
...ndman 1st Class Gagne and Steward's
...te Julius Caesar Alexander display the
...vels they trailed in the oil which came up
...r the prolonged series of depth-charge
...acks on U.190 near Convoy HX.229.

Torpedoed. British merchant seamen
survivors on one of the emergency liferafts
that all merchant ships carried.

Saved. American Coast Guard men clean up
an oil-soaked survivor they have rescued.

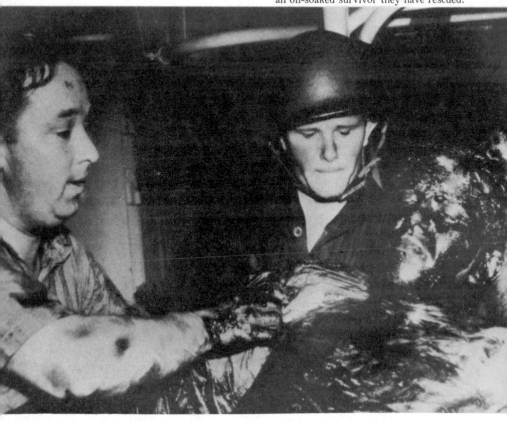

The Liberator. This 120 Squadron aircraft was flown on 18 March 1943 by Flying Officer J. K. Moffat's crew in support of Convoy SC.122 and made a depth-charge attack on U.305.

The Fortresses of 206 and 220 Squadrons, operating from an airfield on remote Benbecula, provided useful cover for the convoys and one of them sank U.384 near Convoy HX.229.

This Sunderland, with Flight Lieutenant N. T. Harvey's crew, flew in support of Convoys SC.122 and HX.229 on 20 March 1943 and was the only aircraft of 201 Squadron not to attack a U-boat on that day.

This particular Sunderland was known affectionately as 'Old Uncle' being 'always in scrapes but always coming back O.K.' but her luck ran out when she disappeared while on a training flight in 1945.

A salvo of depth charges explodes around a
U-boat caught on the surface by a Coastal
Command aircraft; the tail gunner is also
machine-gunning the U-boat.

Survivors of the *Canadian Star* on board the
corvette *Anemone*. Wing Commander
Wrigley, R.A.F., whose wife died after
being rescued, with his daughter Maureen.

Chief Officer Percy Hunt, Lieutenant-Colonel Crouch, Royal Tank Regiment, and Miss Pat Dean (or Deane). This young lady was dragged out of a lifeboat by her long hair.

H.M.S. *Beverley* follows H.M.S. *Highlander* into Londonderry after handing Convoy HX.229 over to local escorts. This may have been the last photograph taken of *Beverley*.

U.S.S. *Babbitt* comes into Londonderry after escorting Convoy HX.229. She is passing H.M.C.S. *Amherst* (K.148) and other escorts of C4 Canadian Group. *Babbitt*'s fo'c'sle party have heaving lines ready for mooring.

American survivors of the Liberty ship, *Walter Q. Gresham*, which was torpedoed in Convoy HX.229, at the Royal Yacht Club, Gourock. At the piano is A.B. Richard Smith from Philadelphia and standing left centre is Wiper Henry Carmichael of St Louis and right centre is A.B. Eugene Swift of Sacramento; the names of the two outside men are not known. They all have recently-issued survivors suits and shirts.

The U-boats come into their Biscay ports. Kapitänleutnant Manfred Kinzel brings U.338 – the Wild Donkey – into St Nazaire after a successful first patrol. His success flags show four merchant ships sunk, all from SC.122, and one Coastal Command Halifax shot down in the Bay of Biscay.

Sergeant Taylor, the Australian flight-engineer of the 502 Squadron Halifax that U.338 shot down, says goodbye to his captors. Kapitänleutnant Kinzel is on the right. The U-boat man with the bandaged hand is Obergefreiter Höhne whose finger had turned septic after being caught in part of the engine-room machinery; he had become known to his shipmates as 'Stinkfinger'.

Kapitänleutnant Heinrich Müller-Edzards of
U.590 also arrives at St Nazaire after the
convoy battle to be greeted by girls and
flowers. His boat had no success against
SC.122 or HX.229 but had claimed two
hits on ships in HX.228 earlier in the patrol.

Crew members of U.618 receive their mail
on arrival at St Nazaire.

The Tower Hill Memorial, London. Here
are commemorated the names of the 37,701
British merchant seamen lost at sea in two
World Wars.

The American Memorial, Cambridge,
England. The names of 1,371 U.S. Navy
men and 201 U.S. Coast Guard men lost in
the Battle of the Atlantic are inscribed on
the memorial wall on the right of this
photograph. The memorial is alongside the
American Military Cemetery near Cambridge.
The greater number of American merchant
seamen lost at sea have no specific memorial
on which their names are recorded.

across the Atlantic. The local convoy went round the north of Scotland and just off Peterhead, during a very dark night, *Carso* rammed the *Atland* which was loaded with iron ore and sank at once with some loss of life. The *Carso* was not seriously damaged.

Convoy HX.229 broke up on 23 March and on the following day Captain S. N. White, Commodore of SC.122, brought *Glenapp* and the other Mersey-bound ships to the Liverpool Bar Light Vessel and ceased to control his convoy when a Liverpool pilot came aboard. Captain White recorded precisely in his report that this was 18 days, 10 hours and 10 minutes after weighing anchor in New York's Hudson River.

The first of the convoy ships to have reached harbour had been the destroyer *Mansfield* which made such an economical run after leaving HX.229 that she was diverted from Londonderry on to Gourock on the Clyde in order to land her 130 survivors. Lieutenant-Commander Hill found orders waiting for him.

We arrived at 03.00 on Sunday 21 March and put the survivors ashore. There was a message there for me to report to the C.-in-C. at Liverpool 'immediately'. I was tired and apprehensive, after all we were not even a Western Approaches ship but Newfoundland-based. I arrived at Liverpool about noon that day and went straight in to Admiral Horton with my Report of Proceedings. I can see his red pencil now underlining the important points. His first comment was about *Mansfield's* defective radar and the lack of spares. He was also concerned at there being no rescue ship but he didn't say much about the convoy losses; he knew we had been short of escorts. I remember his final remark: 'I hope this will be the last picnic the Germans will have. I hope to have the escort carriers in a fortnight.'

Gourock was the main reception port for survivors and most of the shipwrecked sailors from SC.122 and HX.229 were landed there over the next few days. There were many heartfelt farewells and thanks for lives saved. There was a heartrending scene when *Anemone* landed a lady survivor from *Canadian Star*. She had seen her husband drown but hoped that her daughter had been saved by *Pennywort* which also had *Canadian Star* survivors on board. When her daughter was not found on *Pennywort* the woman's last hope went; she stood weeping piteously among the crowd of merchant sailors coming ashore.

The experienced staff of the Gourock Rescue Station soon dealt with the hundreds of survivors. Some were looked after on an old Clyde paddle-steamer, the *Queen Mary II*, and others at a converted warehouse. All were fed, bathed, given money and issued with a complete set of clothes. Many men speak of the kindness with which they were greeted by the people of Gourock. The naval authorities allowed press reporters to interview the civilian survivors from *Canadian Star* but no interviews were allowed with the Merchant Navy men. This resulted in descriptions in the national press on the following morning which stressed the sufferings of the women and children and the violence of the weather but the press reports were surprisingly restrained in their comments on the U-boat men.

A special train was run south from Gourock and the men assembling at the station were ordered not to tell anyone who they were or where they were from but 'as we were 200 men all kitted out with identical suits and white mackintoshes, it was quite obvious we were survivors '. Once aboard the train another man noticed that even though officers and men had shared liferafts in desperate conditions only a few days earlier, 'the return to normal status was immediate' when a coloured fireman came to an officers' compartment and asked his former chief engineer to sign his overtime document, without which he could not be paid what was due for the past voyage. This was done 'very ungraciously indeed'. When the train reached London, one officer had to get to Liverpool Street Station by Underground. 'I still had my uniform stained with salt and vomit and my orange lifejacket. The masses of people using the Underground stations as air raid shelters all wanted to slap my back and shake my hand. It was more of an ordeal than the whole western ocean.'

Some ships put into Londonderry where there was a big strike in the port involving 2,100 dockers and ship-repair workers because of an insufficient wage offer and 'the introduction of dilutees'. *Mathew Luckenbach*'s crew were landed at Londonderry and the following incident, described again by Able Seaman Pat Civitillo, took place there.

> The Captain, through the Agent, put out a draw (money drawn against pay due). We all went out and celebrated. That night, after the pubs closed, we all were feeling no pain.
> We had all turned into our bunks in a large hall and the lights

were out. A conversation sprang up and, before long, it got argumentative. The ex-cons and others of their ilk got abusive. When they referred to us others as 'yellow bastards' – they had slept in the lifeboats while we were under submarine attack – all hell broke loose. Everyone was swinging and punching. Suddenly, the lights were turned on by a person connected with the hall and all went quiet.

The *Mathew Luckenbach* had been a good ship for us and after she had been torpedoed and the crew referred to her as 'Broken-back Luckenbach', I felt the loss of her.

The Merchant Navy men would report to their local Pool offices after their leave and sign on another ship. For many it had been just another bad incident in a bad war, never to be forgotten but not leaving any permanent scars on the mind. The Dutch, Norwegian and Greek survivors were very sorry to lose their ships because this meant that their little communities in exile for the past few years would be broken up. The Americans were all sent home on the *Queen Elizabeth*'s next westbound crossing and arrived safely.

The escort group commanders and captains of individual escorts submitted their Reports of Proceedings, in some cases with a little apprehension. All the reports are available as also are the comments made on them by Western Approaches. Commander Boyle's account of SC.122's actions was received with little comment. A Board of Inquiry was held on the loss of the trawler *Campobello*; the papers of this are not available but it is not thought that *Campobello*'s loss was blamed on other than the stress of weather. Commander Day submitted his report at Londonderry and he expected to be called to Liverpool but was not. Lieutenant-Commander Gordon Luther had cause to be the most nervous of the senior escort officers. He sent his report ashore at Liverpool and waited to be summoned to account for his time in command of HX.229's escort during which twelve merchant ships had been sunk. There was a delay before Luther was called to Derby House, possibly while the HX.229 operation was re-run on the table of the Western Approaches Tactical Unit. Luther was eventually summoned to see Admiral Horton; both men are now dead but it is known that Luther came back to *Volunteer* 'looking fairly relieved'.

The official Western Approaches comments on HX.229 contained four main points: that the escort had sailed without

its senior officer; that the temporary senior officer – Luther
– lacked North Atlantic convoy experience and had no senior
officer's staff; that the re-routing to the east, caused by the
Raubgraf U-boats in the Newfoundland 'bottleneck', and the
good speed made by the convoy during the gale of 15 March
had delayed Commander Day joining in *Highlander*; that the
loss of *Campobello* and the rescue work of the escorts with the
torpedoed merchant ships had weakened the convoy's escort.
The reference to *Campobello*, an SC.122 ship, is either a slip
by the Western Approaches staff officer who compiled these
comments on HX.229 or an implication that, if *Godetia* had
not been forced to stand by the sinking *Campobello* on 16
March, Commander Boyle might have been ordered to give
up some of his escort to reinforce Luther's impoverished
escort with HX.229.

There were neither reprimands nor commendations for
any of the naval officers who sailed with SC.122 and HX.229.
Volunteer returned to her place in B5 Group and Luther and
Day, who had shared the command of the escort of the battered HX.229 convoy, never met again. Commander Day did
meet U.S.S. *Babbitt*'s captain, Lieutenant-Commander Samuel
Quarles, in Londonderry. Quarles remembers the meeting.

I can recall discussing sub attacks with the British escort commander. He stated that he had made a series of good attacks on a
sub on the same date that we had all of our fun in *Babbitt*. The
British commander expressed doubt that he would get credit for a
'Kill' because he said, 'You have to bring back some meat to those
bloody butchers before they will believe you.' The 'bloody butchers'
were those on the various British naval staffs who had to review
action reports and make assessments of results. I thought that was a
very descriptive way of expressing the problem.

Another captain had some hard words for his crew.

After we had landed our survivors, our Canadian captain cleared
lower deck and 'bottled' us over the failure of our depth-charge
attacks during our two contacts on the 17th/18th. He gave full
marks for the rescue operations and our treatment of the survivors
but told us in no uncertain language that we would have to improve
in action. (Wireman R. A. White, H.M.S. *Pennywort*)

The masters or senior surviving officers of the torpedoed
ships also had to submit their reports. Several of these have
already been referred to; the only general comment to be
made is that, on reading the American reports, it is obvious
that there was deep antagonism on some ships between the

merchant crews and the naval Armed Guard. The U.S. Navy recognized the HX.229 battle by awarding all of their men on merchant ships in that convoy a Battle Star to wear on their campaign medal ribbon. It has been recorded earlier in the book how several British Merchant Navy men were awarded medals for their courage when ships were lost, but one man who might have been considered worthy of a decoration did not receive one. This was Captain Albert Hocken who, as master of the *Tekoa*, had stopped his ship and rescued survivors from three torpedoed ships; Hocken received a letter from the Honours and Awards Section of the Admiralty expressing 'their Lordships appreciation of your courage and humanity on that occasion' and also a letter from the Admiralty Trade Division thanking him for picking up survivors 'under difficult and trying conditions' – just these two letters for the four hours spent stopped in such dangerous waters and for the saving of 146 lives!

The ships and men who had sailed in these two convoys were soon scattered throughout the world, little realizing that the ordeal they had just been through would later be judged a turning point in one of the great battles of history. Surgeon-Lieutenant Osmond and another officer from H.M.S. *Volunteer* went on leave to London.

Dexter Gynn, R.C.N.V.R., and I travelled down to London together and were introduced to an M.P. sent to see us by C.-in-C. Liverpool who told him to cheer us up and learn what it was like. Next day he invited us to lunch in the House of Commons. In the lobbies there was much talk about shipping losses and our host indicated that we had been there. We saw Churchill in the distance. It was all very improbable and far, far away from that Liberty ship stricken and foundering in the wild Atlantic. We stayed for Question Time. Dexter was astonished that the main discussion seemed to be about an R.A.F. man whose wife, or maybe his common-law wife, was not getting the proper pay and allowances. I felt proud that even in such difficult times the House felt obliged to deal with such matters. It seemed to me that that was how democracy should work.

Many of the U-boat men involved in the recent battle were also making their way home. There was no fixed time limit at which a U-boat ended its patrol; this decision was made by U-boat Headquarters. The firing of a boat's last torpedoes usually brought orders to return to base and serious depth-charge damage or mechanical defect had the same result. But a crew was expected to deal with its own minor repairs and strain and fatigue were not factors – 'we were always tired'.

The heavy action around SC.122 and HX.229 had left many U-boats unfit for further operations and these were now recalled. It is also possible that Dönitz had decided that a distinct phase of operations was now over, and several boats that still had some torpedoes remaining were surprised to receive their orders to come home too. Another reason for this was that U.119, one of the two tanker U-boats, had almost exhausted its fuel and could no longer supply needy U-boats. The other tanker boat, U.463, was kept busy but only gave each 'customer' just sufficient fuel for a '*sparsamen Rückmarsch*' – a 'thrifty return voyage'. In this, only one of a U-boat's two diesels was run and that on reduced power only. The diesel was clutched in to one of the electric motors which, in turn, ran the second electric motor. This enabled two screws to be driven off one diesel and produced a surface speed of 7 knots, just over half the normal surface speed but giving the most economical use of fuel.

Not all of the SC.122/HX.229 U-boats returned to base; some were ordered north to reinforce Gruppe Seeteufel off Iceland or west to form a new group in the Newfoundland 'bottleneck'. But the larger number of boats now sailing south-east were to spend a pleasant few days in good weather and safe waters. The best rations were found and eaten, the 'paper war' started as officers compiled lists of defects and wrote up their War Diaries. Some men started to worry about the Bay of Biscay crossing with its relentless Allied air patrols but most were 'already at home in their minds'. They were heartened to hear, in the afternoon of 20 March, a long '*Sondermeldung*' – a special broadcast direct from U-boat Headquarters transmitted on the German Allouis radio station. U-boat Headquarters had totalled up all claims and announced that thirty-two ships of 204,000 tons and one destroyer had been torpedoed and sunk. The Germans had somehow got to know of the Atlantic Convoy Conference in Washington and they claimed that this was an act of desperation by the Allies; the recent convoy success was the 'first German reply' to the Conference and would 'paralyse the enemy to the marrow of his bones'. There were detailed descriptions of the first night of the battle and also of the encounter in the early morning of the 19th between a U-boat and a destroyer in which the destroyer (presumably *Highlander*) had been torpedoed while steaming at speed and gone to the bottom with all on board.

One U-boat that had operated in the convoy battle but without success was U.590. She was not ordered to base but her captain, Kapitänleutnant Heinrich Müller-Edzards, wrote a long letter home which he hoped to give to a tanker U-boat for one of the homeward-bound boats. This letter is a useful record of the thoughts of a U-boat officer about both the great and the minor events affecting his life at that time.

On board, 22nd March 1943.

Dear Parents,

In a few hours we shall reach the 'post buoy' and I will take this opportunity to send you a short 'lifeline'. Although we shall be in harbour in a few weeks you will not be sorry to receive this extra letter especially when I tell you that war luck has been with us in the shape of a 6,000-ton steam-freighter which we took out of a convoy (HX.228) and sent down to the fishes within one minute. This sinking and a hit with a torpedo on another ship are the only results of our efforts. Two hard weeks lie behind us. We not only had to fight the convoy ships and the air patrols – a plague on this institution – but also the elements – rain, snow and hail, although these can be a great advantage in some circumstances but when you are standing with water up to your head and the storm sings its grand song then it can all become too much.

We have just passed a field of ruins with a huge oil patch, pieces of wreckage and a body. I have found, to my personal peace of mind, that when I sank a ship in which the loss of the whole crew was certain, it did not affect my nerves even though I caused this catastrophe. One is in such a state of nervous tension when one has overcome all sorts of obstacles, avoiding all the destroyers and crept upon one's victim that one has no other thoughts; just like the enemies who have no moral scruples when they want to take our lives with depth charges and bombs.

The numerous special announcements and the praise of the Ob.d.M. and B.d.U. did our hearts good. Despite, or as a result of, our nerve-wracking activities we are all well; the mood of the men has improved. Only Maschinenmaat Süßenbach has broken his arm. Although it is a complicated fracture he is well and his recovery is not in doubt. I would be grateful if you could tactfully inform his relatives. If you could also inform all the relatives of the crew they would be very pleased.* Apart from that I am not very pleased with my men because it seems that during the last of our rest days in harbour they stole huge quantities of coffee and sold them at shameful prices to the dock workers. So far the culprits have not been

---

* Müller-Edzards's father wrote to the relatives of every man in his son's crew whenever he had news of this kind.

identified. A stoppage of leave would do wonders to make the thieves come forward. Only with such draconian measures could one become 'master of such a pest'.

Apart from the suffering of my piles I feel well; the neuralgic pains have disappeared. Our good doctor can't make any money out of me. The other men are all well apart from a few aches and pains which are unavoidable. As soon as we get into a harbour I will get my piles seen to otherwise the pain will drive me to despair. My birthday – it was not really a celebration, I just lived through it in a north-west gale force 8–9 – I had five pieces left from the Christmas cake which we ate in the afternoon but nobody suspected that it was my birthday. My thoughts were very much of you, dear parents, and doubtless somewhere in the vast Atlantic our thoughts crossed.

The reports from the East Front are very favourable and have been joyfully received here. I hope that all the expectations which we have for the East Front will be fulfilled this year.

In the hope that you my dear ones are all well, I remain with heartfelt greetings and kisses,
Your
Heinrich.*

For those U-boats returning to their bases there was a heightening of tension and an increase in preparedness as they approached the Bay of Biscay. The Metox aerial was fetched up from below and fitted to the conning tower. Metox, often called the Biscay Cross on account of its shape, was a device which gave warning of the approach of a radar-equipped aircraft. It was very effective against the Allied Mark I radar sets but the advanced Mark II sets now coming into service with Coastal Command used a frequency that Metox could not detect. This recent change made the U-boat men very nervous and their visual lookouts were intensified still more. If fuel allowed, the 'thrifty return' was abandoned and U-boats worked up to their top speed of 16 to 18 knots for a dash across Biscay by night.

The Germans would have been even more apprehensive had they known that Coastal Command's 19 Group, with some help from 16 Group, was mounting a special effort for the last eight days of March. Coastal Command realized that the end of the SC.122/HX.229 battle would see an unusual number of U-boats coming into their Biscay bases and 100

---

* Kapitänleutnant Müller-Edzards later left U.590, which was sunk with all hands on its first patrol under a new captain. Müller-Edzards became a schoolmaster after the war and died in 1966. I am grateful to his widow for permission to use this letter.

aircraft had been gathered together for intensive operations during this period.

The very first U-boat to enter the Bay on its return from the recent convoy battle was Oberleutnant Hans-Jürgen Haupt's U.665, a new boat that had only left Germany at the end of the previous month. In the dark early hours of 22 March, U.665 was picked up on the new type of radar fitted in a 172 Squadron Wellington captained by Flying Officer P. H. Stembridge. The Wellington closed to within a mile before switching on its Leigh Light, a powerful light that lit up the surface of the sea. The U-boat was spotted in the act of crash-diving and the depth charges went down. Haupt and all his crew found their graves in Biscay.

The second of the SC.122/HX.229 boats to come into Biscay was Manfred Kinzel's U.338, also a new boat but one which had covered itself in glory by sinking five ships in SC.122 and had then survived attack and damage by escort vessel and aircraft. U.338 was only a few miles behind Haupt's boat and was also picked up by an aircraft. On this occasion the bold Kinzel decided to stay up and fight; his gunners opened fire and it was the aircraft which was destroyed. It was a Halifax of 502 Squadron captained by Flying Officer Leslie McCullock and there was only one survivor, Sergeant J. Taylor from Adelaide who was the aircraft's flight engineer. Taylor was picked up by the U-boat and taken triumphantly into St Nazaire. This incident rounded off a dramatic first patrol for U.338 which had completed the '*grosse Kurve*' – the voyage from Germany out into the Atlantic and thence to Biscay – in the almost record time of just one month from Kiel to St Nazaire.*

There were further encounters in the eight-day air offensive over Biscay. Twenty-seven U-boats were sighted and

---

* This boat's career continued in similar hectic vein. Soon after leaving St Nazaire for the second patrol, it was bombed and damaged and the crew were not pleased to have to return to port, having just spent all their money, and then to unload all the stores in their boat. On the third patrol they were caught on the surface out in the Atlantic by the 120 Squadron Liberator of Flying Officer J. K. Moffat, who had also taken part in the SC.122/HX.229 battle. Kinzel again decided to stay up and U-boat Headquarters received his signal '*bleibe zum Angriff oben*' – 'am remaining up to fight'. But Moffat did not attack at once and, after a long wait, Kinzel did dive. He did not know that the Liberator was carrying a new weapon, an acoustic torpedo, and that Moffat had been waiting for the U-boat to dive. It was the end for the gallant Kinzel and all his crew.

sixteen attacks made but, to Coastal Command's great disappointment, no further U-boats were sunk although another, U.332, was badly damaged. Two more aircraft were lost – a 58 Squadron Halifax shot down by a U-boat and a 59 Squadron Fortress which was shot down by German aircraft during a daylight patrol. The score for Coastal Command's 'Bay Offensive' was thus one U-boat sunk and one damaged against three British aircraft shot down.

The U-boats that had operated against SC.122 and HX.229 had one more casualty to come. The Type IXC boat, U.526 (Kapitänleutnant Hans Möglich), was another on its first patrol. It had achieved no success in the convoy battle and was kept at sea until early April. On the 14th of that month, when almost within sight of Lorient, the U-boat was blown up by a mine and sank. Only one midshipman and eleven ratings were picked up by a German ship. This was a quiet success for R.A.F. Bomber Command's unspectacular but relentless minelaying campaign.

And so the U-boats came into their bases in France. Those who had been successful proudly flew their victory emblems, a triangular white flag for each sinking with the tonnage marked on it. The traditional bouquets of flowers and kisses were bestowed on captains by German Red Cross girls; photographs were taken. After their boat had been handed over to the base staff, the men's first visit was to the local barber, where up to two months growth of beard was removed and, then, off they went to savour all the delights of France or a happy journey home on the *B.d.U. Zug.*

There was much joyous talk in the Officers' Messes at the U-boat ports of the recent huge success – the Germans were never told that the broadcast sinkings were exactly 50 per cent too high – but one captain who had just come in from his fifth patrol decided that, despite the recent sinkings, the war could not now be won. He was given the opportunity of going to a training flotilla and accepted: 'I was pleased to get out of the U-boat war at last.' The captains filed their War Diaries with their flotilla commanders and then went to Angers for an interview with Kapitän Rösing, Flag Officer U-boats West. Those captains with more important items to report were then sent on to Berlin for a meeting with Grossadmiral Dönitz. Dönitz was honoured by Hitler on 6 April with the award of the *Eichenlaube* – the Oak Leaves to the

Knight's Cross – but Dönitz stresses that this was more a recognition for the successful series of recent convoy battles than for this one recent operation.

Dönitz's staff at Am Steinplatz tied up the loose ends of the SC.122/HX.229 battle and compiled their conclusions on it. These contain little that had not already been commented upon, but expressed satisfaction that so many U-boats had found and shared in the attacks on the convoy and that so many boats on their first patrols had done well. The report did show disappointment at the appearance of the heavy air patrols so early in the operation. *Abschlussbetrachtung Geleitzug Nr. 19* – Final Report on Convoy No. 19 – was typed out and filed with the U-boat Headquarters War Diary.

16

An Analysis

Die grösste Geleitzugschlacht aller Zeiten. (The greatest convoy battle of all time.)

German radio broadcast, 20 March 1943.

The German claim that it had been 'the greatest convoy battle' was based mainly on the number and tonnage of ships that were believed to have been sunk. Before proceeding with a study of the implications of this battle and the effect of the losses upon the Allied handling of the U-boat war, it would be useful to summarize the exact losses and show to what extent the German claims were overestimated.

The success signals from the U-boats had claimed the sinking of thirty-two merchant ships with a gross weight of 186,000 tons, and of one destroyer. The U-boat Headquarters radio broadcast claimed the same number of ships but pushed the tonnage up to 204,000. The actual losses of the two convoys were:

HX.229	–	13 ships of 93,502 tons gross weight
SC.122	–	9 ships of 53,094 tons gross weight
Total	–	22 ships of 146,596 tons gross weight

Approximately 161,000 tons of cargo were also lost. Both convoys lost one ship that was sailing out of convoy – the *Clarissa Radcliffe* having straggled and the *Mathew Luckenbach* having romped – and these are included above, but the loss of the *Svend Foyn* in HX.229A is not relevant here and is omitted. The German claim to have sunk one destroyer was without foundation, but the trawler *Campobello* had been lost in bad weather. It had required ninety German torpedoes to sink the twenty-two ships – sixty fired at HX.229 ships and thirty at SC.122 ships. (It is assumed here that *Clarissa Radcliffe* was sunk by four torpedoes from U.663 on 19 March.) The twenty-two lost ships consisted of sixteen cargo ships, five refrigerated ships and one tanker. Nationalities of the lost ships were: British – ten, American – six, Dutch – three, and Norwegian, Greek and Panamanian – one each.

The lost ships had been carrying 1,494 crew and passengers when torpedoed; of these 360 crew members and twelve passengers had lost their lives – almost exactly one quarter of those who had been on board. The dead were made up as follows:

Merchant shipmasters	5
Merchant and radio officers	47
Cadets and apprentices	10
Merchant seamen	230
U.S. Navy Armed Guard	51
Royal Navy gunners	15
Royal Artillery gunners	2
Passengers	12

Two of the dead passengers were women and two were children. The nationalities of the dead were:

American	159
British	125
Dutch	16
Lascars, etc.	38
Others	34

The 'others' came from fifteen different countries, showing the cosmopolitan nature of many merchant-ship crews. The Americans had suffered the heaviest proportional fatal casualties because of the inexperience of so many of the crews in their rapidly expanding merchant fleet.

The system whereby merchant ships drew their crews at a particular port had resulted in heavy casualties in certain communities. The New York area had lost forty-two men, mostly in the *Harry Luckenbach*, and the Savannah area lost eighteen men, mostly when *James Oglethorpe*, Savannah's first Liberty ship, tried to make port on her own after being torpedoed. Of the British seamen, twenty-six from South Wales had died and twenty-three from the Liverpool area. The deaths of two British sailors, both Liverpool men, revealed irregularities. A trimmer in *Coracero* was shown in the Ship's Articles to be 'J. J. Elder', aged eighteen, but was later found to have been a fourteen-year-old boy, Robert Yates, who had run off to sea and signed on under the assumed name. It took some time before his parents could be traced. A fireman in *Port Auckland* who had served as 'T. Williams' turned out to have been a man with another name who had deserted from the Army two years earlier.

The rating washed overboard from H.M.S. *Mansfield* was the only casualty suffered by the naval escorts.

Luck had played its part in this dealing out of death and destruction. Three of the ships sunk in SC.122 – *Kingsbury*, *Zouave* and *Terkoelei* – should not have been in a North Atlantic convoy, having loaded at West African ports, but the suspension of the Sierra Leone convoys had caused them to sail via New York, and a fourth, *Port Auckland*, had been ordered back from HX.228. *Elin K* and *Zaanland* were only in HX.229 because of delay caused by a collision between them in New York harbour; *Irénée du Pont* had put back from her previous convoy, and *Canadian Star* would have sailed as a fast independent but for the accident to her main anti-submarine gun. The American Liberty ships had not been lucky. Out of the six new Libertys sailing with HX.229, three had been sunk, thus contributing to the interesting statistic that one quarter of the 200 Liberty ships lost in the war were lost on their maiden voyages. The New York based Luckenbach Steamship Company had lost both of its ships sailing with HX.229; in fact the company lost four ships during March 1943. Also unlucky at this time was the British company Royal Mail Lines which had lost *Nariva* in HX.229; this company also lost four ships during the month.*

The German casualties do not take long to detail in this one-sided battle. One Type VIIC U-boat, U.384, had been lost. Four officers, two midshipmen and forty-one petty officers and ratings had died. To achieve this the naval and air escorts had expended 378 depth charges – 229 by HX. 229's surface escorts, sixty-nine by SC.122's escorts and eighty by Coastal Command aircraft.

It does not require an expert to appreciate that the German successes had been achieved against two convoys whose escort groups were mainly composed of obsolete escort vessels, that the escorts of one of the convoys had been hopelessly under strength, and that once the battle was joined no effective reinforcements had appeared until the battle was virtually over. It is equally obvious that the failure to provide long-range air escort in the Air Gap had greatly contributed to the German success.

* The other Luckenbach ships lost in March 1943 were *Andrea F. Luckenbach* and *Lillian Luckenbach*; the other Royal Mail Lines losses were *Sabor*, *Loch Goil* and *Nagara*.

The shortage of modern escort vessels was the price that non-aggressive democracies had to pay for their failure to prepare for a war the timing of which was beyond their control. This, and the severe weather of the winter of 1942–3 which caused so many escorts to be put into dock for repairs, was the main reason why Commander Boyle and particularly Lieutenant-Commander Luther had to fight what were really desperate defensive actions when they would so much have liked to have stood and fought it out with the U-boats. However, the failure to provide the support of reinforcing groups of escorts, which would have been of immense value in Boyle's and Luther's battles, and the existence of an Air Gap so late in the war, are more open subjects which will be dealt with later in this chapter. A brief look will first be taken at the way the young commanders on the spot fought their actions.

There is little that need be said about SC.122's battle. Eight ships had been sunk in this convoy, but half of these were the result of one unexpected and sudden encounter with the bold Kapitänleutnant Kinzel in U.338. After this, Commander Boyle's group had conducted a sound and in many ways a routine action with the U-boats. In fact the men who served in Commander Boyle's group often do not remember this particular convoy action and it only later became of historical significance because their action was fought at the same time as HX.229's. An interesting note in Admiralty papers comments on the paucity of R.D.F. (radar) contacts around SC.122 throughout the battle, and concludes that U-boats were being forced to make their attacks 'trimmed well down'.* The truth of the matter is that throughout the battle the Admiralty had erred in assuming that SC.122 was in as great a danger as was HX.229. An analysis shows that twenty-eight U-boats made a total of thirty-seven separate contacts with HX.229, while only six boats made nine contacts with SC.122. The low level of radar activity around SC.122 was simply a result of this lesser U-boat presence. If the Admiralty had known this, they could with some benefit have advised Western Approaches to detach escorts from SC.122 to the aid of HX.229.

The central character concerned with the far more serious battle of HX.229 could not be asked for his opinions of that battle and his memories of that crisis in his career. Gordon

* The report is in Public Record Office ADM199/2019.

John Luther continued in command of H.M.S. *Volunteer* until he suffered a bad fall while ashore in the Azores and broke several bones in his back; this painful injury led to his early death in 1952. Luther had found himself in an almost impossible situation, particularly during the first night of the battle, with only four escort vessels at hand, no regular rescue ship and the last merchant ship in the columns failing to obey orders and pick up survivors. Luther had the terrible choice between deliberately leaving men, and on one occasion women and children, to face certain death and his duty as the senior naval officer of the escort which was to protect the remaining ships in the convoy.

Did Luther make the right or wrong decision when he decided to leave escorts behind for rescue work – a decision which twice left the convoy without any protection whatsoever? Commander Day, the regular commander of Luther's group, was annoyed to find that two out of four escorts were absent on rescue work when he joined HX.229. When pressed to say what his policy would have been in those circumstances, Day was emphatic that he would never have detailed more than one escort vessel to rescue work at any time, and that however many merchant ships were torpedoed, their crews would have had to take their chance on being picked up by that one escort. Commander Day did say that, if things got very bad, he would also have asked the Convoy Commodore to order one merchant ship to stay behind as rescue ship. It would have been interesting to see whether the chosen merchant ship would have obeyed.

Luther's decision to neglect defence for rescue work is the main charge that can be levelled at his handling of HX.229's escorts, but even if valid, and that is open to doubt, it is a charge that can only be levelled in cold blood at his detachment and judgement as a naval officer and, if his reputation is to be that of a man who erred on the side of compassion, he would not have regretted this. It is a fact that Luther died firmly convinced that he had failed to do enough for the crews of those merchant ships.

In other matters Luther had done as well as any officer inexperienced in commanding a group would have done. Communications oversights had left *Mansfield* and then *Pennywort* absent from the convoy during vital periods and this probably did have an effect upon the number of merchant ships sunk. But his decision to stick to a constant course – a decision criticized by many of the merchant ship officers – brought its reward with the appearance of Flying Officer Esler's Liberator at the limit of its range on the second evening of the battle and certainly saved ships from being torpedoed that night.

In the matter of counter-attacks by the escorts there is little to choose between the performance of the two escort groups. HX.229's escorts made eleven separate attacks and damaged five U-boats; the SC.122 escorts made seven attacks damaging two U-boats. These figures do not include *Babbitt's* mayhem with U.190 just before she joined HX.229. The whole affair had demonstrated the facts of life in encounters between escorts and U-boats. To 'kill' a U-boat an escort commander needed one of three factors in his favour – the knowledge that the U-boat was a known singleton, an abundance of spare escorts, or the luck that placed a hurried salvo of depth charges at just the right place. The SC.122 and HX.229 escorts had none of these.

There is no need to dwell long on the performance of the Allied aircraft. The shorter range aircraft from the western side of the Atlantic had done sound work while the convoys had been within their range. Once the convoys had crossed the Air Gap, R.A.F. Coastal Command's aircraft had completely altered the odds in the battle. Fifty-one aircraft had operated in support of the convoys and flown a total of 230 hours either on direct escort of a convoy or in sweeps around the convoys. Only one merchant ship, the *Granville* in SC.122, had been torpedoed while aircraft were with a convoy and that particular aircraft had been attacking another U-boat when *Granville* was hit. Twenty-one depth-charge attacks had been made by aircraft resulting in the sinking of one U-boat and the damaging of six more. Somewhat surprisingly the Fortresses and Sunderlands had been more successful in their attacks than had the more celebrated Liberators, which had only slightly damaged one U-boat during the battle, but it was the early appearance of the Liberators so far out in the Atlantic that really broke up the rhythm of the pack attacks.

In examining the German performance one finds a mixture of technical brilliance and operational disappointment, of wild strokes of bad and good fortune. The B-Dienst had fed a series of potentially decisive decodes to U-boat Headquarters

but several bad assessments of the actual progress of
SC.122 and HX.229 were made before contact was finally
achieved. Perhaps the influence of signals concerning the
extra convoy, HX.229A, really caused this, and the presence
of this convoy, unsuspected for so long by the Germans,
certainly delayed the encounter. Luck had also played its
part; once the boats of the Raubgraf group were avoided, the
two convoys could have expected to sail in safety at least until
the long Stürmer–Dränger line of boats closed in from the
east, but the meeting that would probably then have resulted
would not have taken place until the 17th and the subsequent
action would all have taken place within the range of Coastal
Command's aircraft. Instead, the chance sighting of HX.229
by U.653, which had left its patrol line, brought the action
forward by one day. It was during this vital twenty-four hours
beyond the range of aircraft that the U-boats sank fifteen of
their twenty-two victims.

Once the battle had been joined it was largely left to
individual U-boats to carry out their own attacks and, in most
cases, the results of these rested upon the character, the
determination and the skill of each captain. Examples of the
different types have been met frequently in previous chapters.
Forty-one U-boats had taken part in the operation, of which
thirty-three had made some contact with a convoy and
seventeen had made torpedo attacks. The fourteen new boats
involved in the battle had done very well and were responsible
for torpedoing eleven ships. If any lesson is to be learnt about
the German failure to detect the convoys earlier and then the
failure by some U-boats even to find the convoys at all, it is
that the North Atlantic is a very large place.

In the first twenty days of March 1943, the Germans sank
ninety-seven Allied merchant ships totalling more than
500,000 tons. This was almost twice the rate of new tonnage
being built at that time. Almost two thirds of the losses were
of ships in convoy. During the same period the Germans lost
seven U-boats, which was just half the number of new boats
coming into service. The Monthly History of the Trade
Division of the Admiralty states: 'The import programme
for the U.K. was cut as low as it could be and then seemed
hardly likely to be fulfilled.'* It is known that Lord Hall,

* Public Record Office ADM199/2085.

Financial Secretary to the Admiralty, visited Western Approaches at this time and let it be known that the Admiralty were considering the possibility of abandoning the convoy system. If this had happened all merchant ships would have sailed independently at their best speed and taken their chance with the U-boats. It is not surprising that the naval Official History quotes another Admiralty report which stated that 'the Germans never came so near to disrupting communication between the New World and the Old as in the first twenty days of March 1943'.* How had this crisis in the battle for merchant shipping come about after three and a half years of war and at a time when the Allies were proving successful on every other front?

After reading the chapters covering the battles of HX.229 and SC.122, it will have become evident to the discerning that it might have been within the power of the Allies to counteract at least three of the factors contributing to the German success at this time, and the reader may be entitled to ask the following questions.

If Flying Officer Esler's Liberator could fly from its base in Northern Ireland across 1,000 miles of ocean to meet Convoy HX.229, why could not Liberator aircraft stationed in Newfoundland, which was only 840 miles from that position, have given air cover up to that point? In other words, why did the North Atlantic Air Gap exist at all?

Why, when HX.229 and SC.122 had become embroiled with practically the entire U-boat strength then available in the North Atlantic, had no reinforcement of naval vessels arrived until after the battle was as good as over? In other words, why were no Support Groups available long after the principle of such groups had been accepted and tried?

Why did the Germans seem to be getting so much the better of the signal decoding war without any corresponding success by the Allies?

The Very Long Range version of the American-built B-24 Liberator was an aircraft that was in great demand in early 1943. This was the aircraft best suited for work at extreme ranges over the sea and their allocation was in theory decided upon by joint Allied decisions at the highest level. But the Americans, with their vast naval commitments in the Pacific and the Atlantic, received the greater share of each batch of

* Roskill, op. cit., Vol. II, p. 367.

newly-built V.L.R. Liberators and these were allotted in turn between the Army Air Force and the U.S. Navy, for there was not yet a unified U.S. Air Force on the lines of the Royal Air Force. By March 1943 R.A.F. Coastal Command had a mere eighteen V.L.R. Liberators based in Northern Ireland and Iceland, and their valuable work has been demonstrated in previous chapters. The U.S. Army Air Force had two squadrons operational on similar work, the 1st and 2nd (Antisubmarine) Squadrons, which had been working from south-west England over the Bay of Biscay but had recently been moved to Morocco to cover the eastern end of the American military convoy route to North Africa. The U.S. Navy had 112 V.L.R. Liberators at this time but most of these were operating in the Pacific and not one was based in Canada, Newfoundland or Iceland where they could have given cover to the North Atlantic convoys. The Canadians, who had good bases in Newfoundland, had not been allocated a single Liberator.

The result of this disposition of the Liberator was very simple. The R.A.F. was using its Liberators to give cover over that part of the North Atlantic convoy routes for which the Admiralty was operationally responsible. The two U.S. Army squadrons were working off Morocco. The western half of the North Atlantic, over which the U.S. Navy had operational control, did not have a single V.L.R. aircraft although that Navy possessed 112 of these aircraft. The shadowing of HX.229 by U.653, and the assembly of the Raubgraf U-boats around the convoy before the first night attack, took place in broad daylight in an area that could easily have been covered by a V.L.R. Liberator based in Newfoundland. It is unfortunate that blame for this situation must lie in the office of one of America's famous naval officers – the Commander-in-Chief, Fleet Admiral Ernest J. King. This officer, of such determined character, showed on many occasions two facets of his thinking – that he regarded the Pacific as the prime theatre of operations for the United States, despite the joint Allied decision that victory over Germany should take priority over the Pacific war, and that, in fighting the U-boat, he was not willing to follow British experience hard-won in more than three years of war at the same time as he retained personal control over his navy's anti-submarine operations. It would have been quite feasible to have based two squadrons of U.S. Navy Liberators in Newfoundland many months earlier; they would still have remained under U.S. Navy control, would have flown over an American operational area, and would have saved the lives of American as well as British sailors in the North Atlantic convoys. With the excellent intelligence available to the Submarine Tracking Rooms – about which more later – this modest diversion of effort would have been sufficient to close the Air Gap.

It should be stated, however, that Admiral King was making one contribution to the air side of the U-boat war in the form of an escort carrier, U.S.S. *Bogue*, which with five escorting destroyers was just coming into service at that time in the Atlantic and was soon to be followed by further carriers. These later performed valuable work but they were not a complete answer – each carrier tied up a large number of destroyers in its close escort; there were many days in the North Atlantic on which their aircraft could not fly; the aircraft did not have a great range. Most of the escort carriers' successes were obtained further south and they were not a complete substitute for the land-based Liberators that could so easily have been sent to close the Air Gap from the western side of the North Atlantic.*

The Casablanca Conference of January 1943, in which Admiral King had taken part, had declared that the U-boat threat was the major obstacle to Allied victory in Europe, and had recommended that eighty V.L.R. aircraft should be found at once for what was then called 'the Greenland Air Gap'. Nothing was done. The Atlantic Convoy Conference, of which Admiral King was Chairman, naturally debated the same problem and the daily news received during the conference sittings from SC.122 and HX.229 reinforced the urgency of the situation. A memorandum prepared in advance by Admiral King's staff stated:

> There is urgent need for immediate assignment of more long-radius aircraft to anti-submarine duty in the Atlantic . . . Unless we can augment our A/S defences rapidly, the best we can expect is a loss of ships – particularly tankers – so great as to embarrass planned major operations. At the worst we may suffer serious interruption of supply lines to Great Britain and the Mediterranean . . .

* *Bogue*'s aircraft eventually sank nine U-boats and one Japanese submarine that was caught while carrying scarce materials to Germany, and her score of U-boat kills probably constitutes a record for a single ship which still stands.

The only means for quickly increasing the effectiveness of our AS W measures is to divert heavy bombers from other assignments. In view of the situation in the several theatres of active warfare, it seems that the only possibility is to divert these planes from the bombing of Germany.*

In other words, the aircraft to close the Air Gap in the U.S. Navy's part of the North Atlantic were to come from the U.S. Army Air Force or R.A.F. Bomber Command operating in the European Theater of Operations, rather than from the U.S. Navy's stock of V.L.R. Liberators operating in the Pacific or even from future U.S. Navy allocations of those aircraft, this despite a decision of June 1942 by General George Marshall, Chief of the American Combined Chiefs-of-Staff and therefore Admiral King's superior, that the U.S. Army Air Force should leave the anti-submarine field entirely to the Navy.

Later sittings of the Atlantic Convoy Conference did produce an elaborate time-table by which large numbers of V.L.R. aircraft were to be found for the fight against the U-boats. British delegates to the Conference stressed the seriousness of the current situation and begged for less help but more quickly. It took the interest of President Roosevelt, who intervened on 18 March, the day after HX.229 and SC.122 lost sixteen merchant ships, to get more immediate action; he asked where all the U.S. Navy Liberators had been when the sinkings had taken place and his interest produced results.

Let us leave this high-level debate and go on to examine the action actually taken to close the Air Gap. R.A.F. Coastal Command had already tried a somewhat desperate measure to close part of the gap. There existed in southern Greenland an emergency airstrip known as Bluie West One, operated by the Americans and used mainly by aircraft that found themselves in difficulties while flying across the Atlantic. In January 1943 the British had sent four Hudson aircraft from 269 Squadron in Iceland to fly convoy patrols from Bluie West One. This airstrip, at the end of one of Greenland's many fjords, was very difficult to return to, especially in bad weather. The Hudsons made fourteen sorties on the only four good flying days in five weeks but they provided escort for only one convoy and no U-boats

* From unpublished papers at the Albert F. Simpson Historical Center, U.S.A.F., Maxwell Air Force Base, Alabama.

were found. Two of the Hudsons and their crews were lost when they could not find Bluie West One on the last of these operations and the remainder were then withdrawn to Iceland. American aircraft, probably B-17 Fortresses from Newfoundland, later tried to use Bluie West One for U-boat operations but with no greater success.

The next attempt to close the Air Gap was also R.A.F. inspired. The crews of 120 Squadron's Liberators based in Iceland could see no reason why they should not fly from Iceland to cover a convoy in the Air Gap and then land at Gander in Newfoundland. This 'shuttle' type of operation was well within the Liberator's endurance but the initial suggestion was not approved, the R.A.F. pilot who made it believing that the Americans turned it down. After the disasters of March, the proposal was considered again and the first flight was made by a 120 Squadron crew that had presumably been sent to Newfoundland to collect a new Liberator. At 07.50 on 5 April Flying Officer Smith took off from Gander and made his small contribution to the history of the Second World War when he flew across the Air Gap with a fully operational aircraft. Smith failed to meet Convoy HX.231 but he did spot a surprised U-boat which dived before he could attack. The Liberator then landed safely at Reykjavik. Flying Officer Smith (even his initials are not recorded) and his crew had closed the Air Gap in this somewhat makeshift manner. During the next few weeks 120 Squadron flew seven shuttle patrols, now using Goose Bay in Labrador as the western landing ground, and sighted no less than ten U-boats. These crews had the worst of luck, however, for in every one of the attacks they made on these U-boats there were failures in their depth-charge release mechanisms.

It was left to the U.S. Army Air Force to close the gap in more conventional manner. As a result of President Roosevelt's inquiries and the decisions of the Atlantic Convoy Conference, the Army's 6th (Antisubmarine) Squadron brought the first American-crewed Liberators to Newfoundland. The squadron became operational on 19 April and on that day Lieutenant E. J. Dubeck's crew made a very good attack which damaged a U-boat. At the same time the R.A.F. Coastal Command decided to forgo their next allocation of Liberators, as long as these were given to the Canadians who had crews already trained in anti-U-boat work and experienced in the weather and navigation problems of the North Atlantic. In this way, 10 Squadron R.C.A.F. obtained its first Liberators on 23 April and started operations from Gander on 10 May, the first flight being made by Flight Lieutenant F. J. Green. The first U.S. Navy Liberators (PB4Ys) appeared at Argentia with VB-103 Squadron from San Diego in the middle of May, and their first operational flight was made by Lieutenant H. K. Reese on 20 May.

Earlier in this book it was stated that Admiral Dönitz now considers that two major events settled the outcome of the U-boat war. The first, already referred to, was the delay by Hitler in pressing ahead with U-boat construction in 1940. The second turning point, according to Dönitz, was the appearance in mid-Atlantic of long-range aircraft. This second point was not made with hindsight. The War Diary of U-boat Headquarters contains this passage, written as early as August 1942:

This development must have the consequence of the loss of a large number of U-boats and a reduction in our successes. *There are now no prospects for the successful conclusion of the U-boat war* [my italics].

The subject of provision of aircraft for distant convoy work generates strong international and inter-service feelings. The British blamed the Americans for being slow to close the Air Gap from the west; the R.A.F. blamed the Royal Navy for failing to provide aircraft carriers for convoy work; the Navy pointed at the large force of four-engined bombers being used to bomb Germany and asked why Coastal Command could not have more of these instead of relying on American production. It is a fact that that fine aircraft, the Lancaster, if modified, had the range to become a long-range anti-submarine aircraft; the R.A.F.'s standard post-war maritime reconnaissance aircraft, the Shackleton, was a direct descendant of the Lancaster. Not one Lancaster ever went to Coastal Command; the 'bomb the Germans into defeat' lobby, backed by Churchill, got the lot.

The arguments on this general use of air power all have some validity and have been well thrashed out over the intervening years but, dealing with the specific situation in March 1943, Admiral King had let both his own people and his Allies down badly by insisting on hanging on to operational control of the western North Atlantic but failing to put his V.L.R. Liberators into Newfoundland to plug the Air Gap. Dönitz had been given eight months' grace, but the flights of Flight Lieutenant Smith, Lieutenant Dubeck, Flight Lieutenant Green and Lieutenant Reese and of the other Liberator crews who finally closed the Air Gap settled the outcome of the Battle for Merchant Shipping. Other Allied advances took place in the spring of 1943 but none were as decisive as this one.

It is not possible to extract figures for merchant ships lost in North Atlantic convoys between August 1942 and May 1943, but many Allied merchant seamen paid the price of Admiral King's failure to see that the Air Gap was closed earlier.

The subject of Support Groups was touched upon briefly in Chapter 1. Western Approaches had long realized that such groups of additional escorts, stationed out in the Atlantic, could perform a valuable service by reinforcing the regular escorts of convoys which came under U-boat attack. There were some in Western Approaches who passionately believed that if the necessary ships could have been found earlier in the war the U-boat threat would never have developed to its present critical proportions. The big trouble was where to find the extra ships. In 1942 new sloops and frigates coming into service had been formed into groups which were intended to operate in this support role, but after a brief start in September of that year the Allied landings in North Africa had forced the diversion of these groups to the routine escorting of convoys. The North Atlantic convoys had been forced to continue sailing without Support Groups since that time, although many at Western Approaches and Coastal Command were convinced that extra destroyers could have been transferred from the Home Fleet; there always appeared to be plenty of destroyers at Home Fleet bases being kept back to screen capital ships on their occasional sorties. It was felt that these destroyers could have been better employed as Support Groups especially during this crisis of the Battle of the Atlantic.

There is no doubt that the heavy sinkings during the first twenty days of March 1943 brought the Support Group affair to a head again, and an interesting activity at Western Approaches Headquarters also contributed to the outcome.

One of the enthusiasts for Support Groups had been Captain Neville Lake, a Duty Captain at Derby House. Lake, a First World War submarine officer, took a keen interest in the tactical aspects of the U-boat war and during his time at Western Approaches submitted some fifty papers to Admirals Noble and Horton on various operational subjects. One of these shows that Captain Lake was conducting a 'war game' on the main plot in the Operations Room when HX.229 and SC.122 came under U-boat attack. Lake had 'formed' three Support Groups from various new frigates and sloops recently completed but which had not yet come into service and, since 18 February, had been 'operating' these mythical groups out in the Atlantic in the support role. The movements of actual convoys and U-boats in the subsequent days had been applied to the game, as had the actual weather conditions which so often governed the refuelling of the groups at sea. Lake had broken off the operations of his groups whenever fuel or depth-charge stocks were calculated to have become exhausted and each group was returned to harbour for rest at realistic intervals. Since the commencement of the game it had been found that two of the three groups could always be kept at sea and that no convoy that had actually been attacked by U-boats would have failed to receive help from a Support Group if the game had been a live one. The Wrens working on the main plot were not always enthusiastic for Captain Lake's game – they called it 'Lake's Folly' – as it caused them much extra work, but the other Duty Captain, Captain F. N. Miles, kept the game going when Lake was off duty and it was still being played out on 17 March when Lieutenant-Commander Luther's group was in such desperate need of support.

Captain Lake's papers show that his 'Group C', consisting of the sloops *Whimbrel*, *Wild Goose* and *Woodpecker* and the frigates *Tweed* and *Ness*, was 'sent' to HX.229 at midday on the 16th, soon after U.653's sighting report was picked up, and 'joined' Luther's group that evening two hours and fifteen minutes before the first merchant ship was torpedoed. It requires little imagination to envisage what assistance these five modern escorts might have given to Luther on that first night of the convoy battle. The papers also tell how 'Group C' was able to remain with HX.229 for the remainder of the voyage.

Captain Lake's report goes on:

Admiral Horton was summoned to attend a meeting of the Prime Minister's Anti-Submarine Committee. It suddenly struck us that the question of discontinuing convoys might well be discussed, so the draft of the report on the exercise was hurriedly finished in time to give it to Admiral Horton as he was leaving to catch the night train. On his return Admiral Horton told me that, thanks to the report, he had got his fifteen destroyers. He told me that the Prime Minister in a very sombre mood explained the situation whereby the U-boats were threatening the very prosecution of the war. Oil stocks in the U.K. and North Africa were below danger point and there was a serious shortage of tankers. He turned to Admiral Horton and asked him what he was going to do about it. Horton replied: 'Give me fifteen destroyers and we shall beat the U-boats.' The Prime Minister banged the table and said: 'You Admirals are always asking for more and more ships and when you get them things get no better.' Admiral Horton handed him the report of the exercise and after glancing at it there was a temporary adjournment of the meeting whilst the Prime Minister and Admiral Stark (Commander of U.S. Naval Forces in Europe) studied it. After a short while, Winston turned to Horton and said, 'You can have your fifteen destroyers; we shall have to stop the Russian convoys for the present.' Admiral Horton told me that he had never admired the Prime Minister more than at that moment, weighed down by Russian pressure for a Second Front, he was further embroiling himself in trouble.*

Action quickly followed and by the end of March no less than five Support Groups had been formed or were forming – two from the new sloops and frigates, two from the Home Fleet destroyers and the fifth around the new British escort carrier H.M.S. *Biter*. These five Support Groups, with the American escort carrier *Bogue* and her escorting destroyers, were soon in action. When the next convoys of the SC. and HX. series sailed both were threatened by U-boats, but SC.123's escort was reinforced firstly by *Bogue*'s group and then by a Support Group of Home Fleet destroyers. When SC.123 had passed through the danger area, these destroyers then went to reinforce HX.230. Only one merchant ship, a straggler from HX.230, was torpedoed.

It is so easy to look back now and see how mistakes were made – how the Americans should have put their Liberators into Newfoundland earlier, how the British should have released their Home Fleet destroyers to the Atlantic more quickly – but this writer is very conscious that these solutions

* I am indebted to Captain H. N. Lake for permission to make use of his documents.

were not so obvious under the intense pressure of wartime operations and these notes are written more as historical comment with all the advantages of hindsight than as criticisms of personalities, all of whom served their countries well and many of whom are now long dead.

Again, it would be wrong to attribute these important changes of policy and reallocations of forces solely to the two convoys whose tribulations were described earlier in this book, but HX.229 and SC.122 were certainly the last convoys of a black phase of the shipping war and their losses certainly added urgency to changes that had been long overdue. The men who sailed in these two convoys not only helped to make history but they also helped to change history.

When German resistance was crumbling in May 1945, Admiral Dönitz ordered that Germany's naval records were to be sent to Flensburg near the Danish border to avoid capture by the Russians. Dönitz also ordered that, if a final collapse came, the records should not be destroyed but were to be handed intact to the British or the Americans. Because of this forethought both the B-Dienst documents and the department's key personnel passed safely into British hands, as did the complete German naval records back to 1871. Many of the documents were later handed back to the German state archives and have been open to public inspection for many years.

The early months of 1943, including the time of the HX.229 and SC.122 operations, were found to have marked the high water of German success in the radio war, with the B-Dienst breakthroughs coinciding with both a peak in U-boat strength and a time of Allied weakness. This success was much reduced in May 1943, when the Allies changed their convoy codes again and the B-Dienst had to start once more from the beginning. The sheer volume of signals used eventually resulted in a further partial breaking of the new codes, but by then the war was approaching its end and the Germans no longer had the tactical advantages at sea to make much use of these later decodings.

The German records reveal another side of this story. Although the U-boats found many convoys by following B-Dienst information, other convoys proved impossible to locate even after seemingly reliable decodes had been achieved. It became obvious to the Germans that the Allies were able to divert some convoys away from the U-boat patrol lines almost at the last minute and also that aircraft searches often appeared with uncanny accuracy over new U-boat concentrations. The Germans had looked very carefully into the possibility of a security leak in their own organization and at the possibility that their own codes had been broken by the Allies. They were able to come to no satisfactory conclusion; it was believed that much of the Allied success was attributed to extremely advanced radar equipment, and the French Resistance and the British Secret Service were also credited with greater efficiency than was actually the case. The Germans decided that their own naval codes were still secure.

Post-war books on this aspect of the sea war credit the successful Allied diversions of convoys to a combination of the HF/DF shore stations and shipborne sets that could pick up the direction of U-boat signals, and to the powers of assessment and deduction of the staffs of the various Submarine Tracking Rooms. In particular, Commander Rodger Winn in the London room was credited with great powers of intuition in this respect. Naval authorities did nothing to alter this view and Captain Roskill's Official History, in describing the German decoding work in some detail, does no more than add one sentence on the possibility of similar Allied work – 'The reader should not, of course, assume that we British were meanwhile idle in achieving the opposite purpose.'*

Occasionally a post-war book might hint that British decoding of German signals to U-boats had been a factor but very little detail was given. Security was tight on the U-boat war which had come so near to success, and Churchill had insisted in 1945 that all involved in the intelligence side of the recent conflict should sign a special 'Black Book' which bound them never to write, talk or lecture on the subject. So, for nearly thirty years, the matter rested with readers of naval histories being well informed about the B-Dienst work, with the impression being allowed that the Germans had got the better of this radio war, with such Allied successes as had been achieved being credited to HF/DF and the skill of one outstanding R.N.V.R. officer and his staff.

* Roskill, op. cit., Vol. II, p. 208.

While researching for this book I came upon several pieces of information that did not fit into this picture. A HF/DF operator on H.M.S. *Volunteer* wrote: 'I was amazed by the amount of information we had about German communications and, in particular, the advance notice we received about frequency changes.' A comparatively junior staff officer in the Trade Division was reprimanded for ordering a signal to be sent to a convoy diverting it away from a group of U-boats shown on the Submarine Tracking Room's daily situation report; he was told that his signal might have compromised the source of that information and the convoy should have been left to take its chance. To risk a convoy in order to protect a source of intelligence must indicate that that source was of quite extraordinary value.

The lid was partially lifted on this affair in 1974 with the publication of THE ULTRA SECRET by F. W. Winterbotham. The author had been a wartime R.A.F. officer involved at a high level with radio intelligence, and he revealed the incredibly well-kept secret of how British Intelligence obtained a German Enigma ciphering machine just before the war and of how an intelligence unit at Bletchley Park in Buckinghamshire used this to break top-level codes throughout the war. Much of the results of this work, codenamed Ultra, was shared with Britain's allies, but only a handful of people were informed of the source. Winterbotham's book, written without official cooperation, is not a complete work and, as he himself admits, is based on personal recollections only, but there is enough in it to suggest that, at one stroke, many post-war military, naval and air histories are obsolete or at least incomplete and the full story may one day show that the popular reputations of many high commanders may have to be reassessed. Some were too hesitant or orthodox to use properly the startling Ultra intelligence they were given; others made brilliant use of it.

The Ultra book said little about the U-boat war but it is clear that the U-boat ciphering machine, known to U-boat men as *Schlüssel-M*, had been vulnerable throughout the war. A fresh round of inquiries after the publication of this book produced some interesting reactions. A Professor of History was found at Cambridge who had handled the U-boat side of Ultra at Bletchley Park during the war but this honourable man still felt himself bound by his wartime oath and, with regret, declined his help. But it became obvious that the Admiralty and Coastal Command were fed throughout the war with a stream of reliable intelligence on U-boat dispositions and intentions. The nature of the U-boat pack attacks during the middle period of the war and their need for sighting reports from U-boats and then for detailed orders from U-boat Headquarters to the boats in patrol lines played right into the hands of the Ultra decoders. One naval officer who worked at the Admiralty wrote of how 'we used to wait breathlessly for the electronic machines at Bletchley to get the answers and then we would get a phone call saying they had it out and in no time all the new U-boat positions were plotted on the chart. We had every move and every order direct from Dönitz.'

Looking back to the actions covered by the main chapters of this book, it can be seen that the early diversions of all three convoys because of the Raubgraf U-boats were the result of decoded orders to those U-boats altering the line of their patrol sweep. U.653's sighting report was decoded and Lieutenant-Commander Luther informed that he was being shadowed. It was Ultra decodes that led to the trapping of so many of the German blockade runners, and Ultra also gave information about the movements of supply U-boats and was the cause of the carefully mounted operation by U.S. Navy submarines against the supply U-boats. Finally, when Convoy HX.229A was approaching the area where the Seeteufel group of U-boats was sweeping in their patrol line off Iceland, the Admiralty was able to divert this convoy around the danger area easily. All this was the result of Ultra in just one short period of the naval war.

And so it can be seen that the sailors of both sides, however skilful or brave, were very much at the mercy of the crypto-analysts, and that the Allies were at least as successful as the Germans in this aspect of the war but managed to keep their success a secret for nearly thirty years.

17

The Years That Followed

Late March and early April 1943 were fairly quiet in the North Atlantic, partly because so many U-boats had returned to base after the recent intensive operations and partly because the Allied convoys were again successfully routed around the shorter U-boat patrol lines. But by the end of April the U-boats were back in force with the Germans hoping to resume the slaughter of the merchant ships. The trial of strength came after U-boats gained contact with a slow westbound convoy, ONS.5. By coincidence this convoy contained several of the merchant ships that had survived the SC.122 crossing and had been unloaded and were now on their way back to America. In the fierce battle for ONS.5 thirteen merchant ships, of which two were stragglers, were sunk; two of the losses were the British ship *Dolius* and the American *McKeesport* which had been with SC.122. The convoy's escort was B7 Group, commanded by Commander Peter Gretton who was forced to leave in the middle of the action because his destroyer, *Duncan*, ran short of fuel, but good air cover and the timely arrival of two Support Groups altered the nature of the battle and eight U-boats were sunk, five by escort vessels and three by aircraft. This was too great a loss and the ONS.5 battle was immediately recognized as a defeat for the U-boats.

There is no need to dwell too long on that historic spring and summer of 1943. The combination of Support Groups, escort carriers, the closing of the Air Gap and the changing of the Allied radio codes defeated the U-boats. The month of May 1943 cost Dönitz a staggering forty-one U-boats in exchange for fifty merchant ships sunk – a quite unacceptable balance of losses. Dönitz realized the tide had turned against him and as good as withdrew his boats from the North Atlantic battlefield, at least for the time being. The Germans soon changed their description of the battle involving HX.229 and SC.122 and this became to them 'the last of the great convoy successes'. A further turning point was reached in July 1943 when the launchings of new shipping tonnage, built mainly in American shipyards, finally overtook the figures of tonnage sunk since 1939 and thereafter the issue was never in doubt.

Admiral Dönitz had suffered a personal loss in this period when U.954, on its first patrol, was sunk with all hands while operating against Convoy SC.130 by a 120 Squadron Liberator captained by Flight Sergeant W. Stoves, who had flown in support of SC.122. One of U.954's dead officers was twenty-one-year-old Leutnant Peter Dönitz, the Admiral's son. His other son, Oberleutnant Klaus Dönitz, was to be killed just one year later in his *Schnellboot* in an action in the English Channel.

For followers of the earlier chapters of this book there was also a sad loss, a loss which illustrates again the blow and counter-blow aspect of the U-boat war. The old four-stacker destroyer H.M.S. *Beverley* (formerly the U.S.S. *Branch*) had been the only escort to stay with Convoy HX.229 right through the crossing from Newfoundland to Britain. On 11 April, during her next Atlantic crossing, *Beverley* was torpedoed by U.188 and sank so quickly that only four men were saved. It is ironic that on 4 February *Beverley* had helped to sink U.187, which had been built alongside U.188 at AG Weser's shipyard at Bremen.

The fortunes of the U-boats in the remaining war years have been described many times. Dönitz's command never recovered from the setbacks of the spring and summer of 1943. Large numbers of boats continued to be built and crewed with men whose training periods became steadily shorter until, in the last months, men 'from the farmyards of Bavaria' were going out in operational U-boats only six months after being called up into the navy. Acoustic torpedoes, *Schnorkel* boats and advanced designs of U-boats were produced but these technically brilliant advances came too late. The tactics of the patrol line and the pack attack were abandoned later in 1943 and U-boats were sent to hunt singly in wide spaced areas. The North Atlantic convoys continued to be attacked but there were no more all-out efforts on the lines of early 1943. The morale of the U-boat men was never as high as in the early years but it held up well despite the severe losses. It is probable that the U-boat men never knew the extent of these losses; they were often returned to a different base after each patrol and, possibly by design, could never find out what had happened to their former *Flottille* colleagues. It has been estimated that the life expectation of an operational U-boat fell to one and half patrols.

The Biscay ports continued to function until August 1944 when the R.A.F.'s famous 617 (Dambuster) Squadron started putting their new 'Tallboy' bomb through the roofs of the U-boat shelters, and soon after this the advance of Allied troops in France forced the U-boats out completely. As soon as the U-boats left their concrete shelters they became vulnerable to Allied bombing while in port, and in the last year of the war no less than fifty-seven U-boats were destroyed by bombing compared with only five destroyed by bombers in the previous five years. This shows what might have been achieved had the Allies used their bomber forces to prevent the building of the U-boat shelters in the Biscay ports in 1941 and 1942. In fact air power played the major part in the destruction of U-boats after March 1943. A staggering total of 590 U-boats were destroyed between that month and the end of the war – compared with 194 destroyed in the previous three and a half years – and 290 of these 590 were destroyed by aircraft alone, 174 by ships alone and the remainder by a combination of ships and aircraft or through other causes.* Again this shows what might have been achieved if the Allies had realized earlier how effective aircraft could be against U-boats at sea (or had implemented the realization).

An indication of the hammering taken by the U-boats in the last two years of the war can be seen in the fates of the U-boats that had operated against SC.122 and HX.229. Of the forty-one U-boats that had returned to base from that operation and the two supply U-boats that supported them, twenty were sunk before the end of 1943 and sixteen more in 1944 – a casualty rate of almost 90 per cent! Only five boats survived the war and three of these were no more than scrap at the end.

For the Allied ships and sailors it was a different story. Thirteen of the 149 merchant ships that had sailed in SC.122,

---

* Appendix 6, p. 358, gives details of the 784 U-boats lost by the Germans during the war.

HX.229 and HX.229A were later sunk by enemy action but only the ill-fated *Beverley* of the escort vessels was lost.* H.M.S. *Mansfield* was so weakened in an Atlantic storm in 1944 that she was scrapped in Canada; Lieutenant-Commander Hill and his crew then took over a new American-built destroyer, H.M.S. *Rowley*, at Boston. Most of the other escorts continued to sail the convoy routes to the end. Two exceptions were *Buttercup* and *Godetia*, which paid off late in 1944 so that their Belgian crews could go to help clear mines at the recently liberated Belgian ports. Two of the Belgian sailors were killed at Antwerp and these, with Victor Billet, the founder of the Section Belge who had been killed at Dieppe, were the only fatal casualties of a somewhat fortunate Section Belge.

The officers of the SC.122 and HX.229 escorts were not marked for outstanding success and their names, more by lack of fortunate opportunities than anything else, were not among those who made their reputations in later battles. The only ones who ever achieved flag rank were Lieutenant 'Geordie' Leslie, first lieutenant on H.M.S. *Volunteer*, and two Americans, Lieutenant-Commander E. B. Ellsworth, captain of U.S.S. *Upshur*, and Lieutenant-Commander J. D. Craik, executive officer of the Coast Guard Cutter *Ingham*; all three became Rear Admirals after the war. Admiral Sir Max Horton remained at Western Approaches where, with his own brand of enthusiasm and ruthlessness, he regarded the defeat of the U-boat as his personal crusade. When Admiral of the Fleet Sir Dudley Pound, the First Sea Lord, died late in 1943 it was rumoured that Horton was being considered as Pound's successor for this, the highest position a British naval officer could achieve, but Horton made it clear that 'even if the post had been offered to me officially I should have refused – I had got Dönitz where I wanted him and I intended to keep him there till the war was won!'†

The big surprise when it came to advancement was the decision by Hitler that Admiral Dönitz should be his successor as head of the German State when Hitler committed suicide on 30 April 1945. Dönitz sensibly decided that the war was all but over and spent the last few days trying to save as much

* Details of the later merchant ship sinkings are contained in Appendix 1, p. 332.

† Chalmers, W. S., *Max Horton and the Western Approaches*, Hodder & Stoughton, 1954, p. 211.

as possible of the German forces and the civilian population in the East from falling into the hands of the advancing Russians. The end of the war one week later found forty-nine U-boats still at sea and they were ordered to surrender at Allied ports. U-boat officers at Kiel disobeyed their beloved leader for the first and last time when they ordered that every U-boat in the remaining German-held harbours was to be scuttled on the receipt of the codeword *Regenbogen* (Rainbow), in defiance of the surrender terms that Dönitz was then negotiating with the British and Americans. The codeword was issued while Dönitz was asleep; the U-boats were scuttled and a few hours later the war was over.

The U-boat campaigns had resulted in the destruction of ships and the death of seamen on a massive scale. Captain John M. Waters, U.S.C.G., whose ship *Ingham* sailed briefly with SC.122's escort and whose own book has already been mentioned, makes the very good point that the number of men who lost their lives on both sides was greater than the combined deaths in all the naval battles of the previous 500 years! 5,150 Allied merchant ships were lost during the war and 2,828 of these were sunk by enemy submarines, mostly by German U-boats. The U-boats had also sunk 148 Allied warships, including three aircraft-carriers and two battle-ships. It is very difficult to quote precise Allied losses in men, with so many nationalities involved in several theatres of war and with the loss of ships being due to other causes besides that of U-boat attack, but it is certain that over 50,000 Allied merchant seamen and merchant ship gunners lost their lives and some two thirds of these were probably the victims of U-boat operations.* Platitudes and the 'easy tribute' must be avoided, but history would later decide that the efforts of these men, particularly those who died in the North Atlantic, and of those of their comrades who survived were quite decisive.

The Germans' own losses can be quoted more accurately.

* The following fatal casualties are known to be accurate. Merchant seamen; United Kingdom – 22,490 plus 6,093 Indian Lascars and 2,023 Chinese in British ships; United States – 5,662; Norway – 4,795; Greece – 2,000 (approx.); Holland – 1,914; Denmark – 1,886; Canada – 1,437; Belgium – 893; South Africa – 182; Australia – 109; New Zealand – 72. It is possible that as many as 6,500 merchant seamen from neutral countries also died. Gunners in merchant ships: British – 3,935 (Royal Navy – 2,713, Royal Artillery Maritime Regiment – 1,222); U.S. Navy Armed Guard (to end of 1944 only) – 1,640.

Out of 1,131 U-boats commissioned, 785 were lost. Out of 863 U-boats that actually sailed on operational patrols, 754 were sunk. Out of 39,000 men who went to sea in U-boats, the U-boat Memorial near Kiel records the names of the 27,491 who died; 5,000 became prisoners of war. It is difficult to bring to mind an example of a prolonged campaign producing such a high casualty rate and even more difficult to think of a situation in which morale and discipline held up so well under such fearful casualties.

The end of the war brought a great demobilization of servicemen and a return to civilian life, but for merchant seamen this was not a great landmark; the war had merely been a more hazardous continuation of their peacetime work, although the men from the Occupied Countries enjoyed a happy return to homes and families not seen for five years. Those U-boat ratings who had been taken prisoner were returned to their homes after several months, but many U-boat officers were dismayed to find that they were regarded as a special category of prisoner. They were never told the reasons for their retention but believed that the Allies regarded them as the most fanatical of Nazis (many had certainly been very truculent when captured – 'arrogant' is the word most frequently used by their captors) and that if returned to a recently defeated Germany they might form the core of Resistance groups – '*Werewolf* groups' was the vogue phrase of the period. Many U-boat officers were kept prisoner until well into 1947; when they were released they were more than grateful to settle down to a quiet life with little inclination to take to the woods. Instead they were a little surprised to find that they were soon being courted to serve in the re-born German Navy when Germany was invited to join N.A.T.O.

A harsher fate had befallen Admiral Dönitz. He was put on trial with the German war criminals at Nuremberg and charged on three counts:–

of participating in a conspiracy to commit crimes against peace, war crimes and crimes against humanity;
of participating in planning, preparation and initiation of wars of aggression;
of committing war crimes and being responsible for other persons who committed war crimes.

The charge of conspiracy was not proved and it was accepted that, until September 1939, Dönitz was simply performing his duties as an ordinary naval officer. On the third, or 'War Crimes' charge, he was found guilty of two violations of the pre-war agreement on submarine warfare which Germany had signed in that he had ordered neutral ships to be sunk in certain areas and that he had failed to honour the provisions of the agreement concerning the rescue of survivors. But, because the Royal Navy and the U.S. Navy had also waged what amounted to unrestricted submarine warfare, although after Germany's commencement of such activities, the tribunal decided not to assess Dönitz's sentence on the grounds of his U-boats' activities. Instead, the Allies managed to prove that Dönitz had been guilty of the more vague second charge, that of 'waging aggressive war' and, therefore, 'crimes against peace'. Dönitz was sentenced to ten years in prison and served the full term. Both verdict and sentence were bitterly resented by the former U-boat men, who felt that the Allied wartime propaganda against the U-boats had left them no choice but to punish the leader of the U-boat service. Former U-boat officers say that this imprisonment of Dönitz did much to damp their enthusiasm when they were pressed to serve with the Bundesmarine in N.A.T.O. Karl Dönitz was still alive when this book was prepared, a friendly and approachable man; the great handshaker of the wartime years still shakes a lot of hands at the annual reunions of U-boat men. But the post-war years were not so kind to Dönitz's great adversary. Sir Max Horton was expected to retire from the Royal Navy and become a business tycoon, but he died from a heart attack in 1951 and was given a well-deserved State Funeral Service in Liverpool Cathedral.

The escort vessels that protected SC.122 and HX.229 in March 1943 were mostly consigned to the scrapyard soon after the war. The only one that is still 'alive' is the American ship *Ingham*, which is still in active service with the U.S. Coast Guard. Another SC.122 escort whose name still lives on is the Belgian corvette *Godetia*. The original *Godetia* was scrapped, but when a new Belgian Force Navale was formed for service with N.A.T.O. one of its two new 1,700-ton support ships was named after the old Section Belge's *Godetia* and is the ship now used by the Belgian Royal Family on state voyages.

The surviving seamen of both sides live on and are all immensely proud of having taken part in that great and historic conflict in the Atlantic. What are their thoughts now, those men who took part in the fighting around Convoys SC.122 and HX.229, both on that operation and on the wider scene of their war in the Atlantic?

My lasting impression is not the violence of the enemy but a sense of the omnipotence of Almighty God, whether demonstrated through the awe-inspiring heaving of the ocean during a storm or through the peace and quiet of a starlit night with the ship gently rising and falling in a calm sea. (Sub-Lieutenant E. Walker, H.M.S. *Lavender*)

After one bad trip I remember being a bit dejected and wondering if we could win. Then I'd get ashore and perhaps hear a Winston Churchill speech about the gravity of things, then would come some trips with no losses at all and all would seem cheerful again. (Lieutenant N. T. M. Yeates, H.M.S. *Havelock*)

As a young, impressionable officer less than one year out of the U.S. Naval Academy, the learning process that accompanied SC.122 in particular and the Battle of the Atlantic in general was a tremendous and solid foundation block. I repeatedly and very consciously reviewed both tactical and strategic lessons of those few miserable days throughout my entire career. One of the longest lasting lessons was understanding problems of a minority in a joint command. In this case ours were eased by a very understanding R.N. escort commander. (Ensign D. Lawson, U.S.S. *Upshur*)

I must say that I gained a high respect for these British people in this battle. They were selected from all different types of the population, had received a sound training and fought like lions in a pressing situation. One would almost think they had nerves of steel. While the battle was at its peak, a young seaman came up on the bridge of *Pennywort* with a pot of tea and five or six mugs in one hand. He proceeded to go round to each of the officers, bowed and said, 'Do you want a cup of tea now, Sir?' It was as if he was at home in his mother's kitchen entertaining guests. (Second Officer H. Hansen, M.V. *Elin K*)

You ask about the atmosphere during the battle. Everyone I sailed with just went about his work with a silent contempt; there was nothing to do but press on ahead. Fears were not expressed. Wasn't it the same in all Forces? (Cabin Boy F. Craik, S.S. *Badjestan*)

We were listening to the B.B.C. one evening and one of our fellows said, 'I wonder if those people realize the trouble and danger we face bringing the pork chops over.' Well, probably they did, but most of them would soon forget and maybe sneer at us if we mentioned it. (Fireman P. Murphy, S.S. *Kingsbury*)

I flew co-pilot in those days to captains of quiet and easy demeanour and met men who behaved (in circumstances never presented to me) in the bravest and most exemplary manner. (Sergeant T. E. Kynnersley, 220 Squadron)

The main view of anyone concerned with the business of attacking U-boats from the air is its unreality as compared with anyone who engaged in, and vividly remembers, hand-to-hand fighting whether with fists or bayonets. We did not know how many, or indeed if any, men were consigned to the depths by our depth charges. Nowadays, the so-called 'English bars' in Spanish holiday resorts are popular also with young Germans because of the convivial sing-songs – 'Roll me over', 'Lili Marlene', 'Viva España'. It goes on for hours every night of the season. One can sit, surrounded on all sides by these happy, singing youngsters and reflect, looking at each face in turn, 'did I kill your father?' 'Reflect' is the word, nothing else – no remorse or satisfaction or whatever. (Sergeant R. V. Gurnham, 220 Squadron)

The 'bloody' Atlantic took countless victims from both sides and we bow to their heroism today. May these deeds not be forgotten and may they serve as a warning to the young of today's free countries. (Kapitänleutnant Hansjürgen Zetsche, U.591)

Even to this day, I sometimes wonder what inquisitive fish near Davey Jones' locker somewhere in the grey North Atlantic might be trying to tell the time by my watch hanging from the hook near the head of my bed on the starboard side of *Nariva*. (Second Officer G. D. Williams, S.S. *Nariva*)

I had quite forgotten that it was HX.229 – to me it was just our convoy. There were many others but that one was *the* convoy. (Surgeon-Lieutenant H. F. Osmond, H.M.S. *Volunteer*)

Having taken part in the great Battle of the Atlantic, all else was so tame. (Warrant-Engineer G. H. Ward, H.M.S. *Volunteer*)

Appendices

Appendix 1

The Merchant Ships of Convoys SC.122, HX.229 and HX.229A

Part 1: SC.122

SHIP	TONNAGE	COUNTRY	TYPE	CARGO*	CONVOY EXPERIENCE	LATER CAREER†
New York to Halifax Section						
Asbjörn	4,733	Canada	Freighter	Not known	Arrived safely	Still afloat as *Hoping Er Shi Chiu*
Permian	8,890	Panama	Tanker	Fuel oil	Arrived safely	Broken up 1960
New York to St John's Section						
Livingston	2,140	Newfoundland	Freighter	U.S. military stores	Fell out in storm	Sunk 3 September 1944 by U.541 south of Newfoundland
Polarland	1,591	Norway	Freighter	Not known	Fell out in storm	Sunk 4 January 1945 by U.1232 off Nova Scotia
Sevilla	7,022	Falkland Isles	Tanker	Fuel oil	Arrived safely	Disappeared from Lloyd's Register 1954
New York to Iceland Section						
Alcedo	1,392	Panama	Freighter	General	Fell out in storm and joined HX.229A	Sunk 28 February 1945 by U.1022 off Iceland
Askepot	1,312	Norway	Freighter	General	Arrived but via U.K.	Broken up 1965
Cartago	4,732	U.S.A.	Refrigerated	Meat	Arrived but via U.K.	Disappeared from Lloyd's Register 1950s
Eastern Guide	3,704	U.S.A.	Freighter	General	Fell out in storm	Broken up 1964
Godafoss	1,542	Iceland	Freighter	General	Arrived but via U.K.	Sunk 10 November 1944 by U.300 off Iceland (see *Shirvan*)

New York to United Kingdom

Ship	Tonnage	Flag	Type	Cargo	Convoy outcome	Fate
Alderamin	7,886	Holland	Freighter	Oil, seed, general	Sunk in convoy	—
Atland	5,203	Sweden	Freighter	Iron ore	Arrived safely	Sunk 27 March 1943 in collision with *Carso* in coastal convoy off E. Scotland
Aymeric	5,196	Britain	Freighter	Iron ore	Arrived safely	Sunk June 1943 by U.657 in North Atlantic while in convoy ONS.7
Baron Elgin	3,942	Britain	Freighter	Sugar	Arrived safely	As *Spidola* broken up 1970
Baron Semple	4,573	Britain	Freighter	General	Arrived safely	Sunk 2 November 1943 by U.848 in South Atlantic, no survivors
Baron Stranraer	3,668	Britain	Freighter	Iron ore	Arrived safely	As *Adelfortis II* broken up 1962
Beaconoil	6,893	Britain	Tanker	Fuel oil	Arrived safely	Broken up 1950
Benedick	6,978	Britain	Tanker	Fuel oil	Arrived safely	As *Nonna Mano* broken up 1961
Bonita	4,929	Panama	Freighter	Steel, tobacco	Arrived safely	As *Kongshavn* broken up 1958
Boston City	2,870	Britain	Freighter	General	Arrived safely	As *Cita di Monreale* broken up 1960
Bridgepool	4,845	Britain	Freighter	Linseed	Arrived safely	As *Ramses II* wrecked March 1951 in Bristol Channel
Carras	5,234	Greece	Freighter	Wheat	Sunk in convoy	—
Carso	6,275	Britain	Freighter	Steel, food	Arrived safely	As *Empire Tana* sunk as a blockship November 1944, probably off Normandy
Christian Holm	9,119	Britain	Tanker	Fuel oil	Arrived safely	As *Riza Kaptan* broken up 1959
Clarissa Radcliffe	5,754	Britain	Freighter	Iron ore	Straggled and sunk	—
Dolius	5,507	Britain	Freighter	Bauxite and general	Arrived safely	Sunk 5 May 1943 by U.584 in North Atlantic while in convoy ONS.5
Empire Dunstan	2,887	Britain	Freighter	Sugar	Arrived safely	Sunk 18 November 1943 by U.81 off S. Italy
Empire Galahad	7,046	Britain	Refrigerated	Meat	Arrived safely	As *Ocean Peace* broken up 1967
Empire Summer	6,949	Britain	Freighter	Explosives, general	Fell out in storm	Still afloat as *Hoping Wu Shi Er*

* Deck cargo not listed

† Some ships changed names several times. The one in this column is the last one borne by the ship

SHIP	TONNAGE	COUNTRY	TYPE	CARGO*	CONVOY EXPERIENCE	LATER CAREER†
English Monarch	4,557	Britain	Freighter	Copper, wheat, explosives	Fell out in storm	As Shinwa Maru foundered 23 January 1960 off Taiwan
Filleigh	4,856	Britain	Freighter	General	Arrived safely	Sunk 18 April 1945 by U.245 off North Foreland, Kent
Fort Cedar Lake	7,134	Britain	Freighter	General	Sunk in convoy	–
Georgios P	4,052	Greece	Freighter	Sugar	Returned to New York	Sunk as blockship 1944, probably off Normandy
Glenapp	9,503	Britain	Refrigerated	Cocoa, palm oil, copper	Commodore Ship, arrived safely	As Dardanus broken up 1957
Gloxinia	3,336	Britain	Tanker	Fuel oil	Arrived safely	As Vittorio O broken up 1959
Historian	5,074	Britain	Freighter	General	Arrived safely	As S.R.I. broken up 1960
Innesmoor	4,392	Britain	Freighter	Wheat	Arrived safely	As Weissesee broken up 1959
Kedoe	3,684	Dutch Indies	Freighter	Wheat, zinc, iron ore	Arrived safely	As Havana broken up 1961
King Gruffydd	5,072	Britain	Freighter	Steel, tobacco, explosives	Sunk in convoy	–
Kingsbury	4,898	Britain	Freighter	Soya, timber, bauxite	Sunk in convoy	–
Losada	6,520	Britain	Freighter	General	Arrived safely	Broken up 1952
L.S.T. 305	1,625	R.N. Ship	Tank Landing Ship	Tanks, food, steel	Arrived safely	Torpedoed by E-Boat off Anzio 20 February 1944, sank next day
L.S.T. 365	1,625	R.N. Ship	Tank Landing Ship	Tanks, food, steel	Arrived safely	Returned to U.S. Navy December 1946
McKeesport	6,198	U.S.A.	Freighter	Grain	Fell out in storm	Sunk 29 April 1943 by U.258 in North Atlantic while in convoy ONS.5
Orminster	5,712	Britain	Freighter	Iron ore	Arrived safely	Sunk 25 August 1944 by U.480 in English Channel
Shirvan	6,017	Britain	Tanker	Fuel oil	Arrived safely	Sunk 10 November 1944 by U.300 off Iceland – same day as Godafoss sunk
Vinriver	3,881	Britain	Freighter	Sugar	Arrived safely	As Briikon wrecked 27 November 1949

Name	Tonnage	Registry	Type	Cargo	Voyage outcome	Ultimate fate
[...]oss	1,101		Freighter	Timber, food	Arrived out via U.K.	Still afloat as *Star of Tay*
Selfoss	755	Iceland	Freighter	Timber	Fell out in storm and proceeded independently	Broken up 1955

Halifax to U.K. Section

Name	Tonnage	Registry	Type	Cargo	Voyage outcome	Ultimate fate
Badjestan	5,573	Britain	Freighter	Grain	Arrived safely	As *Somaegra* broken up 1959
Drakepool	4,838	Britain	Freighter	General	Arrived safely	As *Kinugasa Maru* abandoned and foundered in Sea of Japan
Empire Morn	7,092	Britain	Freighter	Grain	Arrived safely	As *Rio Pas* broken up 1972
Franka	5,282	Yugoslavia	Freighter	General	Arrived safely	As *Kordun* broken up 1959
Helencrest	5,233	Britain	Freighter	Grain	Arrived safely	As *Zonguldak* broken up 1969
Ogmore Castle	2,481	Britain	Freighter	Flour	Arrived safely	As *Roodevall* burnt out 23rd September 1961 at Cape Town
P.L.M.13	3,754	Britain	Freighter	Steel, timber	Fell out with engine trouble	As *Dimitrakis* broken up 1961
Parkhaven	4,803	Holland	Freighter	General	Arrived safely	Sunk as blockship 1944, probably off Normandy
Porjus	2,965	Sweden	Freighter	Steel, pulp	Arrived safely	Disappeared from Lloyd's Register 1967
Port Auckland	8,789	Britain	Refrigerated Freighter	Meat, general	Sunk in convoy	—
Zamalek	1,567	Britain	Freighter	None	Rescue ship, arrived safely after saving 165 survivors	Sailed with 64 wartime convoys and picked up 611 survivors. Sunk as blockship by Egyptians 1956 in Suez Canal
Zouave	4,256	Britain	Freighter	Iron ore	Sunk in convoy	—

St John's to U.K. Section

Name	Tonnage	Registry	Type	Cargo	Voyage outcome	Ultimate fate
Reaveley	4,998	Britain	Freighter	Not known	Arrived safely	As *Universal Mariner* broken up 1969

* Deck cargo not listed
† Some ships changed names several times. The one in this column is the last one borne by the ship

Part 2: HX.229

All ships sailed from New York for United Kingdom

SHIP	TONNAGE	COUNTRY	TYPE	CARGO*	CONVOY EXPERIENCE	LATER CAREER†
Abraham Lincoln	5,740	Norway	Refrigerated	Meat, explosives, general	Commodore ship, arrived safely	As *Korsholma* broken up 1962
Antar	5,222	Britain	Freighter	General	Arrived safely	As *Garbeta* broken up 1962
Belgian Gulf	8,237	Panama	Tanker	Lubricating oil	Arrived safely	Broken up 1960
Canadian Star	8,293	Britain	Refrigerated	Meat, butter, cheese	Sunk in convoy	—
Cape Breton	6,044	Britain	Freighter	Linseed	Arrived safely	Broken up 1962
City of Agra	6,361	Britain	Freighter	Copper, military stores	Arrived safely	Broken up 1965
Clan Matheson	5,613	Britain	Freighter	General	Could not maintain speed and fell out	Disappeared from Lloyd's Register 1955
Coracero	7,252	Britain	Refrigerated	Meat	Sunk in convoy	—
Daniel Webster	7,176	U.S.A.	Liberty‡	General	Arrived safely	Broken up 1948
Elin K	5,214	Norway	Freighter	Manganese, wheat	Sunk in convoy	—
El Mundo	6,008	Panama	Freighter	General	Arrived safely	Broken up 1946
Empire Cavalier	9,891	Britain	Tanker	Fuel oil	Arrived safely	Broken up 1959
Empire Knight	7,244	Britain	Freighter	General	Arrived safely	Wrecked 11 February 1944 on coast of Maine
Fort Anne	7,134	Britain	Freighter	Lead, timber, phosphate	Arrived safely	Broken up 1958
Gulfdisc	7,141	U.S.A.	Tanker	Fuel oil	Arrived safely	As *Prelude* broken up 1961
Harry Luckenbach	6,366	U.S.A.	Freighter	General	Sunk in convoy	—
Hugh Williamson	7,177	U.S.A.	Liberty	General	Fell out in storm but proceeded independently	Broken up 1948
Irénée Du Pont	6,125	U.S.A.	Freighter	General	Sunk in convoy	—
James Oglethorpe	7,176	U.S.A.	Liberty	Steel, cotton, food	Torpedoed and presumed to have sunk later	—

Ship	Tonnage	Country	Type	Cargo		
Luculus	6,546	Britain	Tanker	Diesel oil	Arrived safely	As *Pantanassa* broken up 1959
Magdala	8,248	Holland	Tanker	Aviation spirit	Arrived safely	Missing January 1945 between Iceland and U.K., no survivors, no U-Boat claim
Margaret Lykes	3,537	U.S.A.	Freighter	Grain, general	Arrived safely	As *Piauhy* broken up 1967
Mathew Luckenbach	5,848	U.S.A.	Freighter	Grain, general	Romped ahead of convoy and sunk	—
Nariva	8,714	Britain	Refrigerated	Meat	Sunk in convoy	—
Nebraska	8,261	Britain	Refrigerated	Meat	Arrived safely	Sunk 19 April 1944 by U.843 off Gibraltar. Captain B. C. Dodds survived sinking of *Nariva* and of *Nebraska*
Nicania	8,179	Britain	Tanker	Fuel oil	Arrived safely	Broken up 1960
Pan Rhode Island	7,742	U.S.A.	Tanker	Aviation spirit	Arrived safely	As *Amoco Rhode Island* broken up 1961
Regent Panther	9,556	Britain	Tanker	Fuel oil	Arrived safely	Broken up 1959
Robert Howe	7,227	U.S.A.	Liberty	General	Arrived safely	Broken up 1970
San Veronico	8,220	Britain	Tanker	Aviation spirit	Arrived safely	Broken up 1963
Southern Princess	12,156	Britain	Tanker	Fuel oil	Sunk in convoy	—
Stephen C. Foster	7,196	U.S.A.	Liberty	Sugar, general	Fell out with damaged hull	Broken up 1961
Tekoa	8,695	Britain	Refrigerated	Meat, wool	Arrived safely after saving 146 survivors	As *Kyokurei Maru* broken up 1970
Terkoelei	5,158	Dutch Indies	Freighter	Wheat, zinc	Sunk in convoy	—
Walter Q. Gresham	7,191	U.S.A.	Liberty	Sugar, powdered milk	Sunk in convoy	—
William Eustis	7,196	U.S.A.	Liberty	Sugar	Sunk in convoy	—
Zaanland	6,813	Holland	Refrigerated	Meat, wheat, zinc	Sunk in convoy	—

* Deck cargo not listed
† Some ships changed names several times. The one in this column is the last one borne by the ship
‡ The Liberty ships were freighters

Part 3: HX.229A

SHIP	TONNAGE	COUNTRY	TYPE	CARGO*	CONVOY EXPERIENCE	LATER CAREER†
New York to Halifax Section						
Esso Baltimore	7,949	U.S.A.	Tanker	Fuel oil	Arrived safely	As *Amphitrite* broken up 1964
Esso Belgium	10,529	Belgium	Tanker	Fuel oil	Arrived safely	As *Celestina* broken up 1959
Fort Amherst	3,489	Britain	Freighter	Not known	Arrived safely	As *Amherst* broken up 1964
Iris		Norway	Freighter	Not known	Arrived safely	Wrecked October 1944 Frobisher Bay, Canada
New York to St John's Section						
Fairfax	10,351	U.S.A.	Freighter	Not known	Arrived safely	As *Hudson* broken up 1971
New York to Iceland Section						
Empire Airman	9,813	Britain	Tanker	Aviation spirit	Arrived safely	As *San Wenceslao* broken up 1959
North King	4,934	Panama	Freighter	Military stores	Straggled to St John's	Broken up 1958
Shickshinny	5,108	U.S.A.	Freighter	Military stores	Straggled to St John's	As *Chaco* burnt out December 1946
New York to United Kingdom Section						
Clausinia	8,033	Britain	Tanker	Aviation spirit	Arrived safely	As *Metula* broken up 1958
Daphnella	8,078	Britain	Tanker	Aviation spirit	Arrived safely	Broken up 1962
Empire Nugget	9,807	Britain	Tanker	Aviation spirit	Arrived safely	As *Adellen* broken up 1961
Esperance Bay	14,204	Britain	Refrigerated cargo liner	Meat, general	Commodore ship, arrived safely	Broken up 1955
Esso Baytown	7,991	U.S.A.	Tanker	Navy fuel	Arrived safely	Broken up 1955
Franz Klasen	11,194	Panama	Tanker	Aviation spirit	Straggled to Halifax	As *Maggy* broken up 1960
Fort Drew	7,134	Britain	Freighter	Timber	Straggled	Broken up 1961
Henry S. Grove	6,220	U.S.A.	Refrigerated	General	Arrived safely	Broken up 1947
John Fiske	7,250	U.S.A.	Liberty	Steel, general	Arrived safely	Broken up 1971
Michigan	6,419	Panama	Freighter	General	Straggled to St John's	As *Republika* broken up 1962
Orville Harden	11,191	Panama	Tanker	Crude oil	Arrived safely	As *Esso Orinoco* broken up 1962
Pan Florida	7,237	U.S.A.	Tanker	Navy fuel	Arrived safely	As *Peramo* sunk in collision March 1957
Pan Maine	7,237	U.S.A.	Tanker	Fuel oil	Arrived safely	As *Sinchi Roca* broken up 1968

					trouble	Atlantic while in convoy HX.231
Svend Foyn	14,795	Britain	Tanker	Fuel oil	Sunk by iceberg	–
Tactician	5,996	Britain	Freighter	General	Arrived safely	As *Crostafels* broken up 1959
Halifax to St John's Section						
Lady Rodney		Britain	Freighter	Not known	Arrived safely	–
Halifax to United Kingdom Section						
Akaroa	15,130	Britain	Refrigerated cargo liner	Meat, general	Arrived safely	Broken up 1954
Arabian Prince	1,960	Britain	Refrigerated	Sugar, general	Arrived safely	Broken up 1958
Belgian Airman	6,959	Belgium	Freighter	Steel, timber	Damaged by ice, put into Iceland	Sunk 14 April 1954 in North Atlantic
Bothnia	2,407	Britain	Freighter	Nuts	Arrived safely	As *Capetan Manolis* broken up 1961
City of Oran	7,323	Britain	Freighter	General	Arrived safely	Sunk 2 August 1943 by U.196 in Indian Ocean
Fresno Star	7,998	Britain	Refrigerated	Meat	Arrived safely	Broken up 1947
Ganymedes	2,682	Holland	Freighter	Sugar	Arrived safely	Broken up 1955
Lossiebank	5,627	Britain	Refrigerated	Wheat, ore	Arrived safely	Broken up 1962
Manchester Trader	5,671	Britain	Refrigerated	General	Arrived safely	Broken up 1962
Norwegian	6,366	Britain	Freighter	General	Arrived safely	As *Maria Elaine* broken up 1959
Port Melbourne	9,142	Britain	Refrigerated	Meat, general	Arrived safely	Broken up 1947
Ribera	5,598	Britain	Freighter	Flour	Arrived safely	As *Jolanda* broken up 1971
Rosemont	4,956	Panama	Freighter	Steel, general	Arrived safely	As *Ioannis* broken up 1971
Tahsinia	7,267	Britain	Freighter	Sugar, ore	Arrived safely	Sunk 1 October 1943 by U.532 in Indian Ocean
Taybank	5,627	Britain	Refrigerated	Grain	Arrived safely	Broken up 1960
Tortuguero	5,295	Britain	Refrigerated	General	Arrived safely	Broken up 1958
Tudor Star	7,199	Britain	Refrigerated	Meat, general	Arrived safely	Broken up 1949
St John's to United Kingdom Section						
Lone Star	5,101	U.S.A.	Freighter	General, explosives	Diverted to Iceland	As *Huseyin Kaptan* broken up 1953

* Deck cargo not listed

† Some ships changed names several times. The one in this column is the last one borne by the ship

Appendix 2
Local Escort Groups

SHIP AND CAPTAIN	TYPE	DURATION OF ESCORT DUTY
SC.122		
H.M.C.S. *The Pas* (Lt-Cdr E. C. Old, R.C.N.R.)	Flower Class corvette	New York to Halifax
H.M.C.S. *New Westminster* (Lt R. O. McKenzie, R.C.N.R.)	Flower Class corvette	New York to St John's
H.M.C.S. *Rimouski* (Lt R. J. Pickford, R.C.N.V.R.)	Flower Class corvette	New York to St John's
H.M.C.S. *Blairmore* (Lt W. J. Kingsmill, R.C.N.V.R.)	Bangor Class minesweeper	New York to Halifax
H.M.S. *Leamington* (Lt A. D. B. Campbell, R.N.)	Town Class destroyer (ex U.S.S. *Twiggs*)	Halifax to St John's
H.M.C.S. *Dunvegan* (Lt J. A. Tullis, R.C.N.R.)	Flower Class corvette	Halifax to St John's
H.M.C.S. *Cowichan* (Lt K. W. N. Hall, R.C.N.V.R.)	Bangor Class minesweeper	Halifax to St John's
HX.229		
H.M.S. *Chelsea* (Lt-Cdr J. E. R. Wilford, R.D., R.N.R.)	Town Class destroyer (ex U.S.S. *Crowninshield*)	New York to 55 degrees
U.S.S. *Kendrick* (Lt-Cdr C. T. Caufield, U.S.N.)	Benson Class destroyer	New York to St John's
H.M.C.S. *Fredericton* (Lt-Cdr J. H. S. MacDonald, R.C.N.)	Flower Class corvette	New York to St John's
H.M.C.S. *Oakville* (Lt-Cdr C. A. King, D.S.C., R.C.N.R.)	Flower Class corvette	New York to St John's
H.M.C.S. *Annapolis* (Lt-Cdr A. G. Boulton, R.C.N.V.R.)	Town Class destroyer (U.S.S. *Mackenzie*)	55 degrees West to St John's

Ship	Class	Route
H.M.C.S. *Barrie* (Lt R. M. Mosher, R.C.N.R.)	Flower Class corvette	New York to Halifax
H.M.C.S. *Kamsack* (Lt E. Randall, R.C.N.R.)	Flower Class corvette	New York to Halifax
H.M.C.S. *Snowberry* (Lt-Cdr P. J. B. Watts, R.C.N.V.R.)	Flower Class corvette	New York to Halifax
H.M.C.S. *Digby* (Lt S. W. Howell, R.C.N.R.)	Bangor Class minesweeper	New York to Halifax
H.M.C.S. *Gananoque* (Lt E. M. More, R.C.N.R.)	Bangor Class minesweeper	New York to Halifax
H.M.C.S. *Noranda* (Lt J. E. Francois, R.C.N.R.)	Bangor Class minesweeper	New York to Halifax
H.M.C.S. *St Clair* (Lt-Cdr G. O. Baught, R.C.N.R.)	Town Class destroyer (ex U.S.S. *Williams*)	Halifax to St John's
H.M.C.S.s *The Pas* and *Blairmore* (See SC.122)		Halifax to St John's

Appendix 3
Ocean Escort Groups

SHIP AND CAPTAIN	TYPE

SC.122, mainly B5 Group sailing from St John's, 11th March

H.M.S. *Havelock* (Cdr R. C. Boyle, D.S.O., R.N.)	Havant Class destroyer
H.M.S. *Swale* (Lt-Cdr J. Jackson, R.N.R.)	River Class frigate
H.M.S. *Buttercup* (Lt-Cdr J. C. Dawson, R.N.R.)	Flower Class corvette
H.M.S. *Godetia* (Lt A. M. Larose, R.N.R.)	Flower Class corvette
H.M.S. *Lavender* (Lt L. G. Pilcher, R.N.R.)	Flower Class corvette
H.M.S. *Pimpernel* (Lt H. D. Hayes, R.N.R.)	Flower Class corvette
H.M.S. *Saxifrage* (Lt N. L. Knight, R.N.R.)	Flower Class corvette
U.S.S. *Upshur* (Lt-Cdr E. B. Ellsworth, U.S.N.)	Destroyer, class never named
H.M.S. *Campobello* (Lt K. A. Grant, R.N.V.R.)	Isles Class trawler

From 19 March

U.S.C.G. Cutter *Ingham* (Capt. A. M. Martinson, U.S.C.G.)	Secretary Class cutter

HX.229, mainly B4 Group sailing from St John's, 13 March

H.M.S. *Volunteer* (Lt-Cdr G. J. Luther, R.N.)	V and W Class destroyer
H.M.S. *Beverley* (Lt-Cdr R. A. Price, R.N.)	Town Class destroyer

H.M.S. *Pennywort* (Lt O. G. Stuart, R.C.N.V.R.)   Flower Class corvette

From 18 March

H.M.S. *Highlander* (Cdr E. C. L. Day, R.N.)   Havant Class destroyer

From 19 March

U.S.S. *Babbitt* (Lt-Cdr S. F. Quarles, U.S.N.)*   Destroyer, class never named
H.M.S. *Abelia* (Lt-Cdr F. Ardern, R.N.R.)   Flower Class corvette

From 20 March

H.M.S. *Vimy* (Lt-Cdr R. B. Stannard, V.C., R.N.R.)   V and W Class destroyer
H.M.C.S. *Sherbrooke* (Lt J. A. M. Levesque, R.C.N.R.)   Flower Class corvette

---

HX.229A, 40th Escort Group sailing from St John's, 14 March

H.M.S. *Aberdeen* (Cdr J. S. Dalison, D.S.O., R.N.)   Grimsby Class sloop
H.M.S. *Hastings* (Lt-Cdr L. B. Philpott, D.S.O., R.N.R.)   Hastings Class sloop
H.M.S. *Landguard* (Lt-Cdr T. S. L. Fox-Pitt, R.N.)   Banff Class sloop
          (ex U.S.C.G. *Shoshone*)
H.M.S. *Lulworth* (Lt-Cdr R. C. S. Wooley, R.D., R.N.R.)   Banff Class sloop
          (ex U.S.C.G. *Chelan*)
H.M.S. *Moyola* (Lt H. N. Lawson, R.N.R.)   River Class frigate
H.M.S. *Waveney* (Lt-Cdr A. E. Willmott, D.S.C., R.N.R.)   River Class frigate

From 21 March

U.S.C.G. Cutter *Geo. M. Bibb* (Cdr R. L. Raney, U.S.C.G.)   Secretary Class cutter

* *Babbitt* did not join the escort of HX.229 until 20 March but she was in action nearby on the 19th.

Appendix 4
U-boats Involved in the SC.122/HX.229 Operations

Composition of Groups is according to a U-boat's position when HX.229 was sighted on the morning of 16 March 1943.

Note: Rank abbreviations used are KK – Korvettenkäpitan; KL – Kapitänleutnant; and OL – Oberleutnant.

Flotillas were based as follows: 1st and 9th at Brest, 2nd and 10th at Lorient, 3rd at La Pallice, 6th and 7th at St Nazaire, 12th at Bordeaux.

Under 'Convoy Experiences' the following terms are used: 'in contact' with a convoy – had that convoy in sight; 'sank' – was entirely responsible for sinking a merchant ship by torpedo attack; 'damaged' – scored torpedo hits but did not sink the ship; 'finished off' – sank a previously damaged ship; 'attacked without success' – fired torpedoes without scoring hits.

U-BOAT AND CAPTAIN	TYPE	FLOTILLA	CONVOY EXPERIENCES	HOW CAREER ENDED
Gruppe Raubgraf				
U.84 (KL Horst Uphoff)	VIIB	1st	On 7th Patrol. In contact with HX.229 twice but made no attacks. Depth charged by *Anemone* on 19th but not damaged.	24 August 1943, sunk in North Atlantic by aircraft of U.S.S. *Core*, no survivors.
U.91 (KL Heinz Walkerling)	VIIC	9th	On 3rd Patrol. Sank *Harry Luckenbach* of HX.229, probably finished off *Nariva* and *Irénée du Pont*.	25 February 1944, sunk in North Atlantic by H.M.S.s *Affleck*, *Gore* and *Gould*; 16 survivors.
U.435 (KL Siegfried Strelow)	VIIC	1st	On 7th Patrol. Damaged *William Eustis* of HX.229, made one other attack without success.	9 July 1943, sunk off Portugal by Wellington of 179 Squadron; no survivors.
U.600 (KL Bernhard Zurmühlen)	VIIC	3rd	On 3rd Patrol. In contact three times	25 November 1944, sunk off Gibraltar

U-boat (Commander)	Type	Patrol	Operations	Fate
U.603 (OL Hans-Joachim Bertelsmann)	VIIC	1st	On 2nd Patrol. Sank ... and then became official shadower.	... at 16 ... sank ... by U.S.S. Bronstein; no survivors.
U.615 (KL Ralph Kapitsky)	VIIC	3rd	On 3rd Patrol. In contact with HX.229 twice but made no attacks. Depth charged by Anemone and Volunteer on 19th but not damaged.	6 August 1943, attacked off Curacao by nine aircraft from four U.S. squadrons then scuttled; KL Kapitzky went down with his boat but 43 men saved.
U.664 (OL Adolf Graef)	VIIC	9th	On 3rd Patrol. In intermittent contact with HX.229, too low on fuel for high-speed work and made no attacks.	9 August 1943, sunk off Azores by aircraft of U.S.S. Card; 52 survivors.
U.758 (KL Helmut Manseck)	VIIC	6th	On 2nd Patrol. Sank Zaanland and damaged James Oglethorpe of HX.229 then broke off operation, all torpedoes expended.	Severely damaged in March 1945 and not used again.

Gruppe Stürmer

U-boat (Commander)	Type	Patrol	Operations	Fate
U.184 (OL Hans-Günther Brosin)	VIIC	3rd	On 8th Patrol. In intermittent contact with HX.229 but made no attacks. Depth charged and damaged on 19th by Anemone and Volunteer.	24 August 1943, sunk in Bay of Biscay by Wellington of 179 Squadron; no survivors.
U.190 (KL Max Wintermeyer)	IXC	2nd	On 1st Patrol. In intermittent contact with HX.229 but made no attacks. Depth charged and damaged on 19th by Babbitt.	14 May 1945, surrendered at Halifax, Nova Scotia, disposed of by scuttling on 21 October 1947.
U.229 (OL Robert Schetelig)	VIIC	6th	On 1st Patrol. Took part in patrol line sweeps but then ordered to position off Greenland for weather observation duties.	22 September 1943, rammed and sunk off Greenland by H.M.S. Keppel; no survivors.
U.305 (KL Rudolf Bahr)	VIIC	1st	On 1st Patrol. Sank Zouave and Port Auckland of SC.122. Depth charged four times by aircraft on 17th and 18th and forced to break off operation.	17 January 1944, sunk in North Atlantic by H.M.S.s Glenarm and Wanderer; no survivors.

U-BOAT AND CAPTAIN	TYPE	FLOTILLA	CONVOY EXPERIENCES	HOW CAREER ENDED
U.338 (KL Manfred Kinzel)	VIIC	7th	On 1st Patrol. Sank *Kingsbury*, *King Gruffydd* and *Alderamin*; damaged and maybe sank *Fort Cedar Lake* – all of SC.122 – on 17th. Sank *Granville* of SC.122 later on 17th. Depth charged and damaged by *Lavender* and *Upshur* on 18th. Depth charged and damaged again by Sunderland 423 Squadron on 19th.	20 September 1943, sunk in North Atlantic by Liberator of 120 Squadron; no survivors.
U.384 (OL Hans-Achim von Rosenberg-Gruszczynski)	VIIC	Allocated to 3rd Flotilla but never reached base.	On 2nd Patrol. Sank *Coracero* and possibly *Terkoelei* of HX.229. Herself sunk, probably on 19th by Fortress of 206 Squadron. There were no survivors.	–
U.439 (OL Helmut von Tippelskirch)	VIIC	1st	On 7th Patrol. 1st for new captain. In contact with HX.229 twice but made no attacks. Depth charged by Liberator of 86 Squadron on 17th but not damaged. Depth charged and damaged by *Highlander* on 19th.	3 May 1943, sunk in collision with U.659 off Cape Ortegal; 9 survivors.
U.523 (KL Werner Pietzsch)	IXC	10th	On 1st Patrol. Made no contact with a convoy but finished off *Mathew Luckenbach*, romper from HX.229.	25 August 1943, sunk in North Atlantic by H.M.S.s *Wallflower* and *Wanderer*; 37 survivors.
U.526 (KL Hans Möglich)	IXC	Allocated to 10th Flotilla but never reached base.	On 1st Patrol. Made no contact with a convoy. Struck mine and sank when coming into Lorient on 14 April, 12 men rescued by Germans.	–
U.527 (KL Herbert Uhlig)	IXC	10th	On 1st Patrol. Made no contact with a convoy. Damaged *Mathew Luckenbach*. Depth charged and damaged by	23 July 1943, sunk off Azores by aircraft of U.S.S. *Bogue*; 13 survivors.

U-boat (Commander)	Type	Patrol	Notes	Fate
			once and HX.229 twice but made no attacks. Depth charged and damaged by Sunderland of 201 Squadron on 20th.	craft of U.S.N. Patrol Squadron 107; 2 survivors.
U.618 (KL Kurt Baberg)	VIIC	7th	On 3rd Patrol. In brief contact with HX.229 but made no attacks.	14 August 1944, sunk in Bay of Biscay by Liberator of 53 Squadron; no survivors.
U.631 (OL Jürgen Krüger)	VIIC	9th	On 2nd Patrol. Possibly sank *Terkoelei* of HX.229. Depth charged and damaged by Sunderland of 201 Squadron on 20th.	17 October 1943, sunk in North Atlantic by H.M.S. *Sunflower*; no survivors.
U.641 (KL Horst Randtel)	VIIC	7th	On 1st Patrol. In brief contact with SC.122 but made no attacks.	19 January 1944, sunk in North Atlantic by H.M.S. *Violet*; no survivors.
U.642 (KL Herbert Brünning)	VIIC	6th	On 1st Patrol. In contact with SC.122 three times but made no attacks.	5 July 1944, severely damaged at Toulon by U.S.A.A.F. bombing raid and not used again.
U.665 (OL Hans-Jürgen Haupt)	VIIC	Allocated to 1st Flotilla but never reached base.	On 1st Patrol. Fired two torpedoes at ship, possibly finishing off *Fort Cedar Lake*. Sunk by Wellington of 172 Squadron in Bay of Biscay on 22 March; no survivors.	—
U.666 (OL Herbert Engel)	VIIC	6th	On 1st Patrol. Damaged *Carras* of SC.122. Depth charged by Liberator of 86 Squadron and *Godetia* and *Upshur* on 17th; depth charged and damaged by Fortress of 220 Squadron on 19th.	10 February 1944, sunk in North Atlantic by aircraft of H.M.S. *Fencer*.

Gruppe Dränger

U-boat (Commander)	Type	Patrol	Notes	Fate
U.86 (KL Walter Schug)	VIIB	1st	On 6th Patrol. Made no contact with a convoy.	29 November 1943, sunk off Azores by aircraft of U.S.S. *Bogue*; no survivors.

U-BOAT AND CAPTAIN	TYPE	FLOTILLA	CONVOY EXPERIENCES	HOW CAREER ENDED
U.221 (OL Hans Trojer)	VIIC	7th	On 3rd Patrol. Sank *Canadian Star* and *Walter Q. Gresham* of HX.229.	27 September 1943, sunk in Bay of Biscay/North Atlantic by Halifax of 58 Squadron; no survivors.
U.333 (OL Werner Schwaff)	VIIC	3rd	On 6th Patrol. Made no contact with a convoy but finished off a derelict ship, possibly *Carras*.	31 July 1944, sunk off Scilly Isles by H.M.S.s *Loch Killin* and *Starling*; no survivors.
U.336 (KL Hans Hunger)	VIIC	1st	On 2nd Patrol. Made no contact with a convoy.	4 October 1943, sunk off Iceland by aircraft of U.S.N. Bombing Squadron 128; no survivors.
U.373 (KL Paul-Karl Loeser)	VIIC	3rd	On 8th Patrol. Suffering from several mechanical defects but made contact once with HX.229 without attacking. Depth charged by unknown escort on 19th but not damaged.	8 June 1944, sunk off Brest by Liberator of 224 Squadron; 47 men rescued by Germans.
U.406 (KL Horst Dieterichs)	VIIC	7th	On 5th Patrol. In contact with HX.229 once but made no attacks.	18 February 1944, sunk in North Atlantic by H.M.S. *Spey*; 41 survivors.
U.440 (KL Hans Geissler)	VIIC	1st	On 4th Patrol. In one brief contact with HX.229 on 19th but immediately depth charged and slightly damaged, probably by *Anemone*.	31 May 1943, sunk in Bay of Biscay by Sunderland of 201 Squadron during the 'stay up and fight' period; no survivors.
U.441 (KL Klaus Hartmann)	VIIC	1st	On 3rd Patrol. In contact with HX.229 once, attacked without success. Depth charged and damaged by Sunderland of 423 Squadron on 20th.	18 June 1944, sunk off Brittany by Wellington of 304 (Polish) Squadron; no survivors.
U.590 (KL Heinrich Müller-Edzards)	VIIC	6th	On 3rd Patrol. In contact with HX.229 once but made no attacks.	9 July 1943, sunk off Brazil by aircraft of U.S.N. Patrol Squadron 94; no survivors.
U.608 (KL Rolf Struckmeier)	VIIC	6th	On 3rd Patrol. In contact with HX.229 twice, attacked *Highlander* without	10 August 1944, sunk in Bay of Biscay by H.M.S. *Wren* and Liberator of 53

damaged.

U-Boats Not in Groups

U.86 (KL Dietrich Lohmann)	VIIC	9th	On 4th Patrol. Had been in Raubgraf but released for home with engine defects and shortage of fuel. Joined in operation and made contact with HX.229 but depth charged and damaged possibly by *Mansfield* on 17th, made no attacks.	14 May 1943, sunk in North Atlantic by H.M.S.s *Broadway* and *Lagan* and aircraft of H.M.S. *Biter*.
U.119 (KL Alois Zech)	XB	12th	A tanker U-boat, replenished operational boats during convoy action.	24 June 1943, sunk in Bay of Biscay by H.M.S. *Starling*; no survivors.
U.228 (OL Erwin Christopherson)	VIIC	6th	On 1st Patrol. Came from tanker U-boat and made contact with HX.229, made two attacks without success. Depth charged and damaged by *Beverley* on 17th.	4 October 1944, bombed by R.A.F. in Bergen harbour and not used again.
U.230 (KL Paul Siegmann)	VIIC	9th	On 1st Patrol. Came from tanker U-boat but could make no contact with a convoy.	21 August 1943, scuttled at Toulon.
U.463 (KK Leo Wolfbauer)	XIV	12th	A tanker U-boat, replenished operational boats during convoy action.	15 May 1943, sunk in Bay of Biscay by Halifax of 58 Squadron; no survivors. The aircraft was captained by W/Cdr W. Oulton, who sank two other U-boats in the same month. By coincidence the numbers of all three ended in '63'; the others were U.563 and U.663.

U-BOAT AND CAPTAIN	TYPE	FLOTILLA	CONVOY EXPERIENCES	HOW CAREER ENDED
U.616 (OL Siegfried Koitschka)	VIIC	6th	On 1st Patrol. Came from tanker U-boat, made contact with HX.229 and carried out three attacks without success.	17 May 1944, sunk in Mediterranean by eight U.S.N. ships and aircraft of 38 Squadron R.A.F. after a three-day hunt, the longest U-boat hunt of the war; 47 survivors.
U.653 (Kl. Gerhard Feiler)	VIIC	1st	On 6th Patrol. The initial sighter of HX.229 and shadowed until relieved by another boat.	15 March 1944, sunk in North Atlantic by H.M.S.s *Starling* and *Wild Goose* and aircraft of H.M.S. *Vindex*; no survivors.
U.663 (Kl Heinrich Schmid)	VIIC	9th	On 2nd Patrol. Not operating against main convoys but possibly sank *Clarissa Radcliffe*, straggler from SC.122.	7 May 1944, sunk off Brest by Halifax of 58 Squadron; no survivors.

Appendix 5
The Roll of Honour

Here are recorded the names of the men and women who lost their lives in the operations of Convoys HX.229 and SC.122. Torpedoed ships are listed in the order in which they were hit and their owners and ports of registry are shown. The dead of each ship are listed in alphabetical order. The location of their homes is given as accurately as possible, but the home district mentioned may occasionally be that of parents who later moved or of next-of-kin and the dead person may never have lived there. Only the home State of the U.S. Navy casualties can be given.

The following abbreviations are used:

Captain	Capt.
First, Second Officer, etc.	1st, 2nd Off., etc.
Chief, Second Engineering Officer, etc.	Ch., 2nd Eng. Off., etc.
Radio Officer	Rad. Off.
Boatswain	Bosun
Quartermaster	Q.M.
Able Seaman	A.B.
Ordinary Seaman	O.S.
Seaman 1st Class, 2nd Class	S1c., S2c.
Fireman and Watertender	F.W.T.

Convoy HX.229

ELIN K (Owners – A/S Inger. Managers – Jacob Kjøde, A/S. Port of registration – Bergen.)
No fatal casualties.

ZAANLAND (Koninklijke Hollandische Lloyd. Port – Amsterdam.)
No fatal casualties.

JAMES OGLETHORPE (U.S. War Shipping Administration, Savannah.)
A.B. and Q.M. S. I. Bullard, Alamo, Georgia; F.W.T. L. R. Bustin, Savannah; Messman J. D. Carter, Savannah; 3rd Assistant Eng. J. E. Cole, Johnson City, Tennessee; A.B. and Q.M. D. B. Deloach, Savannah; Oiler D. A. Dix, Atlanta; 2nd off. J. L. Duke, Savannah; Messman C. W. Groover, Savannah; Assistant Cook G. W. Hayman, Savannah; Deck Eng. W. W. Huggins, Savannah; F.W.T. L. Jernigan, Savannah; O.S. M. Kiley, Savannah; Rad. Off. H. Kiveit, Jackson Heights, New York; Cadet J. R. Lambert, Del Rio, Texas; Capt. A. W. Long, Fort Lauderdale, Florida; 2nd Cook T. J. McDaniel, Bethlehem, Georgia; 1st Assistant Eng. G. B. Parks, Beaumont, Texas; A.B. C. F. Puckette, Tarborough, North Carolina; Cadet R. M. Record, Oklahoma City; Wiper C. Reed, Savannah; Steward C. F. Salzman, Norfolk, Virginia; A.B. and Q.M. B. G. Schiebold, Savannah; A.B. and Q.M. H. S. Smithson, Rutherford, New Jersey; Wiper J. Thomas, Melbourne, Florida; Ch. Cook T. J. Thomas, Griffin, Georgia; Ch. Eng. Off. W. N. Tiencken, Savannah; Utility A. J. Von Dolteren, Savannah; Oiler T. W. Welsh,

Savannah; Utility C. T. White, Jacksonville, Florida; A.B. and Q.M. F. J.
Williams, Bellaire, Ohio; O.S. R. R. Wilson, Aysleworth, Oklahoma.
U.S. Navy Armed Guard: Lieutenant (j.g.) J. E. Bayne, Michigan; S2c. H. F.
Daggs, Indiana; S2c. J. De Francisco, Illinois; S2c. M. J. Demers, Massa-
chusetts; S2c. K. P. Dreyer, Indiana; Gunner's Mate 3rd Class R. C. Parrish,
Virginia; S2c. C. W. Pelton, State of Washington; S2c. W. P. Sheers, District
of Columbia; S2c. W. J. Smith, Massachusetts; S1c. F. A. Weed, Pennsylvania;
Signalman 3rd Class F. L. Roales, Indiana. Two U.S. Navy enlisted men
passengers cannot be identified by name.

WILLIAM EUSTIS (U.S. War Shipping Administration. Port – Houston.)
No fatal casualties.

HARRY LUCKENBACH (Luckenbach Steamship Company Inc. Port – New
York.)
Ch. Off. C. S. Auetitore, Beechnut, Minnesota; 1st Assistant Eng. R. C.
Bemrick, Baltimore; Messman B. C. Bowers, Jacksonville, Florida; A.B.
J. A. Bryant, British Columbia; Q.M. S. S. Buzzard, Pittsburgh; Cadet L. T.
Byrd, Benson, North Carolina; 3rd Assistant Eng. D. O. Carria, Portugal;
Rad. Off. N. A. Daley, Waterton, South Dakota; A.B. L. P. DaSilva, Bronx;
A.B. L. J. Davis, New York; Watertender J. DeOliveira, Brooklyn; 3rd Cook
H. C. Ferebee, Brooklyn; Deck Eng. R. W. Fleck, Winchester, Virginia;
Storekeeper J. Flood, Philippines; Steward A. S. Gaskin, New York; Fireman
A. Green, Duluth, Minnesota; Messman M. Greenblatt, Brooklyn; A.B.
M. Herbert, Bronx; Wiper J. M. Hotochen, Monessen, Pennsylvania; Ch.
Cook C. Johnson, Roxbury, Minnesota; A.B. and Q.M. S. P. Kellio, Shushan,
New York; 3rd Off. L. W. Kernan, Elgin, Illinois; Oiler G. D. Kline, Somer-
ville, Minnesota; Watertender A. B. Klinga, Bridgeport, Connecticut; A.B.
A. V. Koutavides, New York; O.S. E. B. Lafferty, Elizabeth, New Jersey;
A.B. B. Lane, Staten Island; A.B. M. Levine, Brooklyn; Ch. Eng. Off. R. E.
Lindsay, San Francisco; Fireman J. M. Lopez, New York; Wiper P. Marcin-
kericz, Plymouth, Pennsylvania; Capt. R. McKinnon, Oakland, California;
Cadet W. J. Meyer, Wheaton Place, New Jersey; Utility C. W. Miller,
Martinsville, Indiana; Cadet F. R. Miller, Chesterton, Maryland; Utility
B. C. Mitchell, New York; Oiler J. N. N. Moreira, New York; 2nd Cook F. C.
Mores, Whiting, Indiana; Messman L. H. Nebelett, Brooklyn; Bosun H.
Newell, New York; 4th Assist. Eng. J. H. Nychay, Matapan, Minnesota;
Cadet W. H. Parker, Onancock, Virginia; Fireman S. Pekola, Carteret, New
York; Carpenter P. A. Petersen, Brooklyn; Messman D. Rosado, New York;
Oiler I. Santillon, New York; 2nd Assistant Eng. F. H. Schmidt, Palacios,
Texas; Q.M. M. Soares, Portugal; Messman L. Strong, Babylon, New York;
Watertender J. Trembelas, Washington; Oiler W. A. Vallis, Bayonne, New
Jersey; Oiler J. J. Weldon, Staten Island; O.S. J. Zaia, Winthrop, Minnesota;
Oiler J. A. Zukleia, Plymouth, Pennsylvania.
U.S. Navy Armed Guard: S1c. R. O. Beckwith, Kansas; S1c. A. C. Briede,
Missouri; Ensign W. A. Burr, Arkansas; S1c. J. E. Callahan, New Jersey;
S1c. E. L. Case, Pennsylvania; S1c. D. E. Herzog, Michigan; S2c. H. R.
Howland, Rhode Island; S2c. S. R. Hyde, Pennsylvania; S1c. J. M. Jacaruso,
New York; S1c. A. C. Jackson, North Carolina; S1c. J. C. Karr, Ohio; S2c.
W. A. Klecka, New Jersey; S1c. L. M. Larson, North Carolina; S1c. W. D.
Paton, North Dakota; S1c. T. F. Pucylowski, Pennsylvania; S1c. C. M.
Rothermel, Pennsylvania; S1c. A. M. Roush, West Virginia; S1c. N. L. Seitz,
Missouri; S1c. G. E. Sension, Pennsylvania; S1c. S. O. Shaffer, Missouri;
Coxswain L. J. Shields, New Jersey; S1c. P. C. Shorter, Kentucky; S1c. H. W.
Starkloff, Maryland; S1c. J. A. Stilinovich, Minnesota; S1c. J. R. Thomasson,
Tennessee; Signalman 3rd Class R. A. Topel, Illinois.

IRENEE DUPONT (International Freighting Corporation Inc. Port – Wilmington.)
Ch. Cook L. A. Davis, New York; Messman C. Bruenler, San Francisco; 2nd Cook S. A. Gaynor, Brooklyn; Messman W. Green, New York; Messman J. L. Rivera, New York; Oiler R. Sanguinedo, Brooklyn; O.S. J. F. Valdea, Cuba.
Passenger: Ensign B. Norris, Savannah.
U.S. Navy Armed Guard: S1c. M. E. Borgeson Jr, New York; S1c. H. H. Schwinn, Maryland; S1c. M. E. Stringer, Indiana; S1c. R. J. Trsek, Ohio; Gunner's Mate 3rd Class G. E. Woods, District of Columbia.

NARIVA (Royal Mail Lines, Southampton.)
No fatal casualties.

SOUTHERN PRINCESS (South Georgia Co. Ltd., managed by Christian Salveson and Co. Port – London.)
Messroom Boy P. Scanlan, Greenock; Boy W. A. Wilkins, New York.
Passengers: Greaser T. Cain, Bethnal Green, London and Fireman R. Cooper, Elton, were Distressed British Seamen.

TERKOELEI (N. V. Nederlandsch-Indische Maatschappij voor Zeevaart I. managed by Rotterdamsche Lloyd. Port – Batavia.)
Fireman Ali Sheir Muhammad, India; Greaser Ali Diwan Nadir, India; Trimmer Allafdin Muhammad, India; Steward Amat, Java; Topass Barkat Moontoo, India; Seaman Bhana Ranchord, India; Tindal Bagh Ali Sumari, India; Greaser Bugga Nawab, India; Helmsman P. M. But, Vlissingen, Holland; Seacunny Dayal Soma, India; Tindal Fateh Muhammad Shair, India; Trimmer Fateh Shair Nawab, India; Fireman Fazal Satar Muhammad, India; Steward J. P. Fernandez, Goa; Seacunny Gopal Dhera, India; Helmsman S. Groot, Monnikendam, Holland; Fireman Hamid Ullah, India, Tindal Ismail Roshan, India; Fireman Jarusha, India; Winchman Kewa Muhammad Khan, India; Fireman Ludda Gamma, India; Trimmer Manga Feizoo, India; Seaman Mangal Khan, India; Assistant Purser A. J. Mooij, Zwolle, Holland; Donkeyman Natha Ashmat, India; Seaman Natia Dittia, India; Cassab Parbhu Prag, India; Bhandary Pema Dayal, India; Fireman Pinu Fazaldin, India; Seaman Ramji Seeka, India; Steward Roebaii, Java; Ch. Cook S. I. Schaap, The Hague, Holland; Fireman Shakur Muhammad, India; Fireman Sidiq Nabi, India; Seaman W. Vegter, Wildervank, Holland; Helmsman A. R. Vermeulen, Nijmegen, Holland; A.B. F. J. Warburton, England; A.B. H. Ward, England; Trimmer Zaman Farmanallee, India.

CORACERO (Donaldson Line Ltd. Port – Glasgow.)
4th Eng. Off. A. T. Cherkas, Fairwater, Cardiff; Donkeyman N. Fenlon, Liverpool 5; Fireman R. Gittings, Liverpool 4; Greaser W. D. Snow, Guildford, Surrey (wife – Auckland, New Zealand); Trimmer R. Yates (14 years old and served as J. J. Elder),'Liverpool 5.

WALTER Q. GRESHAM (U.S. War Shipping Administration managed by Standard Fruit and Steamship Co. Port – New Orleans.)
Assistant Cook L. A. Bertrand, Martinville, Louisiana; Ch. Steward J. C. Canney, Sommerville, Minnesota; F.W.T. J. Cebada, home unknown; 3rd Assistant Eng. L. F. Dirks, St. Alban's, Long Island; Rad. Off. C. A. Draper, New Orleans; 2nd Off. W. Farragut, New Orleans; Utility W. Fleet, Hammand, Louisiana; Messman S. Foelak, Forest Hills, Long Island; F.W.T. F. E. Gomez, Spain; Messman M. Hernandez, Spain; 3rd Off. G. N. Johnson, Washington; 1st Assistant Eng. G. R. King, Detroit; Oiler J. Larrea, Spain; Deck Eng. S. J. Levandosky, Staten Island; Deck Maintenance R. Madland, Brooklyn; Clerk L. A. Miller, Ridgeway, Missouri; O.S. D. Milliorn, Fort

Worth, Texas; Oiler H. D. Oker, Minneapolis; Bosun J. H. Peters, New Orleans; Messman G. Snaith, New York; 2nd Assistant Eng. H. H. Steelman, Brooklyn; Utility H. F. Winger, Dayton, Ohio.
U.S. Navy Armed Guard: S1c. J. M. Burns, Illinois; S1c. W. D. Butler, Indiana; S1c. R. E. Cantrell, Texas; S1c. N. E. Cetnar, Michigan; Gunner's Mate 3rd Class D. League, Alabama.

CANADIAN STAR (Union Cold Storage Co. Ltd. managed by Blue Star Line Ltd. Port – London.)
Assistant Steward W. E. Bevan, St Thomas, Swansea; Ch. Eng. Off. E. G. Buckwell, Crosby, Liverpool; Donkeyman T. Christie, Liverpool 5; Cadet J. Coghlan, Liverpool 13; Fireman D. Connor, Liverpool; Donkeyman F. Edwards; 2nd Refrigeration Eng. J. A. Forbes, Bluff, Southland, New Zealand; Junior Eng. Off. J. Gee, Blaydon-on-Tyne, Co. Durham; Greaser W. Grieve, Rush Green; Donkeyman T. Hughes, Treharris, Glamorgan; Cook L. F. Humphries, Lower Peacedown St John, Somerset; Donkeyman R. S. Jones, Caernarvon; Ch. Cook H. Mack, Wallsend-on-Tyne and Australia; Ch. Refrigeration Eng. C. Marsh, Sunderland, Co. Durham; Capt. R. D. Miller, Penybont, Radnorshire; Assistant Cook J. B. O'Reilly, Old Trafford, Manchester; Junior Eng. Off. A. I. Towers, Bolton, Lancashire; 4th Off. V. Trillo, Wellington, New Zealand; Electrician K. St C. Vincent, Leigh-on-Sea, Essex; Carpenter J. G. Watson, Whitehills, Banffshire; Assistant Steward S. Williams, Swansea.
R.N. Gunners: A.B. E. H. Hayward, Andover, Hants; A.B. S. Slater, Hyde, Cheshire.
Passengers: Colonel A. C. Craighead and Miss A. E. Craighead (12 years old) of Simla, India and Forfar, Scotland; Mr H. T. W. Early, Melbourne; Colonel and Mrs J. H. Ord and Master J. H. Ord (2 years old), Duntroon Staff College, New South Wales and London; Mrs M. Wrigley, Newcastle-on-Tyne.

MATHEW LUCKENBACH (Luckenbach Steamship Co. Inc. Port – New York.)
No fatal casualties.

H.M.S. MANSFIELD lost one man overboard: A.B. A. Thomas, Brompton, Derbyshire.

Convoy SC.122

KINGSBURY (Alexander Shipping Co. Ltd. managed by Capper, Alexander and Co. Port – London.)
Fireman D. W. Cox, Llandaff, Glamorgan; Ch. Eng. Off. J. Nicoll, Aberdeen; A.B. S. A. Ward, Hamnavoe, Shetland Isles.
Passenger: Mr S. Fairey, Barry, Glamorgan.

KING GRUFFYDD (King Line Ltd. managed by Dodd, Thompson and Co. Ltd. Port – London.)
A.B. L. A. Chitham, Garston, Liverpool and New Zealand; Fireman J. P. Clune, Liverpool; Ch. Eng. Off. C. E. Cox, Liverpool; A.B. W. H. Diggens, St Paul's Cray, Kent; Fireman A. Edwards, Liverpool; O.S. J. Feenan, Liverpool; 3rd Eng. Off. T. A. Fuge, Liscard, Cheshire; Capt. H. Griffiths, Burnside, Lanarkshire; Fireman C. Heron, Bargaddie, Glasgow; Fireman R. Hughes, Bangor, Caernarvonshire; Assistant Cook G. Jones, Dingle, Liverpool; A.B. H. Jones, Wallasey, Cheshire; Fireman J. Little, Liverpool; A.B. S. McDonough, Liscard, Cheshire; Messroom Boy S. Martin, Cannon Hill, Birmingham; 2nd Rad. Off. W. Mayall, Leeds; Donkeyman W. J. Miller,

Walton, Liverpool; O.S. C. G. Pringle, Kensington, Liverpool; 3rd Rad. Off.
J. J. Ramsay, Shettleston, Glasgow; Fireman T. C. Reilly, Liverpool.
Gunners: Naval – A.B. G. E. Boardman, High Wycombe, Bucks. Army
Gunner F. C. Roberts, St Osyth, Essex.

ALDERAMIN (Van Nievelt Goudriaan. Port – Rotterdam.)
Seaman J. den Bakker, Brielle, Holland; Seaman J. Bogaart, Holland; 2nd
Eng. Off. D. van Brandwijk, Dordrecht, Holland; Fireman J. Inguanez, Sliema,
Malta; Greaser A. J. M. de Landes, Paramaribo, Dutch Guiana; Fireman
H. B. Lems, Rotterdam, Holland; Fireman F. van der Maath, Groningen,
Holland; Trimmer A. Mizzi, Malta; 2nd Cook E. Peper, Rotterdam, Holland;
Trimmer J. G. Press, Motherwell, Lanarkshire; Boy W. Shuttleworth,
England; Fireman J. Spanjersbergh, Vlaardingen, Holland; Trimmer S.
Summerfield, Dagenham, Essex; Messroom Boy A. Unwin, St Pancras,
London; 2nd Helmsman A. Vader, Vlissingen, Holland.

FORT CEDAR LAKE (Ministry of War Transport managed by Port Line Ltd. Port
– none.)
No fatal casualties.

GRANVILLE (U.S. Maritime Commission. Port – Panama.)
Oiler J. Aransolo, Spain; Ch. Eng. Off. H. Christjansen, Denmark; Fireman
N. Ekkholm, Finland; Oiler J. Graham, U.S.A.; 3rd Eng. Off. M. Mand,
Estonia; Fireman Martin, Boston, U.S.A.; 2nd Off. C. Micallef, Sliema,
Malta; Fireman J. Miranda, Montevideo, Uruguay; Coalpasser J. Sellers,
U.S.A.; 2nd Eng. Off. B. Vahtra, Estonia.
U.S. Navy Armed Guard: S1c. R. Colebank, West Virginia; S3c. Gregoire,
Oregon; S3c. A. C. Kapel, State of Washington.

PORT AUCKLAND (Port Line Ltd. Port – London.)
Fireman P. Collins, New Cross, London; 4th Eng. Off. J. C. Doran, South
Shields, Co. Durham; Fireman P. B. Gallantrey, Southampton; Fireman J. A.
Huton, Dagenham, Essex; Fireman H. Huyton (served as T. Williams),
Bootle, Liverpool; Greaser J. S. Mansbridge, Shirley, Southampton; Fireman
A. E. Williams, Southampton; Assistant Steward R. C. Yalden, Havant,
Hampshire.

ZOUAVE (Zinal Steamship Co. Ltd. managed by Turner, Brightman and Co.
Port – London.)
Bosun A. Azzopardi, Cardiff and Malta; Ch. Eng. Off. Halliday, O.B.E.,
Sunderland, Co. Durham; 2nd Eng. Off. J. H. Kinnon, Bishopbriggs, Glasgow;
Fireman C. Leuben, Cardiff; Fireman E. Lynch, Cardiff; Fireman A. C. Lyttle,
Kingston, Jamaica; Fireman J. F. Payne, Barry Dock, Glamorgan; O.S.
L. Pereira, Oporto, Portugal; Cook W. Smith, Cardiff; A.B. P. Xerri, Cardiff
and Malta; A.B. E. Xicluna, Cardiff and Gozo, Malta; Carpenter A. Sahra,
Cardiff and Malta.
Army Gunner: Gunner W. T. Clegg, Burnley, Lancs.

CARRAS (C. J. Carras. Port – Chios.)
No fatal casualties.

CLARISSA RADCLIFFE (Owned by Wynnstay Steamship Co. Ltd. and W. L.
Radcliffe Steamship Co. Ltd., managed by E. Thomas Radcliffe and Co. Ltd.
Port – London.)
2nd Eng. Off. H. M. Bennie, Edinburgh; A.B. W. R. Black, Greenock,
Renfrewshire; A.B. B. L. Blackie, Aberdeen; Fireman B. Brown, Cardiff;
Fireman R. J. Brown, Liverpool 7; A.B. R. A. Bruce, Burravoe, Isle of Yell,
Shetland Isles; 2nd Rad. Off. F. C. Budd, London S.W.9; 4th Eng. Off. C.

Clark, Gateshead, Co. Durham; Apprentice C. A. Cook, London E.15; Ch. Off.
C. M. J. De Mathelin, Château de Missanur, Belgium; Ch. Eng. Off. D. T. P. L.
Evans, Rumney, Cardiff; Shipwright I. B. Evans, Mosman Park, Western
Australia; Capt. S. G. Finnes, Greasby, Cheshire; 3rd Eng. Off. J. R. Greany,
Dublin; Cook J. Hammerson, Sierra Leone and Liverpool 8; Fireman J.
Harding, Freetown, Sierra Leone; 1st Rad. Off. G. Hargreaves, Burnley,
Lancashire; A.B. C. B. Henderson, Cullivoe, Shetland Isles; Fireman T. James,
Cardiff; O.S. C. T. G. Jones, Cathays, Cardiff; Fireman B. Koroma, Freetown,
Sierra Leone; Fireman P. Lopez, Barry, Glamorgan; 3rd Rad. Off. J. J. Nunan,
Ballysheedy, Co. Limerick; Boy F. O'Driscoll, Cardiff; O.S. D. S. Palmer,
Liverpool 13; Apprentice G. H. Pembridge, Penllyne, Glamorgan; Fireman
T. Peter, Cardiff; Fireman P. A. Phipps, Barry, Glamorgan; Fireman T. Pratt,
Freetown, Sierra Leone; Assistant Cook R. S. A. Prendergast, St Andrew,
Jamaica; Assistant Steward W. H. Rance, Chelsea, London; A.B. T. W.
Robertson, Burravoe, Isle of Yell, Shetland Isles; Greaser T. Rogers, Free-
town, Sierra Leone; Fireman Samuel R. Arose, Lagos, Nigeria; Apprentice G.
Shepperson, Ramsey, Huntingdonshire; Fireman F. H. Smart, Liverpool 8
and Freetown, Sierra Leone; Greaser S. R. Sonkoh, Bathurst, Gambia; 3rd
Off. E. J. V. Thomas, Pontypridd, Glamorgan; 2nd Off. L. R. Vickery, Sheffield;
Fireman J. Williams, Cardiff; Bosun O. Williams, Port Dinorwic, Caernarvon-
shire; Ch. Steward R. Williams, Liverpool 11; O.S. A. E. Williamson, Walls,
Shetland Isles.

Naval Gunners: A.B. P. Black, Brentwood, Essex; A.B. H. Burns, Highbury,
London; A.B. E. W. Clarke, Reading, Berks.; A.B. E. Clewlow, Stafford;
A.B. J. Hitchenor, Stafford; A.B. R. Jones, Coventry, Warwicks.; A. B.S.
Lawrence, Stoke-on-Trent, Staffs; A.B. L. Lomas, Failsworth, Manchester;
A.B. H. McCart, Belfast; A.B. G. Miller, Pudsey, Yorks.; A.B. C. Opie,
Altrincham, Cheshire; A.B. I. B. Todd, Glasgow.

U-Boats

U.384 (Launched at Howaldtswerke, Kiel, on 28 May 1942, designated for
3rd U-boat Flotilla at La Pallice but lost on first patrol.)

Oberleutnant (Ing.) Max Bähnge, Kiel; Maschinengefreiter Hermann
Blessing, Vöhrenbach, Black Forest; Funkmaat Erwin Blömer, Neumarkt,
Bavaria; Bootsmaat Franz Blut, Essen; Maschinenmaat Willi Bogatz, Lötzen,
East Prussia; Fähnrich Elfried Eckert, Pauske, Latvia (now U.S.S.R.);
Mechanikergefreiter Heinz Fischer, Weissenfels, near Leipzig; Obermaschinist
Paul Fischer, Leuna, near Leipzig; Bootsmaat Erich Fittkau, Weitmar, near
Bochum; Obermaschinenmaat Walter Friedmann, Hamburg; Matrosenge-
freiter Erich Giebel, Sorau, near Görlitz; Maschinenobergefreiter Richard
Gimpel, Unterweiden, near Krefeld; Matrosengefreiter Ludwig Gras, Gross-
höflein; Matrosengefreiter Karl Grosse, Aken, near Calbe; Matrosengefreiter
Josef Grund, Ehrenfeld, Cologne; Matrosengefreiter Günter Hau, Bochum;
Maschinenmaat Karl Herold, Limbach; Mechanikermaat Karl Hornenboich,
Cologne; Maschinengefreiter Karl Huf, Reifenthal, near Regensburg;
Maschinengefreiter Johannes Jakobi, Warburg, near Kassel; Maschinenge-
freiter Heinz Jeschke, Grossenhain, near Bremerhaven; Matrosengefreiter
Walter Kästner, Grünstedt, Weissensee; Matrosengefreiter Georg Keppel,
Essen; Maschinenmaat Jakob Kerberg, Krefeld; Leutnant Kurt Krumscheid,
Kiel; Obersteuermann Paul Landgraf, Meingen, Thuringia; Funkgefreiter
Heinrich Leinweber, Frankfurt; Matrosenobergefreiter Heinz Lücking,
Mettmann, Rhineland; Funkgefreiter Martin Mainzer, Kassel; Maschinen-
maat Heinrich Marschinke, Bochum; Maschinengefreiter Peter Martinett,

Cologne; Maschinengefreiter Albert Munderich, Münsingen, Württemberg; Matrosengefreiter Heinz Nehring, Karlsfeld, near Munich; Obermaschinist Karl Oppermann, Düsseldorf; Funkmaat Karl-Heinz Orpick, Bübingen, Saar; Matrosengefreiter Werner Recknagel, Zeitz, near Leipzig; Mechanikergefreiter Herbert Richert, Brambauer, near Dortmund; Maschinengefreiter Johannes Richter, Hopfgarten, near Chemnitz; Maschinengefreiter Kurt Riege, Berlin; Oberleutnant Hans-Achim von Rosenberg-Gruszczynski, Krefeld; Maschinengrefreiter Bernhard Schadt, Berlin; Maschinenobergefreiter Kurt Sinholz, Köslin; Obermaschinenmaat Günter Sobetzko, Ratibor, Upper Silesia (now Poland); Leutnant Rudolf Scherdel, Schwarzenbach, near Hof; Fähnrich Kurt Schrader, Brunswick; Bootsmaat Heinz Streit, Langenbielau, Schleswig; Maschinengefreiter Gerhard Weckopp, Krefeld.

Appendix 6

German U-boats Destroyed September 1939–May 1945

Part 1: Cause of U-boat Loss

1,170 U-boats were completed, of which 863 were commissioned as operational boats. 784 of these operational U-boats were lost at sea or completely put out of action while in port. The following tables give the best post-war estimate of the cause of loss. The war is divided into two parts, to show the influence of air power following the provision of more aircraft and the closing of the North Atlantic Air Gap in the spring of 1943.

Notes: The total sunk by aircraft includes forty-three U-boats sunk by carrier-borne aircraft. Forty-nine U-boats destroyed by the joint efforts of ships and aircraft have been allocated evenly under both headings. The lack of U-boats destroyed by bombing in the early years of the war shows how well the U-boat pens in the Biscay ports served the Germans, although Air Force leaders would claim, with justification, that bombers destroyed some partially-built U-boats in German shipyards and, by bombing inland factories, prevented prefabricated U-boat sections reaching shipyards, but the results of this cannot be quantified.

The decline in the number of U-boats 'killed' by aircraft on sea patrol in the last few months of the war is a result of the introduction by the Germans of *Schnorkel* equipment which allowed U-boats to breathe while submerged.

| | DESTROYED BY ALLIED SHIPS | | | DESTROYED BY ALLIED AIRCRAFT | | | | |
	Surface Vessels	Submarines	Mines laid by ships	Aircraft on sea patrols	Bombing at ports	Mines laid by aircraft	Other Causes and Unknown	Total
1939 September–December	5	1	3	0	0	0	0	9
1940	12	2	2	3	0	0	4	23
1941	26	1	0	4	0	0	4	35

January–March	14	2	0	21	0	0	3	40
Total in first 43 months of war	92½	8	5	68½	0	3	17	194
1943 April–December	51½	4	0	125½	2	1	13	197
1944	79½	7	1	79½	24	8	43	242
1945 January–May 8	47	3	3	41	36	4	17	151
Total in last 25 months of war	178	14	4	246	62	13	73	590

Total destroyed by Allied ships	301¼
Total destroyed by Allied aircraft	392½
Other causes and unknown	90
	784

Part 2: Nationality of Ships and Aircraft Sinking German U-boats

Only U-boats sunk either by ships or by aircraft are shown here; U-boats sunk by the joint efforts of ships and aircraft are not included.

	BRITISH	AMERICAN	JOINT 'KILLS'
Sunk by surface vessels	206	37	3
Sunk by submarines	19	3	
Sunk by shore-based aircraft	195	48	2
Sunk by carrier-based aircraft	14	29	
Sunk by bombing	22	40	

Acknowledgements

Before all others I would like to thank the following men and women, both Allied and German, who sailed in the ships of Convoys SC.122, HX.229 and 229A, who flew in the Coastal Command aircraft covering those convoys, who sailed in the U-boats involved in the operation against those convoys or who performed various duties in the shore-based headquarters of both sides. Without their generous and willing help this book would not have been written.

(The contributors are listed ship by ship in alphabetical order of surname. Ranks shown are those held in March 1943.)

CONVOY SC.122, Merchant Ships

Badjestan: Cabin Boy F. Craik. *Baron Elgin*: Captain J. S. Cameron. *Baron Stranraer*: Chief Officer M. Maclellan. *Carso*: Chief Officer A. H. Tarr. *Cartago*: Cadet W. E. West. *Dolius*: Able Seaman W. Carmody. *English Monarch*: Captain T. N. Dobbie, Chief Officer W. D. Morton. *Fjallfoss*: Chief Engineer B. Jonsson, Captain H. Olafsson, Second Officer G. Sigurdsson. *Fort Cedar Lake*: Captain C. L. Collings. *Glenapp*: Chief Engineer R. W. Douglas, Midshipman C. H. F. Hill, Midshipman M. Lees, Second Officer C. G. Ross. *Granville*: Bosun H. B. Starckx. *Gudvor*: Able Seaman Karsten Moe. *King Gruffydd*: Second Officer F. R. Hughes. *Kingsbury*: Fireman D. E. B. Evans, Able Seaman A. Garriock, Fireman P. Murphy (died May 1974). *Livingston*: Second Officer W. K. Blackmore. *Losada*: Chief Electrician J. R. Fortune, Chief Officer G. H. Hobson, Captain W. M. Horsfall. *L.S.T.365*: Sub-Lieutenant J. H. H. Bayley. *McKeesport*: Engineer M. P. Linkowich. *Parkhaven*: Radio Officer P. H. den Engelsen, Fourth Officer H. Meuldijk. *Port Auckland*: Chief Officer V. G. Battle, Third Officer P. J. Chrisp, Second Officer J. G. A. Dunn, Able Seaman A. N. Johnson, Seventh Engineer W. P. Shevlin, Sergeant J. B. Ware, R.A.A.F. (passenger). *Selfoss*: First Officer E. Thorgilsson. *Zouave*: Cook/Baker S. Banda, Cabin Boy J. S. Kelly, Second Officer J. D. Sharp.

CONVOY SC.122, Naval Escorts

WESTERN ATLANTIC LOCAL ESCORTS. H.M.C.S. *Blairmore*: Lieutenant T. MacDonald. H.M.C.S. *Cowichan*: Ordinary Seaman J. R. Graham. H.M.C.S. *Dunvegan*: Signalman G. Nisbet, Leading Telegraphist G. A. Stevens. H.M.S. *Leamington*: Lieutenant A. D. B. Campbell. H.M.C.S. *New Westminster*: Lieutenant W. McCombie, Sub-Lieutenant T. C. Marshall. H.M.C.S. *Rimouski*: Lieutenant R. J. Pickford.

OCEAN ESCORTS. H.M.S. *Buttercup*: Able Seaman G. Goormachtig, Lieutenant R. Jonckheere, Lieutenant W. A. M. J. Libert, Able Seaman E. Moeyaert, Sub-Lieutenant P. M. J. van Schoonbeek, Able Seaman L. Vandekerckhove, Lieutenant P. H. V. M. van Waesburghe. H.M.S. *Campobello*: Lieutenant G. B. Rogers. H.M.S. *Godetia*: Lieutenant J. C. O. Böting, Able Seaman R. Crudenaire, Lieutenant J. A. Delforge, Sub-Lieutenant C. M. L. De Pierre, Lieutenant M. A. F. Larose. H.M.S. *Havelock*: Commander R. C. Boyle,

Signalman A. H. Dossett, Leading Seaman R. Freatly, Lieutenant A. I. Hughes, Surgeon-Lieutenant C. S. Savage, Lieutenant N. T. M. Yeates. U.S.C.G.C. *Ingham*: Lieutenant-Commander J. D. Craik, Lieutenant J. M. Waters. H.M.S. *Lavender*: Sub-Lieutenant E. Walker. H.M.S. *Pimpernel*: Lieutenant H. D. Hayes, Leading Seaman J. G. Owen. H.M.S. *Saxifrage*: Lieutenant R. G. Cropp, Lieutenant A. M. Hodge, Lieutenant M. C. C. F. Shaw. H.M.S. *Swale*: Sub-Lieutenant L. A. Spicer. U.S.S. *Upshur*: Lieutenant-Commander E. B. Ellsworth, Lieutenant H. Gravely, Ensign D. Lawson, Lieutenant J. G. Prout, Lieutenant J. J. White.

CONVOY HX.229, Merchant Ships

Abraham Lincoln: Radio Officer H. Andersen, Able Seaman A. Ekedal, Chief Cook T. Saga, Refrigeration Engineer E. Solberg. *Canadian Star*: Second Officer W. A. Clarke-Hunt, Flying Officer and Mrs W. B. Dobree (passengers), Chief Officer P. H. Hunt, Ordinary Seaman J. Jones, Third Officer R. H. Keyworth, Chief Radio Officer A. E. Phillips. *City of Agra*: Chief Officer W. Howel. *Coracero*: Engineer J. Crawford, Third Officer R. McRae. *Daniel Webster*: Ordinary Seaman F. W. Corey. *Elin K*: Seaman First Class L. T. Engebretsen, Motorman J. A. Johannessen, Second Engineer T. Tobiassen. *Fort Anne*: Second Officer H. Gravell, Captain S. T. Roach. *Irénée du Pont*: Ensign F. N. Pilling (passenger). *James Oglethorpe*: Fireman-Watertender W. J. Brantley, Able Seaman T. Napier. *Luculus*: Captain W. Luckey. *Mathew Luckenbach*: Ordinary Seaman P. Civitillo, Seaman Third Class J. O. Jackson. *Nariva*: Engineer D. Anthony, Able Seaman H. J. Brinkworth, Second Engineer H. W. Brophy, Third Officer J. A. P. Matthews, Second Officer G. D. Williams. *Nebraska*: Second Officer A. G. Maxwell. *Pan Rhode Island*: Chief Steward R. P. Morgan. *Southern Princess*: Able Seaman T. Craigie, Chief Engineer C. K. Crockett. *Tekoa*: Captain A. Hocken, Chief Steward R. L. Keen. *Walter Q. Gresham*: Wiper H. E. Carmichael, Able Seaman R. J. Smith, Able Seaman E. P. Swift. *Zaanland*: Chief Officer P. G. van Altveer, Cadet G. H. Baird-Jones, Captain G. J. J. M. Franken, Gunner L. E. French, Steward J. Steenhart, Second Officer J. Wassenaar.

CONVOY HX.229, Naval Escorts

WESTERN ATLANTIC LOCAL ESCORTS. H.M.S. *Chelsea*: Sub-Lieutenant E. McMillan, Leading Torpedoman C. Nowell, Sub-Lieutenant A. D. Powell, Lieutenant-Commander J. E. R. Wilford.

OCEAN ESCORTS. H.M.S. *Abelia*: Lieutenant F. N. Salisbury. H.M.S. *Anemone*: Lieutenant D. C. Christopherson, Lieutenant-Commander P. G. A. King, Stoker R. V. Procter. U.S.S. *Babbitt*: Chief Watertender J. J. Haddad, Lieutenant H. D. Hirschler, Ensign O. Jensen, Lieutenant-Commander S. F. Quarles. H.M.S. *Beverley*: Petty Officer C. G. Braillard, Able Seaman G. Truswell. H.M.S. *Highlander*: Surgeon-Lieutenant W. A. B. Campbell, Commander E. C. L. Day, Lieutenant D. G. M. Gardner, Lieutenant-Commander R. M. Miller, Lieutenant G. E. Milner, Gunner (T) G. H. Page. H.M.S. *Mansfield*: Able Seaman R. Brett, Leading Seaman A. D. Cornelius, Sub-Lieutenant A. R. Guy, Lieutenant-Commander L. C. Hill, Lieutenant J. K. Laughton, Engineroom Artificer G. T. Smith, Engineroom Artificer C. Wilburn. H.M.S. *Pennywort*: Sub-Lieutenant L. M. Maude-Roxby, Able Seaman P. W. Maxim, Steward C. A. Mitton, Lieutenant O. G. Stuart, Wireman R. A. White. H.M.C.S. *Sherbrooke*: Leading Seaman R. Hatfield. H.M.S. *Vimy*: Petty Officer H. F. Hopcliffe. H.M.S. *Volunteer*: Ordinary Telegraphist R. T. Brown, Stoker First Class E. C. Claxton, Sub-Lieutenant R. G. Goudy, Ordinary Telegraphist D. Greenhouse, Petty Officer R. V.

Jackson, Lieutenant G. C. Leslie, Communication Rating E. Norman, Surgeon-Lieutenant H. F. Osmond, Chief Petty Officer R. H. Potts, Warrant Engineer G. H. Ward, Able Seaman W. R. Wilson. H.M.S. *Witherington*: Lieutenant Sir William Blunden, Bart, Able Seaman L. W. Davis, Able Seaman W. Ellis.

CONVOY HX.229A, Merchant Ships

Akaroa: Third Radio Officer J. Whiten (died 1975). *City of Oran*: Third Radio Officer F. C. J. Lockett. *Esperance Bay*: Assistant Steward F. R. Canham. *John Fiske*: Captain J. P. Chiles.

CONVOY HX.229A, Naval Escorts

H.M.S. *Hastings*: Lieutenant-Commander L. B. Philpott. H.M.S. *Landguard*: Lieutenant-Commander T. S. L. Fox-Pitt. H.M.S. *Waveney*: Lieutenant-Commander A. E. Wilmott.

CANADIAN EASTERN AIR COMMAND

5 (B.R.) Squadron R.C.A.F: Flying Officer G. E. Fletcher.

COASTAL COMMAND

86 Squadron: Flight Sergeant J. H. Bookless, Sergeant W. H. Bryan, Flying Officer C. W. Burcher, Sergeant C. C. Cooper, Sergeant H. R. Derrick, Sergeant J. Pheasey. 120 Squadron: Sergeant B. H. Harvey, Flight Sergeant T. J. Kempton, Squadron Leader J. R. E. Procter, Flying Officer R. T. F. Turner, Sergeant V. C. S. Wilson. 201 Squadron: Sergeant J. H. Bunton, Flight Lieutenant N. T. Harvey, Flight Lieutenant D. G. T. Hayes. 220 Squadron: Pilot Officer N. Carpenter, Sergeant R. V. Gurnham, Sergeant T. E. Kynnersley. 228 Squadron: Flight Sergeant L. Ledingham. 423 (Canadian) Squadron: Sergeant A. R. Caterham, Flying Officer J. B. Donnett, Flying Officer N. V. Martin, Sergeant M. N. Trowbridge, Sergeant J. H. Wright.

U-BOATS

U.91: Kapitänleutnant Heinz Walkerling. U.190: Oberleutnant Hans-Edwin Reith, Kapitänleutnant Max Wintermeyer. U.228: Oberfunkmaat Gustav Klütz, Oberbootsmaat Walter Kürten, Oberleutnant Carl Schauroth. U. 230: Kapitänleutnant Paul Siegmann. U.305: Maschinengefreiter Hans Bethke, Fähnrich Wolfgang Jacobsen. U.338: Oberleutnant Herbert Zeissler. U.373: Kapitänleutnant Paul-Karl Loeser. U.406: Obermaschinist Artur Kolbe, Oberleutnant Rudi Toepfer. U.439: Machinenobergefreiter Gerhard Schmeling. U.523: Leutnant Alexander Grau, Oberleutnant Walter Lorch. U526: Fähnrich Wilhelm Daude. U.527: Kapitänleutnant Herbert Uhlig. U.530: Funkgefreiter Werner Hess, Maschinenobergefreiter Heinz Koch, Bootsmaat Hermann Lawatsch. U.603: Oberleutnant Rudolf Baltz, Funkobergefreiter Erich Klein, Mechanikerobergefreiter Max Zweigle. U.615: Leutnant Claus von Egan-Krieger, Matrosenobergefreiter Hans Krohn. U.616: Oberleutnant Siegfried Koitschka. U.618: Kapitänleutnant Kurt Baberg. U.641: Oberleutnant Hans-Arend Feindt. U.642: Kapitänleutnant Herbert Brünning. U.653: Kapitänleutnant Gerhard Feiler, Obersteuermann Heinz Theen. U.664: Oberleutnant Adolf Graef. U.758: Kapitänleutnant Helmut Manseck.

The U-boats of the following men were all in the vicinity of the convoy battle; although they did not take part in the action they did help me by answering general questions about that period.

U.409: Kapitänleutnant Hanns-Ferdinand Massmann. U.415: Oberleutnant Kurt Neide. U.448: Oberleutnant Halmut Dauter. U.566: Oberleutnant Hans Hornkohl. U.591: Kapitänleutnant Hansjürgen Zetsche.

OTHERS AT SEA

First Officer G. van de Vuurst was a prisoner from the Dutch ship *Madoera* aboard U.591. Lieutenant-Commander J. F. Holm was captain of H.M.S. *Crocus* involved in the rescue of survivors from the *Empress of Canada*.

Shore-based Staffs

U.S. NAVY HEADQUARTERS, WASHINGTON

Convoy and Routing Division: Captain W. C. Wickham.
Operational Intelligence Division: Captain K. A. Knowles.

ADMIRALTY, LONDON

Trade Division: Commander H. C. Burton, Lieutenant-Commander J. H. Dundas-Grant, Commander R. A. Hall, Captain C. M. Leggatt, Captain B. B. Schofield. Submarine Tracking Room: Lieutenant P. Beesly, Signals Division: Captain G. R. Waymouth (died January 1975).

WESTERN APPROACHES, LIVERPOOL

Commander C. D. Howard-Johnston, Captain H. M. Lake, Captain R. W. Ravenhill, Captain G. Roberts, Sub-Lieutenant G. S. Shobrook, Commander H. M. Wilson (died April 1976), Lieutenant-Commander J. B. Wright. W.R.N.S.: Leading Wren Mary Carlisle (now Mrs Hall), Second Officer Mary Leveson.

COASTAL COMMAND

Air Marshal Sir John Slessor (Commander-in-Chief), Wing Commander J. R. S. Romanes (Commanding Officer 206 Squadron), Flight Lieutenant E. H. Sneath (R.A.F. Benbecula).

OTHER ALLIED

Captain J. M. Rowland (Captain (D), St John's), Captain A. Stanford (U.S. Navy Headquarters, London).

GERMAN

U-Boat Headquarters: Grossadmiral Karl Dönitz, Korvettenkapitän Hans Meckel, Kapitänleutnant Adalbert Schnee, Kapitänleutnant Herbert Schultze. B. Dienst: Kapitän Heinz Bonatz.
U-Boat Headquarters West: Kapitän Hans-Rudolf Rösing.

Personal Acknowledgements

I would like to express my thanks to many people and organizations in many countries for their most generous and friendly help with the preparation of this book.

Of the many private individuals who have helped me, I would particularly like to mention two ladies: Mrs Annemarie Lamb, a German-born neighbour who translated so patiently my large German correspondence, and Mrs Audrey Luther of Sherborne St John, Hampshire, the widow of Lieutenant-Commander Gordon John Luther, who, with great kindness and understanding, answered my questions about her husband. I think it would be better not to place my other private helpers in any order of merit but will record them by countries and in alphabetical order; I am grateful to every one of them.

Australia: P. W. Stanbridge of Kangaroo Island. Belgium: M. Vandekerckhove of Ostend. Britain: Mrs Jeanette Croft, Captain A. T. Harris, Mike Hodgson, Clement Poole, Mrs Denise Sagar and Mrs Thorunn Tunnard, all of Boston and district; Bob Childs of Rochester, Michael Cooper of Chesham (for particular help with U-boat research), H. W. F. Johnson of Eastbourne, Neville Mackinder of Finchley, R. C. McNeill of Bootle, Mrs J. C. Mansfield of Norwich, Patrick Mahoney of Chadwell Heath, E. J. Nevin of Wrexham (formerly of the U.S. Merchant Marine), Mrs Wendy Perrin of Cambridge, Alfred Price of Uppingham, David Robinson of Thurton, Norfolk, and Cliff Vincent of Bristol. Germany: Karl-Egbert Houy of Schiffweiler, Kapitän Claus Korth of Kiel, Norbert Krüger of Essen, Frau Martha Müller-Edzards of Hermannsburg, Major Ulrich Neck of Goch and Kapitän Reinhart Reche of Bonn-Bad Godesberg. Holland: Bastian van Os of Vlaardingen. Iceland: Mrs Kristin Thorbjarnardottir. Norway: Mrs Hans-Hendrik Hansen, who sent documents belonging to her late husband who was an officer on the *Elin K* but did not send her address, and Mrs Bjørg Natvig of Kristiansand. U.S.A.: Peter Danielak of Mineral City, Ohio, Rev. Lynn Smith of Boston, Mass. and Nicholas Wynnick of Ansonia, Connecticut.

I would also like to thank Janet Mountain who typed so well much of my correspondence and two drafts of the manuscript, also Jean Smith, Cherry Robinson, Bridget Bemrose and Frances Grebby who also contributed to the typing. My appreciation also to my wife, Mary, and my daughters, Jane, Anne and Catherine, and to my brother, Gregory, for their help in a variety of ways. I wish to thank the staff of the Boston Public Library for their efficient assistance.

Of the official organizations with whom I have been in contact, I would like to thank in particular the Naval Historical Branch in London (Mr J. D. Lawson and Mr J. D. Brown) and the Naval Historical Center (Mr D. C. Allard) in Washington for their most reliable and diligent help with my numerous requests. Many other organizations gave reliable help to less demanding requests and it might be fairer again to list these by countries and alphabetically. Argentine: Armada Argentina. Australia: Records Section, R.A.A.F. Belgium: Association des Anciens de la Force Navale and Service de l'Historique des Forces Armées. Britain: Ministry of Defence – Air Historical Branch, Naval Home Division, Department of Naval Secretary, Departments AR8b and AR9 (R.A.F.) and OS10 (Navy), R.A.F. Personnel Management Centre and 120 Squadron, Kinloss. Commonwealth War Graves Commission, Federation of Shipping (Information Division), Imperial War Museum, Lloyd's Register of Shipping (particularly Derek Eastaway) and the Corporation of Lloyds, Registrar General of Shipping and Seamen, Royal Naval Association, R.N.R. Officers Club (Liverpool and London), R.N.V.R. Officers Association, The Navy Club, United States Information Service of U.S. Embassy London (Miss Nancy Phillipps). Canada: Directorate of History, Canadian Forces Headquarters. France: L'Archiviste de la Ville of Brest and also of Lorient. Germany: Bundesarchiv-Militärarchiv, Freiburg; Deutsche Dienststelle and Verband Deutscher U-Bootfahrer e. V. Holland: Federatie Van Werknemersorganisaties Ter Zeevaart, Historical Section and the Public Information Department of the Royal Netherlands Navy, Nederlandse Vereniging van Kapiteins Grote Vaart and Oorlogsgravenstichting, Rijksinstitut voor Oorlogsdocumentatie. Norway: Norges Krigsseilerforbund and Norsk Sjöfartsmuseum. U.S.A.: Albert F. Simpson Historical Research Center U.S.A.F., American Battle Monuments Commission (Colonel William E. Ryan Jr), Board of Commissioners of Pilots of the State of New York (R. H. Martin), Military Personnel Records Center, United States Atlantic Fleet Public Affairs Section (Captain H. E. Padgett), United States Coast Guard (Mr Strobridge, Commander W. E. Whaley and Chief Warrant Officer H. M. Kern). Officials of the following port cities kindly helped in my search for American seamen participants in the convoy battle: Boston, Houston, Los Angeles, Mobile, New Haven, Norfolk, San Diego, Savannah.

I would also like to thank the following shipping companies for help given in a variety of ways: Blue Star Line, London; Booth Steamship Co. Ltd, Liverpool; Christian Salveson Ltd, Edinburgh; Cunard Steam-Ship Company, London; Evan Thomas Radcliffe and Co. Ltd, Cardiff; Furness Withy Group, London; Glen Line Ltd, London; Jacob Kjøde A/S, Bergen; Koninklijke Nedlloyd B.V. Rotterdam; Anton von der Lippe, Tonsberg; Luckenbach Steamship Company Inc., New York; Goudriaan van Nievelt, Rotterdam; Ocean Group, Liverpool; Fred Olsen and Company, Oslo; P. and O., London; Pacific Steam Navigation Company, Liverpool; Shaw Savill Line, London; Shell Tankers B.V., Rotterdam.

Acknowledgements for permission to include quotations from *The Second World War* by Winston S. Churchill are gratefully made to Cassell & Company, and from Crown copyright records in the Public Record Office and from the Official History of the Second World War to the Controller of H.M. Stationery Office.

The Press

Many newspapers and magazines published my appeals for participants in the convoy battle; this help was always given without charge and I am pleased to be able to record my gratitude for this valuable help.

BELGIUM: *De Zeewacht, Lloyd Anversois.*

BRITAIN: *Belfast Telegraph, Birmingham Post, Bradford Telegraph and Argus, Bristol Evening Post, Brixham News, Cardiff Evening Post, Eastern Daily Press, Edinburgh People Journal, Falmouth Packet, Glasgow Evening Times, Glasgow Herald, Grimsby Evening Telegraph, Hampshire Telegraph, Lincolnshire Echo, Lincolnshire Standard, Liverpool Daily Post, London Evening News, Manchester Evening News, Middlesbrough Evening Gazette, Newcastle Evening Chronicle, Nottingham Evening Post, Paignton News, Poole Evening Echo, Shetland Times, Shields Gazette, Southern Evening Echo, Sunderland Echo, Western Morning News, York Evening Press.*

Air Mail, Air Pictorial, East Coast Shipping News, Flight International, Journal of Commerce (Liverpool), *Lloyd's List, M.M.S.A. Reporter, Navy News, Ocean Wave, R.U.S.I. Journal, R.A.F. News, Sea Breezes, The Naval Review, The Telegraph, Wave.*

CANADA: *Cape Breton Post, Halifax Chronicle-Herald, Halifax Mail-Star, Montreal Gazette, New Brunswick Telegraph-Journal, St John's Daily News, St John's Telegram, Vancouver Sun, Victoria Colonist, Canadian Shipping and Marine Engineering, Legion, Starshell, Wings at Home.*

GERMANY: *Flensburger Tageblatt, Nordsee Zeitung, Schiff und Hafen* (Hamburg). *Marine, Schaltung Küste, Schiffahrt International.*

HOLLAND: *Alle Hans, Dagblatt Scheepvaart, De Blauwe Wimpel, De Zee, Ons Zeewezen, Stam Tijdschriften, Unieschakel.*

ICELAND: *Timinn.*

NEW ZEALAND: *The Windmill Post* (Dutch language publication).

NORWAY: *Bergens Tidende, Faedrelandsvenden, Haugesunds Avis, Krigssei-leren, Norges Handels og Sjøfartstidende, Risoer, Vi Menn.*

U.S.A.: *Galveston Daily News, New Haven Register, New Orleans States-Item, Portland Press Herald.*

Coast Guard Reservist, Journal of Armed Forces, Journal of the American Society of Naval Engineers, Master Mate and Pilot, MEBA Journal, Navy Pictorial News, Navy Time, Seafarer's Log, The NMU Pilot, The Sea Breeze (Norwegian Veterans), The Retired Officer, United States Naval Institute Proceedings, Veterans of Foreign Wars Magazine.

Bibliography

OFFICIAL HISTORIES

Roskill, Captain S. W., *The War at Sea*, H.M.S.O., 1954–61.
Morison, S. E., *History of United States Naval Operations in World War II*, Oxford University Press, 1948–56.
Behrens, C. B. A., *Merchant Shipping and the Demands of War*, H.M.S.O. and Longmans, Green, 1955.
Craven, W. F., and Cate, J. L., *The Army Air Forces in World War II*, University of Chicago Press, 1949.
Webster, Sir Charles, and Frankland, N., *The Strategic Air Offensive against Germany 1939–1945*, H.M.S.O., 1961.

OTHER PUBLICATIONS

Bekker, Cajus, *Hitler's Naval War*, Macdonald, 1974.
Busch, Harald, *U-Boats at War*, Putnam, 1955.
Chalmers, Rear Admiral W.S., *Max Horton and the Western Approaches*, Hodder & Stoughton, 1954.
Creighton, Rear Admiral Sir Kenelm, *Convoy Commodore*, William Kimber, 1956.
Delmer, Sefton, *Black Boomerang*, Secker & Warburg, 1962.
Dönitz, Grossadmiral Karl, *Ten Years and Twenty Days*, Weidenfeld & Nicolson, 1959.
Frank, Wolfgang, *Die Wölfe und der Admiral*, Gerhard Stalling, 1953.
Gretton, Vice Admiral Sir Peter, *Convoy Escort Commander*, Cassell, 1964.
Herzog, Bodo, *60 Jahre Deutsche U-Boote 1906–1966*, Lehmanns Verlag, 1968.
Jacobsen, Dr Hans-Adolf, and Rohwer, Dr Jürgen, *Decisive Battles of World War II*, André Deutsch, 1965.
Kahn, David, *The Codebreakers*, Weidenfeld & Nicolson, 1966.
Lenton, H. T., and Colledge, J. J., *Warships of World War II*, Ian Allen, 1964.
Macintyre, Captain Donald, *The Battle of the Atlantic*, Batsford, 1961.
McLachlan, Donald, *Room 39*, Weidenfeld & Nicolson, 1968.
Marder, Arthur, 'The Influence of History on Sea Power: The Royal Navy and the Lessons of 1914–1918', an article in *Pacific Naval Review*, November 1972.
Price, Alfred, *Aircraft Versus Submarine*, William Kimber, 1973.
Rayner, Commander D. A., *Escort*, William Kimber, 1955.
Rohwer, Dr Jürgen, *Die U-Boot-Erfolge der Achsenmächte 1939–1945*, Lehmanns Verlag, 1968.
Sawyer, L. A., and Mitchell, W. H., *The Liberty Ships*, David & Charles, 1970, and *The Oceans, The Forts and the Parks*, Sea Breezes, 1966.
Schofield, Vice Admiral B. B., and Martyn, Lieutenant-Commander L. F., *The Rescue Ships*, William Blackwood, 1968.
Slessor, Marshal of the R.A.F. Sir John, *The Central Blue*, Cassell, 1956.
Waters, Captain John M., *Bloody Winter*, D. Van Nostrand, 1967.
Waters, Sydney, *Ordeal by Sea*, New Zealand Shipping Co., 1949.
Winterbotham, Group Captain F. W., *The Ultra Secret*, Weidenfeld & Nicolson, 1974.

Index

Admiralty
and convoys in First World War, 1
introduces convoys in 1939, 9
asks R.A.F. for long-range aircraft, 11
and American Naval Mission, 16
arms merchant ships, 23
control of convoys, 45–6
Operational Research Section, 54, 92
codes, 75–6
in planning of SC.122 and HX.229, 90–94
in SC.122 and HX.229 convoy action, 114, 151–2, 164–5, 203–4, 212, 227, 247, 282, 305
meanness in award of decorations, 295
in developments of Spring 1943, 308–10, 314–21 passim
Aldergrove, 50, 53, 206, 212, 242, 245, 272, 275
Allen, Capt. A. M., 90
Allied and neutral merchant ships*
Abraham Lincoln, 95, 193, 202, 254
Akaroa, 114
Alderamin, 196, 198–9
Alpherat, 199
Andrea F. Luckenbach, 247, 304n.
Atland, 290–91
Athenia, ix, 3n., 4, 56, 63, 72
Avondale Park, ix
Badjestan, 329
Baron Stranraer, 202
Benedick, 103, 147, 196
Bonneville, 115
Brandt, 173n.
Calabria, 23n.
Canadian Star, 82, 96, 149, 247–53, 255, 259, 290–92, 304
Carras, 262–3, 271
Carso, 83, 290–91
Cartago, 96, 214

Christian Holm, 147
City of Agra, 227
Clan Matheson, 108
Clarissa Radcliffe, 107, 282–3, 302
Coracero, 81, 89, 108, 150, 223–7, 271, 303
Daniel Webster, 80n., 171
Dettifoss, 31
Dolius, 322
Elin K, 84, 171–4, 183, 202, 277, 304, 329
Empire Galahad, 147
Empire Morn, 244, 272–3
Empress of Canada, 285
English Monarch, 83, 107
Esperance Bay, 95
Fjallfoss, 113, 214
Fort Anne, 82–3, 196
Fort Cedar Lake, 82–3, 100, 196, 198, 200, 211
Georgios P, 105
Glenapp, 81, 87–8, 95, 101, 102–4, 107, 166, 214, 291
Glittre, 173n.
Godafoss, 31, 214
Granville, 214–15, 217, 307
Gulf Disc, 163
Harry Luckenbach, 171, 184–7, 303
Hekla, 31
Historian, 103
Hugh Williamson, 150
Irénée du Pont, 96–7, 188–92, 218–20, 304
James Oglethorpe, 80n., 88, 174, 176–8, 184, 217, 220–21, 251, 303
John Fiske, 8n.
Kedoe, 107
King Gruffydd, 100, 196–8
Kingsbury, 196–8, 304, 329
Lillian Luckenbach, 304n.
Loch Goil, 304n.
Losada, 25

* Only those merchant ships mentioned in the chapters are indexed; Appendix 1 lists all other ships in SC.122, HX.229 and HX.229A.

Luculus, 84, 98, 193, 202
McKeesport, 322
Madoera, 142
Magdala, 165, 193
Mathew Luckenbach, 100–101, 109, 255, 266–9, 277, 292–3, 302
Nagara, 304n.
Nariva, 81, 166, 188–92, 218–20, 252, 290, 304, 330
Normandie, 81, 88
Parkhaven, 214
Pierre Soule, 80n.
Port Auckland, 234–7, 242, 243n., 303–4
Queen Elizabeth, 284, 293
Queen Mary, 97
Queen Mary II, 292
Robert E. Peary, 27
Robert Howe, 80n., 192
Sabor, 304n.
San Veronico, 84
Selfoss, 114, 148
Southern Princess, 100, 188–93, 218–19
Southern Sun, 108
Stephen C. Foster, 80n., 114–15
Sunoil, 108
Stigstad, 173n.
Svend Foyn, 97, 280–81, 302
Tekoa, 82, 165, 190–92, 194, 222, 287, 290, 295
Terkoelei, 82, 172, 174, 177, 224–7, 304
Vinriver, 107
Walter Q. Gresham, 80n., 150, 248, 250–51, 253, 255
William Eustis, 80n., 114–15, 181–183, 185, 218, 220, 249
Zaanland, 30, 84, 86, 174–5, 177–8, 304
Zamalek, 111–13, 198–200, 211, 237, 262, 287, 290
Zouave, 22, ·100, 167, 235, 237, 242, 243n., 304
American merchant seamen or ships, 19n., 25–9, 99–101, 176–7, 182, 185–6, 188–9, 214, 217–18, 220–21, 251, 253, 266–9, 287, 294–5, 302–4, 326n.
American naval seamen or ships, 36, 42–3, 96, 213, 216, 243, 264–6 (*see also* United States Navy)

Anderson, Petty Officer Tel., 151
Angers, 64, 74
Argentia, 12, 39, 126
Asdic, 7, 9–10, 14*
Atlantic Convoy Conference, 203, 290, 296, 311–13
Australian servicemen, 36, 52, 78, 96, 171, 212–14, 236–7, 245, 299, 326n.
Auten, Capt. H., 95

B.d.U. Zug, 66, 300
Baberg, KL. K., 146, 159, 222n.
Bahr, KL. R., 217, 235n., 238, 242–3
Baird-Jones, Cadet G. H., 175
Baltimore, 26, 83
Baltz, OL. R., 170–71
Banda, Cook S., 22, 100, 235
Bayley, Sub-Lt J. H. H., 86, 106
Belgian merchant seamen or ships, 29, 101, 326n.
Belgian naval seamen or ships, 36, 43–5, 148–9, 216, 237–9, 263, 325, 328
Benbecula, 51, 53, 269–70, 276
Berge, Ch. Off., 171
Berlin, 74–6, 135, 206, 210, 300
Bertelsmann, OL. H.-J., 169–71, 173, 277
Beuren, Coxswain R., 263
Billet, Lt V., 43, 325
Birnie, Commodore R. C., 115
Blackett, Prof. P. M. S., 92
Blohm and Voss, 61, 225
Bonatz, Kapitän H., 76, 118, 209
Bonatz, Lt H., 209
Bookless, Flight Sergeant J. H., 245
Bordeaux, 63, 66, 285
Borden, Capt. A. H., 267
Bortoft, Warrant Officer M., 270
Boston (U.S.A.), 26, 41, 134, 325
Boyle, Mrs G., 203
Boyle, Cdr R. C., 120–27 *passim*, 131, 147–9, 164–5, 195, 203, 205, 217, 234, 237, 242, 255, 257, 267, 293–4, 305
Bradley, Flight Lieutenant C. L., 272
Bremen, 61, 66, 323
Brest, 63–6, 152, 283
Brinkworth, A.B. H. J., 166
British merchant seamen or ships, 19–25, 326n.†

* Passing references to technical devices in the 'action' chapters are not indexed.

† Many other passing references to British men are not indexed.

British naval seamen, 36–40, 326n.*
British airmen, 49–54*
Brooklyn, 27, 41, 86
Brophy, 2nd Eng. H. W., 218
Brosin, OL, H.-G., 259
Brown, Ordinary Tel. R. T., 128
Brünning, KL, H., 243, 274, 276n.
Bunton, Sergeant J. H., 276
Burcher, Flying Officer C. W., 212–14

Cambridge, Capt. W. H., 235
Cambridge, Sub-Lt J., 235n.
Campbell, Lt A. D. B., 113, 127
Canadian airmen, 52, 272, 277 (*see also* Royal Canadian Air Force)
Canadian seamen or ships, 31, 33–4, 36, 40–41, 110–11, 205, 326n. (*see also* Royal Canadian Navy)
Cardiff, 20–21, 197, 282
Carlisle, Leading Wren M., 204
Casablanca Conference, 203, 311
Casey, Commodore D. A., 95
Castle Archdale, 53, 269, 276
Christopherson, Lt D. C., 252–3
Christopherson, OL. E., 142, 194
Church, Flight Lieutenant G. A., 272
Churchill, Winston
  fear of U-boat danger, 1
  support for Bomber Command, 11, 314
  honoured by Dutch, 30
  his Anti-Submarine Committee, 205, 317
  radio broadcasts, 286, 329
  in Parliament, 295
  wartime secrecy, 319
Civitillo, A.B. P., 101, 109, 255, 292–3
Clark, Pilot Officer L. G., 270–71
Clarke-Hunt, 2nd Off. W. A., 250
Collings, Capt. C. L., 198
Convoys (other than SC.122 and HX.229)
  HS.224, 79
  HX.228, 93, 115–17, 126, 135, 142, 247, 297, 304
  HX.230, 317
  HX.231, 313
  ONS.1, 209–10, 282, 284
  ONS.5, 322
  ON.168, 121, 126
  ONS.169, 117, 121
  ON.170, 139–40, 284
  ONS.171, 152, 210, 284
  ON.172, 138–9, 152, 284
  ON.173, 152, 284
  PQ.17, 78, 113
  SC.118, 79, 99
  SC.121, 115, 135, 142, 264
  SC.123, 317
  SC.130, 323
  SC.143, 246n.
  SL. Convoys, 81, 122, 304
  TM.1, 120n.
  UGS.6, 119, 284
  WS. Convoys, 89n.
Craighead, Colonel A. C., 249
Craik, Cabin Boy F., 329
Craik, Lt-Cdr J. D., 43, 325

Dalison, Cdr J. S., 122, 127
Danish merchant seamen or ships, 29, 215, 326n.
Dauter, OL. H., 63
Dawson, Lt-Cdr J. C., 262–3
Day, Cdr E. C. L., 121, 152, 246, 254–6, 258, 260–61, 270, 293–4, 306
Delforge, Lt J. A., 149
Depth charges, 9, 14, 179–80, 228–9
Dieterichs, KL. H., 258
Dodds, Capt. B. C., 189, 218–19
Dönitz, Grossadmiral K.
  in opening stages of war, 4, 15–17
  and tactical changes, 7, 79, 110, 115, 314, 322
  his character and attitude, 57–76 *passim*, 233
  in SC.122 and HX.229 action, 118, 126, 135, 137, 154–9, 206–9, 212, 264, 269, 274–5, 282, 296
  decorated by Hitler, 300–301
  loses sons, 323
  at end of war and after, 318, 325–8
Dönitz, OL. K., 323
Dönitz, Lt P., 323
Dossett, Signalman A. H., 39–40
Douglas, Ch. Eng. R. W., 87, 166
Dubeck, Lt E. J., 313, 315
Duke, 2nd Off. J., 176
Dunn, 2nd Off. J. G. A., 236
Dutch seamen or ships, 20n., 29–30, 101, 174–7, 198–9, 214, 224–6, 287, 293, 302–3, 326n.

Edser, Squadron Leader W. E., 270
Eglinton, 213, 230
Ellsworth, Lt-Cdr E. B., 42, 216, 325
Endrass, KL. E., 8
Engebretsen, S1c L., 172
Engel, OL. H., 257, 262, 271

* Many other passing references to British men are not indexed.

Engelmann, Fregattenkapitän K.-E., 285
Esler, Flying Officer S. E., 230-32, 241, 274-5, 307, 309

Farragut, 2nd Off. W., 253
Feiler, KL. G., 152, 156, 163n.
Finnes Capt. S. G., 283
Flensburg, 66, 318
Fortress, Boeing (B.17), 51-2, 54, 64, 264, 269-71, 273, 276, 300, 307, 313
Franken, Capt. G. J. J. M., 175, 177
Fraser, Pilot Officer A. W., 242-3, 275n.

Gardener, Lt D. G. M., 130, 259
Geissler, KL. H., 259
German merchant ships
  Doggerbank, 285
  Karin, 285
  Regensburg, 156, 284-5
German Navy
  B. *Dienst*, 75-6, 79, 118-19, 126, 134-40, 149, 151, 158, 209, 212, 284, 307, 318-19
  Ships:
    Bismarck, 235n.
    Erich Giese, 58
    Graf Zeppelin, 2, 59
    Hela (1914-18), 48
    Prinz Eugen, 235n.
  (All U-boats and their units are indexed under 'U-boats')
Glasgow, 6, 20
Godt, Konteradmiral E., 75
Goodfellow, Flying Officer R., 246
Goudy, Sub-Lt R. G., 166, 182
Graef, OL. A., 160, 181, 221
Gravely, Lt H., 134, 213
Greek seamen or ships, 20n., 30, 101, 262-3, 293, 302, 326n.
Green, Flight Lieutenant F. J., 314-15
Greenhouse, Ordinary Tel. D., 151
Greenock, 34, 40, 47, 284
Gretton, Cdr P. W., 34, 322
Gurnham, Sergeant R. V., 330
Guy, Sub-Lt A. R., 226
Gynn, Lt A. D., 295

Halifax, Handley Page, 283, 299-300
Halifax (Nova Scotia), 34, 41, 76, 80, 83, 89n., 93, 101, 107-8, 111-19 *passim*, 135, 137, 203, 266n.
Hall, Cdr R. A., 45, 203
Hall, Lord, 308-9

Hamburg, 61, 66, 225, 229
Hammond, Flying Officer E. C. 230
Hansen, 2nd Off. H., 329
Harris, Air Chief Marshal Sir Arthur, 64
Hartmann, KL. K., 259
Harvey, Sergeant B. H., 53
Haupt, OL. H.-J., 220, 299
Hayes, Squadron Leader D. G. T., 278
Hedgehog, 179-80
Hess, Funkgefreiter W., 5, 59
Hessler, Fregattenkapitän G., 75
Hewitt, Flight Lieutenant D., 276-7
Heyme, Ch. Off. W., 109, 268
High Frequency Direction Finder, 13, 112, 318-19
Hill, Apprentice C. H. F., 87-8, 101, 107
Hill, Lt-Cdr L. C., 160, 163, 185, 222-3, 225-6, 244, 291, 325
Hirschler, Lt L. D., 42
Hitler, Adolf
  and rules of submarine warfare, 3-4, 72
  and U-boat building programme, 15, 314
  orders blockade of Britain, 30
  support among U-boat men, 71, 74
  promotes Dönitz, 73
  attempt on his life, 78-9
  decorates Dönitz, 300
  commits suicide, 325
Hobson, 1st Off. G. H., 25
Hocken, Capt. A., 82, 165, 190, 295
Holtorf, KL. G., 200, 277
Horton, Admiral Sir Max
  his character and attitude, 47-9, 325
  and tactical changes, 108, 316-17
  in SC.122 and HX.229 action, 204-5, 291, 293
  his death, 328
Howard-Johnston, Cdr C. D., 10
Howell, Flying Officer A. B., 277
Hudson, Lockheed, 54, 312-13
Hughes, 2nd Off. F. R., 197-8
Hunger, KL. H., 209n.
Hunt, Ch. Off. P. H., 249-50, 254
Hvalfjordur, 42, 244-5

Icelandic seamen or ships, 31, 101, 214
Isted, Squadron Leader D. J., 246, 275n.

Italian submarines, 6
  Leonardo da Vinci, 285
  Nani, 180

Jackson, Petty Officer R. V., 182
Jacobsen, Fähnrich W., 59, 143, 216, 234
Johannessen, Capt. R., 171
Johannessen, Motorman J., 172
Jonsson, 2nd Eng. B., 113

Kaiser, Henry J., 27
Kapitsky, KL. R., 70n., 160
Kempton, Flight Sergeant T. J., 231, 274
Kersley, Capt. L. W., 102
Keyworth, 3rd Off. R. H., 247, 250, 252-4
Kiel, 57, 59-62, 299, 326-7
King, Rad. Off. B., 197
King, Admiral E. J.
  and U.S.N. involvement in N. Atlantic convoys, 46-7, 90, 290
  at Atlantic Convoy Conference, 203
  attitude to U.S. aircraft cover for convoys, 310-12, 314-15
King, Lt-Cdr P. G. A., 178-81, 218-220, 261
Kinzel, KL. M., 146, 195-6, 199, 208, 213, 215-16, 244, 272-3, 299, 305
Klusmeier, KL. E., ix, 3n.
Knowles, Flying Officer H., 271
Knowles, Capt. K. A., 47, 137
Koitschka, OL. S., 58, 187, 218-19
Kolbe, Obermaschinist A., 60
Kretschmer, KK. O., 5, 8
Krüger, OL. J., 274, 278
Krushka, OL. M., 282-3
Kynnersley, Sergeant T. E., 51, 270, 330

Laidler, Capt. W., 197
Lake, Capt. H. N., 316-17
Lancaster Avro, 314
Lange, KL. K., 228-9
La Pallice, 63, 65-6
Larose, Lt M.A.F., 148, 238
Lascar seamen, 21, 23, 224, 226, 281, 287, 303, 326n.
Lawatsch, Bootsmaat H., 60
Lawson, Ensign D., 134, 329
Leggatt, Capt. C. M., 45
Leigh Light, 299
Lemp, KL, F.-J., ix, 4, 63, 72
Leslie, Lt G. C., 181-3, 239-40, 325
Liberator, Consolidated (B.24), 18, 50-52, 54, 64, 206, 212-13, 216-17, 230-32, 242-3, 245-7, 269, 271, 273-5, 299n., 307, 309-15, 317, 323
Liberty ships, 26-7, 80, 303-4
Liverpool, 6, 20-21, 34, 40, 47, 198, 204-5, 210, 237, 291, 293, 303, 328
Lloyd, Sergeant J., 212
Lohmann, KL. D., 164, 180
London, 6, 20, 114, 135, 210, 292
Londonderry, 34, 40, 122, 194, 213, 241, 245, 263, 281, 290, 292-4
Long, Capt. A. W., 176, 178, 221
Lorch, OL. W., 59
Lorient, 63-6, 74, 300
Luckey, Capt. W., 84, 98, 201-2
Lundon, Flying Officer G., 144
Luther, Lt-Cdr G. J., 124-5, 127, 149, 151-2, 159-64, 167-8, 170, 172-3, 178, 181-2, 185-7, 192, 201, 203, 205, 221-3, 225-7, 230-31, 239-40, 245, 247, 253-4, 274, 293-4, 305-6, 316, 321

MacDonald, Lt T., 111
MacLellan, Ch. Off. M., 202
McCullock, Flying Officer L., 299
McKinnon, Capt. R., 186
McRae, 3rd Off. R., 89, 108, 150, 223
Manseck, KL. H., 143, 160, 174
Marder, Prof. A., 9
Maritime Regiment, Royal Artillery, 24, 238, 303, 326n.
Marshall, General G., 312
Marshall, Sub-Lt T. C., 105
Martin, Lt J. A., 268
Martinson, Capt. A. M., 268
Matzen, Capt. F., 215
Maude-Roxby, Sub-Lt L. M., 173, 251
Mayall, Commodore M. J. D., 95, 99, 162-3, 165, 192-3, 223, 227, 286, 290
Mazavinos, Capt. D. C., 262
Meckel, Kapitän H., 13
Merchant Navy: *see* British merchant seamen or ships *and* Allied and neutral merchant ships
Metcalf, Rear Admiral M. K., 46
Metox, 298
Micallef, 2nd Off. C., 217
Miles, Capt. F. N., 316
Miller, Capt. R. D., 250, 254
Mitchell, North American (B.25), 105

Moffat, Flying Officer J. K., 242–3, 299n.
Möglich, KL. H., 300
Morris, Capt. O. C., 263
Morton, Ch. Off. W. D., 83
Müller-Edzards, KL. H., 274, 297–8
Murphy, Fireman P., 196, 329
Murray, Rear Admiral L. W., 114, 123
Mussolini, Benito, 6

Nancollis, Capt., 227
Napier, A.B. T., 88, 251
Neide, KL. K., 57
New Jersey, 104–5
New Orleans, 26, 253
New York, ix, 26, 34, 41, 77–109 *passim*, 218, 284, 287, 303–4
New Zealand servicemen, 212, 252–3, 326n.
Noble, Admiral Sir Percy, 48–9, 316
Norris, Ensign B., 96–7, 192
Norwegian seamen or ships, 19n., 29, 101, 171–3, 202, 281, 287, 293, 302, 326n.
Nuremberg Trials, 72, 327–8

Old, Lt-Cdr E. G., 94, 102, 104–5, 107
Olsen, Hysing, 29
Osmond, Surgeon-Lt H. F., 241, 248, 295, 330
Ostend, 43–4, 263

Panamanian ships or crews, 101, 214–17, 302
Paris, 66, 74–5
Philadelphia, 26, 100
Pietro Orseolo (Italian), 285
Pietzsch, KL. W., 269
Pilcher, Lt L. G., 217
Pilling, Ensign F. N., 96–7, 188
Philpott, Lt-Cdr L. B., 280–81
Powell, Sub-Lt A. D., 98, 111
Prien, KK. G., 3, 8, 73n.
Price, Lt-Cdr R. A., 124, 184–5, 193–4, 228–9
Procter, Stoker, R. V., 166

Quarles, Lt-Cdr S. F., 264–5, 294

Radar, 12–13, 298
Raeder, Grossadmiral E., 4, 73
Ravenhill, Capt. R. W., 49, 206
Reese, Lt H. K., 314–15
Reinicke, Capt. F. G., 92–3, 98
Reith, OL. H.-E., 266

Reykjavik, 50, 53, 113, 148, 164, 206, 214, 242, 313
Robertson, Flying Officer W. C., 278
Rogers, Lt G. B., 148, 237–8
Roosevelt, President, 26, 29, 312–13
Rösing, Kapitän H.-R., 64, 300
Roskill, Capt. S. W., 256, 319
Rowland, Capt. J. M., 124
Royal Air Force
    Bomber Command, 11, 52, 54, 65–6, 78, 300, 312
    Coastal Command, 10–11, 50–54, 64, 69, 203–4, 212, 269, 273, 276, 298, 300, 307, 310, 312–15, 321
    15 Group, 50, 52–3, 206
    16 Group, 298
    18 Group, 54
    19 Group, 54, 298
    50 Squadron, 237n.
    58 Squadron, 300
    59 Squadron, 300
    86 Squadron, 50–51, 212, 213n., 230, 245, 269
    120 Squadron, 50, 53–4, 206, 216, 230, 243, 245, 269, 299n., 313, 323
    172 Squadron, 144, 299
    179 Squadron, 261n.
    196 Squadron, 237n.
    201 Squadron, 52, 261n., 276, 278
    206 Squadron, 51, 269–70
    220 Squadron, 51, 269, 330
    228 Squadron, 52, 272, 277n.
    246 Squadron, 52
    269 Squadron, 312
    502 Squadron, 299
    617 Squadron, 324
Royal Canadian Air Force, 310, 313–14
    10 Squadron, 314
    423 Squadron, 52, 246n., 272, 277
Royal Canadian Navy
    provides convoy escorts, 11, 40–41
    Western Local Escort Force, 34, 41, 94, 110–11
    its Submarine Tracking Room, 47
    in planning of SC.122 and HX.229, 94
    and times to be used in convoys, 99
    in SC.122 and HX.229 action, 114, 123
    Ships:
        Annapolis, 114
        Blairmore, 102, 104–5, 111, 113
        Cowichan, 113

Ships: *(cont)*.
  Dunvegan, 113
  New Westminster, 102, 104–5
  Rimouski, 102, 104, 114
  St Croix, 243n.
  Sherbrooke, 121–2, 205, 254, 290
  The Pas, 94, 102, 104, 107, 113,
    282–3
Royal Naval Reserve, 37–8, 49
Royal Naval Volunteer Reserve, 37–8,
  49
Royal Navy
  Western Approaches Command,
    10, 36, 38, 40, 47–9, 93, 112,
    120, 203–6, 227, 232, 293–4,
    305, 315–16, 325
  Escort Groups:
    2nd, 163n.
    40th, 120–27 *passim*
    B4, 120–27 *passim*
    B5, 44, 120–26 *passim*, 294
    B7, 34, 322
  Ships:
    Abelia, 121–2, 186, 205, 246–7,
      254, 273
    Aberdeen, 122, 127, 279–80
    Aboukir (1914), 57n.
    Aconit, 247
    Active, 285
    Anemone, 121–2, 166, 174, 178–
      81, 183, 186, 191–2, 218–
      20, 222, 227, 240, 245, 248–
      9, 251, 253, 258–9, 261,
      287, 290–91
    Arbutus, 8n.
    Aubretia, 3n.
    Barham, 3n.
    Beverley, 121–2, 124–5, 170,
      174, 178, 187, 192–4, 222,
      227–30, 232, 240, 245, 290,
      323, 325
    Biter, 317
    Boreas, 285
    Broadway, 3n.
    Bulldog, 3n.
    Buttercup (Belgian), 44, 121,
      126, 211, 263, 325
    Burza (Polish), 35n.
    Camilla, 8n.
    Campbeltown, 32
    Campobello, 123, 126, 131, 134,
      148–9, 164–5, 195, 211,
      237, 293–4, 302
    Chelsea, 94, 98, 111, 114
    Corinthian, 285
    Courageous, 3
    Cressy (1914), 57n.

    Crocus, 285
    Daring, 8
    Deptford, 8n.
    Dinosaur, 44
    Duncan, 322
    Glasgow, 285
    Glenarm, 243n.
    Godetia (Belgian), 44, 121, 126,
      148–9, 164–5, 195, 211,
      214, 216, 237, 287, 294,
      325, 328
    Harvester, 33n., 117, 126
    Hastings, 122, 280, 282
    Havant, 33n.
    Havelock, 39, 120, 126–7, 131,
      147, 164–5, 211–12, 217,
      234, 239, 242, 261, 274–5,
      329
    Highlander, 121, 123–4, 130, 152,
      204–5, 244, 246–7, 254,
      259–60, 297
    Hogue (1914), 57n.
    Hood, 235n.
    Hurricane, 33n.
    Jonquil, 263
    Keppel, 284n.
    L.S.T. 305, 83–4 106n.
    L.S.T. 365, 83–4, 86, 106
    Languard, 122
    Laurentic, 5
    Lavender, 121, 126, 147, 211,
      215–17, 243–4, 261, 272,
      329
    Leamington, 113–14, 127
    Londonderry, 122
    Lulworth, 122
    Mansfield, 32, 123, 131, 150,
      160–63, 168, 174, 183–5,
      187, 190–93, 222–3, 225–7,
      239–40, 244–5, 287, 291,
      304, 306, 325
    Ness, 285, 316
    Orkan (Polish), 246n.
    Pennywort, 121, 149, 170, 173–4,
      178, 184, 186, 191, 222–3,
      227, 240, 245, 248–9, 251,
      258–9, 287, 290–91, 294,
      306, 329
    Petunia, 285
    Pimpernel, 121, 126, 211, 234
    Rowley, 325
    Royal Oak, 3, 8
    Samphire, 8n.
    Saucy, 263
    Saxifrage, 121, 126, 167, 198–
      200, 211, 244

Spey, 258n.
Starling, 163n.
Stockforce (1918), 95n.
Swale, 121, 126, 196, 211, 234, 261–2
Tweed, 243n., 316
Vanoc, 8n., 124n.
Vimy, 121, 162, 205, 227, 254, 274
Vindex, 163n.
Volunteer, 121, 123–5, 127–8, 131, 151, 159–60, 163–4, 166–7, 172, 174, 181–2, 185–7, 192, 202, 218, 222, 225–6, 230, 232, 239–41, 244–8, 259, 261, 287, 290, 293–5, 306, 320, 325, 330
Wallflower, 269n.
Walker, 8n.
Wanderer, 243n., 269n.
Warwick, 121
Waveney, 122, 279–80
Whimbrel, 316
Wild Goose, 163n., 316
Winchelsea, 121
Witherington, 123, 131, 149
Woodpecker, 316

Saga, Chief Cook T., 202
St John's, 34, 39–41, 93, 115–28 *passim*, 149, 152, 203, 205, 244, 273, 280, 290
St Nazaire, 63, 65–7, 74, 178, 299
San Francisco, 26–7
Savannah, 176, 303
Schaap, Chief Cook S.I., 226
Schauroth, OL. C., 142
Schepke, KL. J., 8, 124n.
Schetelig, OL. R., 156, 284
Schmeling, Maschinenobergefreiter G., 260
Schmid, KL. H., 283
Schnee, KL. A., 75, 209
Schofield, Capt. B. B., 45
Schubert, Mechanikermaat K., 171
Schuhart, OL. O., 3
Schütze, Korvettenkapitän V., 23n.
Schwantke, OL. H.-J., 285
Sharp, 2nd Off. J. D., 167
Shaw, Lt M. C. C. F., 167
Shetlands, 20, 283
Shevlin, Eng. W. P., 235–6
Simonsen, Capt. C., 96
Slatter, Air Vice-Marshal Sir Leonard, 205–6

Slessor, Air Marshal Sir John, 50
Skillen, Pilot Officer E., 237n.
Smith, Engine Room Artificer, G. T., 225
Smith, Flying Officer, 313, 315
'Snowflake' flares, 12, 99
Spicer, Sub-Lt L. A., 261
Stannard, Lt-Cdr R. B., 227n.
Stapleton, Signalman, 111
Starshell, 12
Staten Island, 41, 102
Stembridge, Flying Officer P. H., 299
Stoves, Flight-Sergeant W., 216–17, 234, 323
Strelow, KL. S., 181, 185
Struckmeier, KL. R., 231, 260
Stuart, Lt O. G., 178, 186
Süssenbach, Maschinenmaat, 297
Sunderland, Short, 51, 246n., 261n., 269, 271–3, 276–8, 307
Swedish merchant seamen or ships, 20n., 30, 101

Tait, Cdr A. A., 117
Taylor, Sergeant J., 299
Theen, Obersteuermann H., 152–3
Thompson, Capt. F., 281
Tickner, Flying Officer R., 237n.
Toepfer, OL. R., 59, 67, 258
Torpedoes, 170
Trefflich, Obersteuermann, 196
Trojer, OL. H., 231, 247, 254
Turner, Flying Officer R. T. F., 242–3

U-boats
in First World War, 1
in 1939–40, 2–6
'pack' tactics, 6–10, 13
in 1941–2, 11–18
crews' background and morale, 56–74 *passim*, 273–4, 323–4
different Types, 60–62, 67–9, 143–4
organization, 62–7, 72–5
casualties, 70, 324, 326–7
in early 1943, 79–80
War Diary system, 105n.
tanker U-boats, 142–4, 156, 296
achievements, 326
U-boat Headquarters (B.d.U.), 7, 62, 72–5, 117–19, 135–40, 143–5, 150, 153–60, 163, 169, 173, 196, 206–9, 240, 242, 270, 273, 275, 295–6, 299n., 301–2, 307, 321

U-boat Groups:
Gruppe Burggraf, 142
Gruppe Dränger, 135, 137, 140,
   146, 151–2, 156–8, 164,
   280–89, 223, 233, 368
Gruppe Neuland, 117
Gruppe Ostmark, 142
Gruppe Raubgraf, 117, 120,
   135–40, 146, 150, 152,
   156–8, 169, 194, 208, 223,
   279, 294, 308, 310, 321
Gruppe Seeteufel, 210, 233,
   282–3, 296, 321
Gruppe Stürmer, 135–8, 140,
   146, 151–2, 156–8, 164–5,
   195, 200, 208–9, 223, 233,
   308
Gruppe Westmark, 142
Individual U-boats:
1914–18:
U.9, 57n
U.80, 95n.
1939–45:
U.29, 3
U.30, ix, 63
U.43, 285
U.46, 8n.
U.47, 3, 8n.
U.84, 156, 162
U.89, 164, 180–81, 233
U.91, 150, 156, 160, 185, 191,
   220–21, 233
U.99, 8n.
U.100, 8n., 124n.
U.103, 23n.
U.110, 3n.
U.119, 142–3, 296
U.130, 284
U.134, 156, 259, 261n.
U.163, 285
U.187, 227, 323
U.188, 323
U.190, 156, 265–6, 274, 307
U.221, 231, 247, 254
U.225, 35n.
U.228, 142, 156, 184, 191,
   193–4, 220
U.229, 156, 233, 282, 284
U.254, 247
U.305, 59, 143, 214, 216–17,
   233–5, 238–9, 242–3
U.332, 300
U.333, 144, 263
U.336, 209, 269
U.338, 146, 156, 195–6, 200,
   208, 213–16, 244, 257, 261,

272, 299, 305
U.358, 211
U.384, 156, 225, 270–71, 304
U.406, 59–60, 67, 257–8, 266
U.415, 33n., 57
U.419, 213n.
U.432, 33n.
U.435, 156, 181, 183, 185, 233
U.439, 146, 156, 211–12, 239,
   260, 261n.
U.440, 259, 261n.
U.441, 259
U.448, 63
U.456, 79
U.461, 285
U.463, 142–3, 184, 296
U.468, 233
U.523, 59, 143–4, 269
U.526, 300
U.527, 159, 267–9, 277
U.530, 5, 59, 60, 156, 228–9, 233
U.567, 8n.
U.590, 274, 297–8
U.591, 142, 330
U.597, 50
U.598, 156, 200, 269, 277
U.600, 156, 162–3, 187, 229, 233
U.603, 156, 164, 169–71, 173–4,
   183, 191, 214, 223, 232, 277
U.606, 35n.
U.608, 231, 260, 272
U.610, 246
U.615, 61, 70, 156, 160, 163,
   184, 191, 258–9
U.616, 58, 156, 184, 187, 191,
   218, 233
U.618, 69, 146, 156, 159, 222n.,
   269
U.621, 282–3
U.626, 35n.
U.631, 156, 225, 274, 276–8
U.632, 213n.
U.633, 271
U.638, 233
U.641, 156, 200
U.642, 243, 256–7, 261, 274, 276
U.653, 152–63, 169, 208, 233,
   263, 308, 310, 316, 321
U.661, 231
U.663, 283, 302
U.664, 156, 160, 181, 221, 233
U.665, 156, 220, 233, 299
U.666, 211, 213–14, 216, 257,
   261–2, 271
U.758, 143, 156, 160, 174, 178,
   181, 233

U.954, 323
U.2336, ix
U.3518, 276n.
Uhlig, KL. H., 159, 267–8, 277
'Ultra', 320–21
Underhill, Flight Sergeant J., 237n.
United States Army Air Force, 65–6, 78, 310, 312–13
Anti-Submarine Squadrons
1st, 310
2nd, 310
3rd, 105
6th, 313
11th, 105
20th, 54
United States Coast Guard, 17, 35, 42–3, 85–6, 119, 264, 267–8, 325–6, 328
Cutters:
Aivik, 281
Algonquin, 281
Campbell, 35
Duane, 35
Frederick Lee, 281
Geo. M. Bibb, 35, 281–2
Ingham, 35, 43, 164–5, 205, 264, 267–8, 275, 290, 325–6, 328
Modoc, 281
Spencer, 35
United States Navy
sends fifty destroyers to Britain, 6
Neutrality Patrol, 12
faces U-boat attacks off American coast, 16–17
in Western Local Escort Force, 35, 111
ships and men in SC.122 and HX.229 action, 36, 42–3, 96, 213, 216, 243, 264–6
Convoy and Routing Section, 46–7, 89–95, 114, 118, 123, 137–9, 151–2, 160, 162, 164, 203
control of convoys, 80, 89
in planning of SC.122 and HX.229, 89–95
Eastern Sea Frontier, 90–94, 96, 118
Tenth Fleet, 90n.
Port Director, New York, 92, 118–19
Submarine Squadron Fifty, 285
withdraws escorts from North Atlantic, 290
personnel casualties, 303, 326n.
attitude towards V.L.R. Liberators, 309–15

VB-103 Squadron, 314
Ships:
Babbitt, 42, 164–5, 205, 254, 264–6, 274, 290, 294, 307
Bogue, 277n., 311, 317
Bronstein, 173n.
Card, 222
Champlin, 284
Cowie, 94, 111
Herring (submarine), 285
Kearney, 12
Kendrick, 111
Reuben James, 12
Upshur, 42, 123, 126, 131, 134, 140, 211, 213, 215–16, 243, 267, 272, 290, 325, 329
Uphoff, KL. H., 162

van Altveer, Ch. Off. P. G., 174, 177
van Dongen, Rad. Off. J. F. J., 166
van Os, Capt. C. L., 199
van Waesburghe, Lt P. H. V. M., 263
von Egan-Krieger, Lt C., 61, 70, 184, 258
von Freyberg, KL. W., 246
von Rosenberg-Gruszczynski, OL. H.-A., 271
von Tippelskirch, OL. H., 146, 211, 239, 260

Walker, Sub-Lt E., 329
Walker, Capt. F., 37, 163n.
Walkerling, KL. H., 160, 185, 220
Ward, Warrant Eng. G. H., 241, 330
Ward, A.B. S. A., 197
Ware, Sergeant J. B., 236–7
Washington, 46–7, 90, 93, 114, 135, 137, 203, 210, 296
Wassenaar, 2nd Off. J., 30, 86, 175, 177
Waters, Lt. J. M., 267, 326
Weddigen, Otto, 57n.
Weller, Lt W. F., 215
Wellington, Vickers, 144, 261n., 299
Wermuth, OL. O., 229
White, Lt J. J., 215–16
White, Wireman R. A., 251, 294
White, Capt. S. N., 95, 102, 104–5, 107, 214–15, 291
Whiting, Lt-Cdr C. J., 94
Wilford, Lt-Cdr J. E. R., 94
Wilhelmshaven, 58, 61, 74
Williams, 2nd Off. G. D., 189, 218, 252, 330
Willmott, Lt-Cdr A. E., 279
Winn, Cdr R., 45–6, 204, 319

Winterbotham, Gp Capt. F. W., 320
Wintermeyer, KL. M., 265–6
Women's Royal Naval Service, 40,
    49, 204, 316
Wright, Noel, 96, 253

Yates, Trimmer R. (served as J. J.
    Elder), 303
Yeates, Lt N. T. M., 329

Young, Apprentice, 224
Yugoslav merchant seamen or ships,
    30, 101

Zeissler, OL. H., 146, 196, 199–200
Zetsche, KL. H., 142, 330
Zurmühlen, KL. B., 162, 187
Zweigle, Mechanikerobergefreiter
    M., 169

Also available from Quill

The United States Navy in World War II
The One-Volume History, from Pearl Harbor to Tokyo Bay—by Men Who Fought in the Atlantic and the Pacific and by Distinguished Naval Experts, Authors, and Newspapermen
S. E. Smith, ed.

A superb narrative history of mammoth conflict on the oceans of the globe.
0-688-06274-1

The Mighty Endeavor
The American War in Europe
Charles B. MacDonald

The only single-volume history containing the *entire* story of the American armed forces in Europe and the Mediterranean during World War II.
0-688-06074-9

The Pacific War
John Costello

Thorough and in vivid detail, the first comprehensive one-volume account of the causes and conduct of World War II in the Pacific. "Readable and enlightening"—*Publishers Weekly*
0-688-01620-1

Barbarossa
The Russian-German Conflict, 1941–45
Alan Clark

A vivid history of one of the most brutal campaigns in military history in which the victor, the Red Army, lost seven million lives. This has become a classic of military history.
0-688-04268-6

The German Generals Talk
B. H. Liddell Hart

As professional soldiers, the German generals talk over the war and what led up to it.
0-688-06012-9

The Marauders
Charlton Ogburn, Jr.

A war story turned war legend, as dramatic today as when it was first published. "One of the noblest and most sensitive books by an American about his experience in the war. It is good history, eyewitness reporting at its most accurate best."—*The New York Times Book Review*
0-688-01625-1

A Short History of World War II
James L. Stokesbury

The only objective, comprehensive one-volume survey of the entire war.
0-688-08587-3

At your local bookstore